Coral Reefs of the World

Volume 17

Coral Reefs of the World is a series presenting the status of knowledge of the world's coral reefs authored by leading scientists. The volumes are organized according to political or regional oceanographic boundaries. Emphasis is put on providing authoritative overviews of biology and geology, explaining the origins and peculiarities of coral reefs in each region. The information is so organized that it is up to date and can be used as a general reference and entry-point for further study. The series will cover all recent and many of the fossil coral reefs of the world.

Prospective authors and/or editors should consult the Series Editors B.M. Riegl and R.E. Dodge for more details. Any comments or suggestions for future volumes are welcomed:

Dr. Bernhard M. Riegl/Dr. Richard E. DodgeNova Southeastern UniversityDania Beach, FL 33004USA

e-mail: rieglb@nova.edu and dodge@nova.edu

Ichiro Takeuchi • Hideyuki Yamashiro
Editors

Coral Reefs of Eastern Asia under Anthropogenic Impacts

Springer

Editors
Ichiro Takeuchi
Graduate School of Agriculture
Ehime University
Matsuyama, Ehime, Japan

Hideyuki Yamashiro
Sesoko Station, Tropical Biosphere Research Center
University of the Ryukyus
Motobu, Okinawa, Japan

ISSN 2213-719X ISSN 2213-7203 (electronic)
Coral Reefs of the World
ISBN 978-3-031-27559-3 ISBN 978-3-031-27560-9 (eBook)
https://doi.org/10.1007/978-3-031-27560-9

This Springer imprint is published by the registered company Springer Nature Switzerland AG
The registered company address is: Gewerbestrasse 11, 6330 Cham, Switzerland

Preface

When we started the studies on marine biodiversity and ecosystems from the 1980s to the early 1990s, coral reefs composed of *Acropora* spp. and other hermatypic corals thrived like thick bushes along the coasts of the Ryukyu Archipelago in southern Japan. Although the outbreak of the crown-of-thorns starfish was reported there in the 1970s, to the best of our knowledge, no one had attempted a rapid survey of coral reef coverage around 1990. Without detailed analysis, it was difficult to identify specific scientific significance in the measurement of coral coverage. However, coral reefs have been diminishing due to the rapid economic growth of countries around the East China Sea during that period.

Ichiro Takeuchi started his career as a marine environmental scientist by studying the natural history of amphipod crustaceans inhabiting macro-algal and seagrass communities in central Japan. In addition, he has been studying the behavior of anthropogenic chemicals in shallow-water ecosystems since the mid-1990s. Since the 1990s, he joined several field surveys to study amphipod crustaceans from Maine on the east coast of the USA to Lützow-Holm Bay in Antarctica. In most of these surveys, he collected the amphipod crustaceans by directly observing the specific habitats of these invertebrates using SCUBA diving. In particular, field surveys at New Caledonia and along the east coast of Australia from Lizard Island, Queensland to Jervis Bay, New South Wales have enabled him to recognize the various intense anthropogenic pressures on the shallow waters along the coasts of East Asia around the East China Sea. Ichiro Takeuchi acknowledges the late Dr. Jim Lowry, Dr. Penny Berents, and Dr. Bertrand Richer de Forges for their invitation to participate in field surveys and various discussions on natural history and anthropogenic impacts on shallow water ecosystems. These experiences led him to shift his main research focus to the anthropogenic impacts on *Acropora* spp., which is one of the most declined hermatypic corals in East Asia, from the mid-2010s.

Hideyuki Yamashiro, who was born and grew up in Okinawa, started studying corals at University of the Ryukyus, Japan. He began to focus on mushroom corals (solitary Fungiidae), especially skeletal dissolution by coral itself and phototaxis, under the late Dr. Kiyoshi Yamazato and Dr. Moritaka Nishihira. Over more than 40 years, he experienced coral degradation processes around Okinawa Island caused by land-based overdevelopment, repeated predation by crown-of-thorns starfish, and mass mortality due to severe bleaching events in 1998 and later. His interests turned to coral infectious diseases in 1998 and coral-killing sponges from late 2010, which have emerged and spread over time throughout coral reefs in Japan.

Various recent reports, including the IPCC, have mentioned the possibility that corals will be replaced by algae and sponges under climate change and various anthropogenic impacts in the near future. Thus, we sometimes feel that we are living witnesses to drastic changes in coral reefs in the Ryukyu Archipelago, East China Sea, and around the world. Even in such a difficult era, we hope that the present book contributes in several ways to the conservation of coral reefs in East Asia. Finally, we would like to thank editors of SpringerNature, for their help and encouragement.

Matsuyama, Ehime, Japan — Ichiro Takeuchi
Motobu, Okinawa, Japan — Hideyuki Yamashiro
14 October 2022

Contents

1 Introduction: The Relevance of Anthropogenic Factors to Coral Reef Conservation in the Coastal Areas of the East China Sea

Ichiro Takeuchi

Abstract

Although human activities on the Earth are geologically recent phenomena, the variety, scale, and impact of these activities are so great that they have led to our present era being termed the Anthropocene. The rapid economic growth of East Asia since the 1960s pushed the countries around the East China Sea to be the most economically active regions in the world, along with Central Europe. Thus, in addition to climate change, various anthropogenic pressures such as coastal development, plastic run-off, nutrients, and anthropogenic chemicals are thought to have been extensive, particularly around the East China Sea. As a result, coral reefs in the East China Sea are thought to have been exposed to various anthropogenic impacts earlier than reefs in other parts of the world.

Keywords

Anthropocene · Anthropogenic Impacts · Coastal Environments · East China Sea · Eastern Asia

1.1 Age of the Anthropocene

Scleractinian coral reefs emerged approximately 65 million years ago (Ma) in the early Paleogene and developed into coral reefs with extant form and function by the early Neogene, approximately 24 Ma (Bellwood et al. 2017; Williams and Graham 2019).

Although various human activities on Earth are geologically recent phenomena, the variety and extent of those actions are transforming terrestrial surfaces and changing the chemical composition of the atmosphere, which has led us to refer to the present as the Anthropocene Epoch (Lewis and Maslin 2015). Human impacts on the environment surged in the mid-twentieth century as a result of rapid increase in water consumption and fertiliser application; this period is known as the Great Acceleration (Monastersky 2015). The 1950s marked by extraordinary escalation in energy consumption and productivity due to the increase in human population have been proposed as the beginning of the Anthropocene (Syvitski et al. 2020). Starting from the mid-1960s, which marks the period of notable economic progress for in Japan (Abegglen 1970), China, and major East Asian countries, including Korea, the per capita GDP steadily grew from 1960 to 1994 at the average rate of over 4% (Collins and Bosworth 1996). East Asia was considered the only region in the developing world where living standards have been catching up to those of industrial countries (Collins and Bosworth 1996). Currently, East Asian countries bordering the East China Sea, along with Central Europe, are among the regions with the highest economic activities globally (Takeuchi 2004) (Fig. 1.1).

1.2 Present Status of Anthropogenic Impacts on Earth

The global mean surface temperature (GMST) increased by 0.87 °C in 2006–2015 relative to that in 1850–1900 (IPCC 2018). Without strong and urgent mitigation initiatives to drastically decline greenhouse gas emissions by 2030, global warming will surpass 1.5 °C in the following decades, driving irreversible loss of most ecosystems on the Earth (IPCC 2018). The increase in water temperature will cause various biotas to relocate to higher latitudes and the emergence of novel ecosystems; however, reef ecosystems, such as kelp forests in temperate regions and coral reefs in tropical regions, are relatively less able to migrate and will experience

I. Takeuchi (✉)
Graduate School of Agriculture, Ehime University, Matsuyama, Ehime, Japan

Center of Advanced Technology for the Environment, Graduate School of Agriculture, Ehime University, Matsuyama, Ehime, Japan
e-mail: takeuchi@agr.ehime-u.ac.jp

I. Takeuchi, H. Yamashiro (eds.), *Coral Reefs of Eastern Asia under Anthropogenic Impacts*, Coral Reefs of the World 17,
https://doi.org/10.1007/978-3-031-27560-9_1

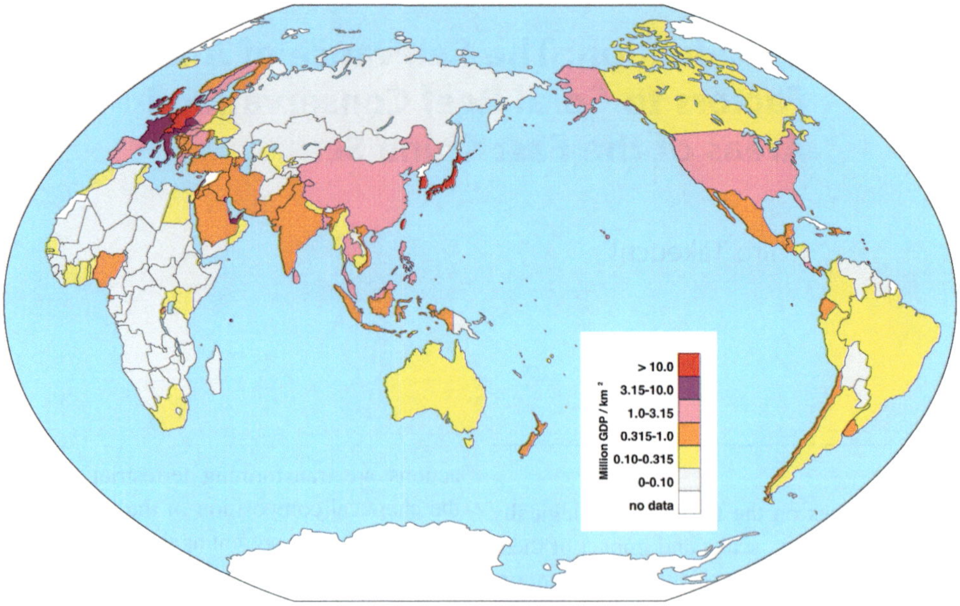

Fig. 1.1 GDP (2019) in US dollars per square kilometre. GDP values of each country are based on IMF (https://www.imf.org/external/datamapper/NGDPD@WEO/OEMDC/ADVEC/WEOWORLD accessed on 12-September-2021). The area of each country is based on "the country area" in FAO (http://www.fao.org/faostat/en/#data/RL/visualize accessed on 06 September 2021)

high mortality rates and loss of dominant biota (IPCC 2018). The oceans have absorbed approximately 30% of the anthropogenic carbon dioxide, leading to a decrease in pH, known as ocean acidification (IPCC 2018). Climate change due to anthropogenic activities has already affected many weather extremes worldwide (IPCC 2021). The increase in global surface temperature is projected to exceed 1.5 °C and 2 °C in the twenty-first century unless significant reductions in greenhouse gas emissions are realised in the coming decades (IPCC 2021). The increase in temperature of 1.5 °C has caused bleaching and long-term degradation of approximately 70–90% of coral reefs worldwide, and further rise in temperature of 2.0 °C will affect 99% of the reefs (IPCC 2018).

Anthropogenic mass, mainly comprising concrete, aggregates, bricks, asphalt, and metals, has doubled approximately every 20 years, and it is estimated to surpass the present global living biomass (approximately 1.1 tera tons) in 2020 ± 6 (Elhacham et al. 2020). With the development of various industrial and agricultural activities, the number of anthropogenic chemicals registered in the Chemical Abstracts Service increased from 20 million in 2002 to 156 million in 2019 (Escher et al. 2020). Of these, over 350,000 chemicals and mixtures of chemicals have been registered for production and use (Wang et al. 2020), and approximately 400,000 are somewhat regulated in international markets (Lara-Martín et al. 2020).

Plastics are widely used in various industry sectors, including packaging, toys, textiles, household items, electronics, automotive, and construction, with a record global production of over 350 million tons per year (Wiesinger et al. 2021). Thus, countries worldwide are struggling to manage the current volume of plastic waste accumulated in freshwater and marine ecosystems from the poles to the deep ocean (Borrelle et al. 2020). The plastic outflow from Asian rivers is estimated at ~93,000 metric tons (MT) year^{-1}, accounting for 69% of the global outflow (Mai et al. 2020). The amount of plastics flowing into the East China Sea from the Yangtze River is estimated at ~17,000 MT year^{-1}, which is the largest amount released by any of the world's rivers (Mai et al. 2020). Of the more than 10,000 chemicals potentially used in plastic production as plastic monomers and additives, more than 2400 substances are considered to potentially meet at least one of the criteria of persistence, bioaccumulation, and toxicity in the European Union (Wiesinger et al. 2021). Approximately half of them are not regulated in the most of the world (Wiesinger et al. 2021).

Agricultural activities occupy 37.5% of the land area on earth, with an average of 35.6% of land area being used for agriculture in Organisation for Economic Co-operation and Development (OECD) countries (Sud 2020). The land use by agricultural sectors provides a range of benefits, including the production of food, fodder, and fibre, input for industry, and various ecosystem services (Sud 2020). Although nutrient inputs in the agricultural sector are necessary for maintaining crop growth, nutrient loss from agricultural land is considered a significant source of pollution for freshwater and coastal ecosystems (Carpenter et al. 1998; West et al. 2014). Approximately 48% to 60% of nitrogen and phosphorus inputs are estimated to exceed the recommended amounts for 17 major crops worldwide (West et al. 2014), with China,

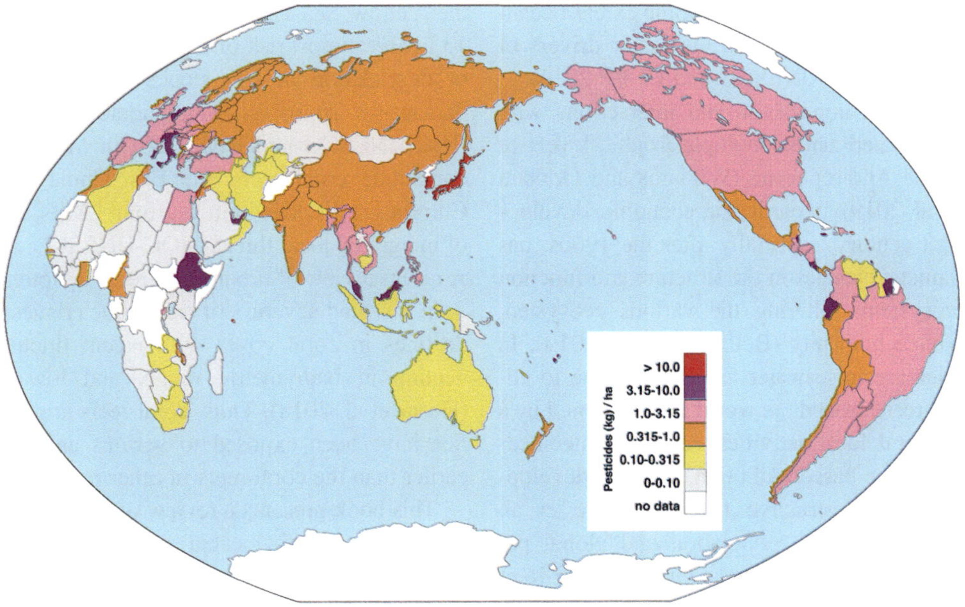

Fig. 1.2 Pesticide usage (2019) per square kilometre of agriculture land. Pesticide usage is based on FAO (http://www.fao.org/faostat/en/#data/RL/visualize accessed on 06-September-2021). Among countries with relatively small areas, countries with high pesticide usage (>10.0 kg/ha) include Bermuda, 210.8 kg/ha; Samoa, 55.8 kg/ha; Tonga, 36.7 kg/ha; Trinidad and Tobago, 21.7 kg/ha; Saint Lucia, 19.7 kg/ha; Maldives, 19.4 kg/ha; Barbados, 16.9 kg/ha; Bahamas, 12.7 kg/ha; Seychelles, 12.1 kg/ha. These countries are located in tropical coral reef areas of the Pacific, the Indian Ocean, or the Caribbean Sea

India, and the USA together accounting for approximately 64% to 66% of excess nitrogen and phosphorus globally (West et al. 2014).

The coastal region of the East China Sea has the highest frequency of pesticide use in the world. Among OECD countries, pesticide sales per 1000 ha of agricultural land area in 2013–2014 were fairly high in Japan and Korea, both recorded at 11.7 tons, followed by the Netherlands at 5.8 tons and Belgium at 4.9 tons (Sud 2020). According to recent Food and Agriculture Organization of the United Nations (FAO) data, pesticide use in Japan, Korea, and Taiwan surrounding the East China Sea exceeded 10.0 kg/ha of agricultural land in 2019 (Fig. 1.2). Besides, pesticide production and usage in China has been rapidly increasing since 1990, positioning China among the top producers and consumers of pesticides in the 2010s (Grung et al. 2015). Habitat loss induced by land conversion for intensive agriculture, urbanisation, and pollution mainly by synthetic pesticides and fertilisers are major drivers of species decline compared to other biological factors, including pathogens, alien species, and climate change (Sánchez-Bayo and Wyckhuys 2019). The abundance of a small number of species is increasing, occupying the niches left by the wiped-out species (Sánchez-Bayo and Wyckhuys 2019). The worldwide decline of insect biodiversity, with a dramatic extinction of 40% of the world's insect species, is expected to take place over the next few decades (Sánchez-Bayo and Wyckhuys 2019). The amount of hexachlorocyclohexanes (HCHs; 4.9 million tons) and dichlorodiphenyltrichloroethanes and its metabolites (DDTs; 0.4 million tons) used in China accounted for 33% and 20%, respectively, of the world's total usage in 1950–1983 (Bao et al. 2012; Ya et al. 2019). Both organochlorine pesticides are listed in the initial 12 persistent organic pollutants (POPs) of the Stockholm Convention (Lallas 2001). The seasonal fluctuation of HCH and DDT concentrations in the East China Sea and northern South China Sea suggests long-range oceanic transport into the Western Pacific from these marginal seas (Ya et al. 2019).

1.3 Present Status of Coastal Environments

Nearly 40% of the global population has been estimated to dwell within 100 km of the coastline at the beginning of the twenty-first century (Burke et al. 2011). The ecosystems in the coastal zone are threatened by anthropogenic activities, including settlement and infrastructure development, industry, and aquaculture, which alter the coastal zone either through direct physical damage, such as dredging and landfilling, or indirectly through increased runoff of sediments, pollutants, and sewage (Burke et al. 2011).

A recent analysis of the cumulative impact of approximately 20 different types of anthropogenic stressors on the global marine ecosystem revealed that human impact has pervaded every part of the global ocean, and nearly the entire ocean (97.7%) has been affected by multiple stressors

(Halpern et al. 2015). Land-based and to some extent ocean-based human activities are considered the major drivers of coastal water pollution (IOC-UNESCO and UNEP 2016).

Thus, various human impacts on and interactions with coral reefs have escalated across multiple trophic levels of the ecosystem in the Anthropocene (Williams and Graham 2019; Williams et al. 2019). In particular, economic development over the past century, especially after the 1960s, has brought about dramatic changes in the structure and function of coral reef ecosystems, altering the various ecosystem services they provide humanity (Bellwood et al. 2019). In addition to the increase in seawater temperature due to climate change, coral reefs around the world are threatened by a combination of several local activities, such as watershed-based pollution, marine-based pollution, coastal development, overfishing, and destructive fishing (Burke et al. 2011). With increasing human population and global per capita GDP, coral coverage, even in the Great Barrier Reef in Australia, which is located far from industrialised countries in the Northern Hemisphere, has been declining since 1960 (Bellwood et al. 2019). On a local scale, enhanced terrestrial runoff has been considered one of the significant stress factors on coastal and inshore coral reefs since the 2000s because of the elevated input of inorganic nutrients and particulate organic matter (POM), turbidity-related light limitation, and sedimentation (Fabricius 2005). In the early 2000s, to protect the Great Barrier Reef, The State of Queensland (2009) established management practices to reduce the runoff of photosystem II-inhibiting herbicides (Ametryn, Atrazine, Diuron, and others), decrease nitrogen and phosphorus (in dissolved or particulate form) and total suspended solids from agricultural land, and promote improved land usage.

1.4 Anthropogenic Impacts in the East China Sea and Adjacent Waters

Climate change and other anthropogenic drivers of biodiversity change are unequally distributed across terrestrial and marine ecosystems (Bowler et al. 2020). The highest intensities of multiple drivers in the marine realm were found in the East China Sea of the Northern Pacific, Central Indo-Pacific, and North Sea of the Northern Atlantic (Bowler et al. 2020). Among these areas, high cumulative intensities of human activities, human population, pollution, and invasions of alien species have been reported in the East China Sea and North Sea (Bowler et al. 2020).

IOC-UNESCO and UNEP (2016) separated the world's coastal areas into 66 large marine ecosystems (LMEs), covering more than 200,000 km^2 and extending from the shoreline to the edge of continental shelves or to the outer margins of major coastal currents. The East China Sea and adjacent the Yellow Sea and the South China Sea are listed in the 13 LMEs at most risk of environmental threats including loss or degradation of fish stocks and ecosystem health, and damage from climate-related extreme events (IOC-UNESCO and UNEP 2016). Moreover, of 24 LMEs containing coral reefs, the East China Sea and Kuroshio Current are listed within the four LMEs at the highest risk of integrated local threats (IOC-UNESCO and UNEP 2016). A recent review of urban coral reefs in Southeast and Eastern Asia revealed several critical characteristics, including early declines in coral cover with recent fluctuating periods, a decline in bathymetric range, and low reef complexity (Heery et al. 2018). Thus, coral reefs around the East China Sea have been exposed to various anthropogenic impacts earlier than the coral reefs in other regions of the world.

This book presents a review of the anthropogenic impacts on the coral reefs located along the coastline of the East China Sea. The recent succession and present status of coral reefs from Taiwan to southern Japan and Korea are reviewed. Additionally, the current spread of coral diseases in Taiwan and Okinawa, effects of ocean acidification on corals, and the changes in gene expression of corals induced by anthropogenic chemicals in hermatypic corals are summarised. A new, compact rearing system that can be used to study various anthropogenic effects on corals is presented. In the last chapter, I offer a perspective on coral reef conservation under anthropogenic impacts.

References

Abegglen JC (1970) The economic growth of Japan. Sci Am 222:31–37

Bao L-J, Maruya KA, Snyder SA, Zeng EY (2012) China's water pollution by persistent organic pollutants. Environ Pollut 163:100–108

Bellwood DR, Goatley CHR, Bellwood O (2017) The evolution of fishes and corals on reefs: form, function and interdependence. Biol Rev 92:878–901

Bellwood DR, Pratchett MS, Morrison TH, Gurney GG, Hughes TP, Álvarez-Romero JG, Day JC, Grantham R, Grech A, Hoey AS, Jones GP, Pandolfi JM, Tebbett SB, Techera E, Weeks R, Cumming GS (2019) Coral reef conservation in the Anthropocene: confronting spatial mismatches and prioritizing functions. Biol Conserv 236: 604–615

Borrelle SB, Ringma J, Law KL, Monnahan CC, Lebreton L, McGivern A, Murphy E, Jambeck J, Leonard GH, Hilleary MA, Eriksen M, Possingham HP, De Frond H, Gerber LR, Polidoro B, Tahir A, Bernard M, Mallos N, Barnes M, Rochman CM (2020) Predicted growth in plastic waste exceeds efforts to mitigate plastic pollution. Science 369:1515–1518

Bowler DE, Bjorkman AD, Dornelas M, Myers-Smith IH, Navarro LM, Niamir A, Supp SR, Waldock C, Winter M, Vellend M, Blowes SA, Böhning-Gaese K, Bruelheide H, Elahi R, Antão LH, Hines J, Isbell F, Jones HP, Magurran AE, Cabral JS, Bates AE (2020) Mapping human pressures on biodiversity across the planet uncovers anthropogenic threat complexes. People Nat 2:380–394

Burke L, Reytar K, Spalding M, Perry A (2011) Reefs at risk revisited world. Resources Institute, Washington, DC, x + 114 pp

Carpenter SR, Caraco NF, Correll DL, Howarth RW, Sharpley AN, Smith VH (1998) Nonpoint pollution of surface waters with phosphorus and nitrogen. Ecol Appl 8:559–568

Collins SM, Bosworth BP (1996) Economic growth in East Asia: accumulation versus assimilation. Brookings Pap Econ Act 1996:135–203

Elhacham E, Ben-Uri L, Grozovski J, Bar-On YM, Milo R (2020) Global human-made mass exceeds all living biomass. Nature 588: 442–444

Escher BI, Stapleton HM, Schymanski EL (2020) Tracking complex mixtures of chemicals in our changing environment. Science 367: 388–392

Fabricius KE (2005) Effects of terrestrial runoff on the ecology of corals and coral reefs: review and synthesis. Mar Pollut Bull 50:125–146

Grung M, Lin Y, Zhang H, Steen AO, Huang J, Zhang G, Larssen T (2015) Pesticide levels and environmental risk in aquatic environments in China—a review. Environ Int 81:87–97

Halpern BS, Frazier M, Potapenko J, Casey KS, Koenig K, Longo C, Lowndes JS, Rockwood RC, Selig ER, Selkoe KA, Walbridge S (2015) Spatial and temporal changes in cumulative human impacts on the world's ocean. Nat Commun 6:7615. https://doi.org/10.1038/ncomms8615

Heery EC, Hoeksema BW, Browne NK, Reimer JD, Ang PO, Huang D, Friess DA, Chou LM, Loke LHL, Saksena-Taylor P, Alsagoff N, Yeemin T, Sutthacheep M, Vo ST, Bos AR, Gumanao GS, Hussein MAS, Waheed Z, Lane DJW, Johan O, Kunzmann A, Jompa J, Suharsono TD, Bauman AG, Todd PA (2018) Urban coral reefs: degradation and resilience of hard coral assemblages in coastal cities of east and Southeast Asia. Mar Pollut Bull 135:654–681

IOC-UNESCO and UNEP (2016) Large marine ecosystems: status and trends. In: A global comparative assessment of baseline status and future trends. United Nations Environment Programme (UNEP), Nairobi, xliii + 299 pp

IPCC (2018) In: Masson-Delmotte V, Zhai P, Pörtner H-O, Roberts D, Skea J, Shukla PR, Pirani A, Moufouma-Okia W, Péan C, Pidcock R, Connors S, Matthews JBR, Chen Y, Zhou X, Gomis MI, Lonnoy E, Maycock T, Tignor M, Waterfield T (eds) Global warming of 1.5° C. An IPCC special report on the impacts of global warming of 1.5° C above pre-industrial levels and related global greenhouse gas emission pathways, in the context of strengthening the global response to the threat of climate change, sustainable development, and efforts to eradicate poverty. Cambridge University Press, Cambridge, xi + 616 pp

IPCC (2021) Summary for policymakers. In: Masson-Delmotte V, Zhai P, Pirani A, Connors SL, Péan C, Berger S, Caud N, Chen Y, Goldfarb L, Gomis MI, Huang M, Leitzell K, Lonnoy E, Matthews JBR, Maycock TK, Waterfield T, Yelekçi O, Yu R, Zhou B (eds) Climate change 2021: the physical science basis. Contribution of working group I to the sixth assessment report of the intergovernmental panel on climate change. Cambridge University Press, Cambridge, pp 3–31

Lallas PL (2001) The Stockholm convention on persistent organic pollutants. Am J Internat Law 95:692–708

Lara-Martín PA, Chiaia-Hernández AC, Biel-Maeso M, Baena-Nogueras RM, Hollender J (2020) Tracing urban wastewater contaminants into the Atlantic Ocean by nontarget screening. Environ Sci Technol 54:3996–4005

Lewis SL, Maslin MA (2015) Defining the anthropocene. Nature 519: 171–180

Mai L, Sun X-F, Xia L-L, Bao L-J, Liu L-Y, Zeng EY (2020) Global riverine plastic outflows. Environ Sci Technol 54:10049–10056

Monastersky R (2015) The human age. Nature 519:144–147

Sánchez-Bayo F, Wyckhuys KA (2019) Worldwide decline of the entomofauna: a review of its drivers. Biol Conserv 232:8–27

Sud M (2020) Managing the biodiversity impacts of fertiliser and pesticide use. Overview and insights from trends and policies across selected OECD countries. OECD Environment Working Papers No. 155, 60 pp

Syvitski J, Waters CN, Day J, Milliman JD, Summerhayes C, Steffen W, Zalasiewicz J, Cearreta A, Gałuszka A, Hajdas I, Head MJ, Leinfelder R, McNeill JR, Poirier C, Rose NL, Shotyk W, Wagreich M, Williams M (2020) Extraordinary human energy consumption and resultant geological impacts beginning around 1950 CE initiated the proposed Anthropocene Epoch. Commun Earth Environ 1:32. https://doi.org/10.1038/s43247-020-00029-y

Takeuchi I (2004) Anthropogenic chemicals, environmental policy and fisheries production. In: Takeuchi I, Tanabe S, Hino A (eds) Fisheries Series 140. Biological monitoring for anthropogenic chemicals in low doses. Koseisha-Koseikaku, Tokyo, pp 9–23. (In Japanese)

The State of Queensland (2009) Reef water quality protection plan 2009. In: For the Great Barrier Reef World Heritage area and adjacent catchments. The Reef Water Quality Protection Plan Secretariat, Brisbane, QLD, 30 pp

Wang Z, Walker GW, Muir DCG, Nagatani-Yoshida K (2020) Toward a global understanding of chemical pollution: a first comprehensive analysis of national and regional chemical inventories. Environ Sci Technol 54:2575–2584

West PC, Gerber JS, Engstrom PM, Mueller ND, Brauman KA, Carlson KM, Cassidy ES, Johnston M, MacDonald GK, Ray DK, Siebert S (2014) Leverage points for improving global food security and the environment. Science 345:325–328

Wiesinger H, Wang Z, Hellweg S (2021) Deep dive into plastic monomers, additives, and processing aids. Environ Sci Technol 55: 9339–9351

Williams GJ, Graham NA (2019) Rethinking coral reef functional futures. Funct Ecol 33:942–947

Williams GJ, Graham NA, Jouffray JB, Norström AV, Nyström M, Gove JM, Heenan A, Wedding LM (2019) Coral reef ecology in the Anthropocene. Funct Ecol 33:1014–1022

Ya M, Wu Y, Wu S, Li Y, Mu J, Fang C, Yan J, Zhao Y, Qian R, Lin X, Wang X (2019) Impacts of seasonal variation on organochlorine pesticides in the East China Sea and northern South China Sea. Environ Sci Technol 53:13088–13097

Transitional Coral Ecosystem of Taiwan in the Era of Changing Climate

2

Chao-Yang Kuo, Shashank Keshavmurthy, Ya-Yi Huang, Ming-Jay Ho, Hernyi Justin Hsieh, An-Tzi Hsiao, Wei-Cheng Lo, Yi-Chia Hsin, and Chaolun Allen Chen

Abstract

The transitional coral ecosystem of Taiwan features tropical and temperate scleractinian coral species and suboptimal environmental conditions. As the largest continental island in the region, Taiwan is proposed as a steppingstone for range expansion to higher latitudes of East Asia. In this chapter, we synthesize the literature and highlight the importance of the transitional role of Taiwanese coral ecosystem in sustaining its high-latitudinal counterparts in this era of changing climate. The boundary line separating tropical coral reefs and subtropical non-reefal coral communities stretches from south Penghu towards the Sandiao Cape in northeastern Taiwan. Between 1948 and 2020, the average seawater warming trend around Taiwan was 1.58 °C. This trend was not homogeneous; the region with non-reefal coral communities experienced a higher warming rate owing to gradually increasing winter sea surface temperatures. However, studies on the effect of typhoons, ocean acidification, and bioerosion on this transitional coral ecosystem are limited. To comprehensively understand how coral ecosystems in Taiwan respond to climate change, we recommended four research directions: (1) biodiversity formation, (2) the role of Kuroshio Current in reef formation and in connecting coral ecosystems of East Asia, (3) long-term ecological research of these coral ecosystems, and (4) the conservation and governance of transitional coral ecosystems using novel socio-ecological paradigms.

Chao-Yang Kuo, Shashank Keshavmurthy and Ya-Yi Huang contributed equally with all other contributors.

C.-Y. Kuo · S. Keshavmurthy · Y.-Y. Huang
Biodiversity Research Centre, Academia Sinica, Taipei, Taiwan

M.-J. Ho
Biodiversity Research Centre, Academia Sinica, Taipei, Taiwan

Green Island Marine Research Station, Marine Science Thematic Centre, Biodiversity Research Centre, Academia Sinica, Taitung, Taiwan

H. J. Hsieh
Penghu Marine Biological Research Centre, Taiwan Fishery Research Institute, Magong, Penghu, Taiwan

A.-T. Hsiao
Biodiversity Research Centre, Academia Sinica, Taipei, Taiwan

Institute of Bioinformatics and System Biology, National Yang Ming-Chiao Tung University, Hsinchu, Taiwan

W.-C. Lo
Institute of Bioinformatics and System Biology, National Yang Ming-Chiao Tung University, Hsinchu, Taiwan

Y.-C. Hsin
Research Center for Environmental Changes, Academia Sinica, Taipei, Taiwan
e-mail: ychsin@gate.sinica.edu.tw

C. A. Chen (✉)
Biodiversity Research Centre, Academia Sinica, Taipei, Taiwan

Department of Life Science, National Taiwan Normal University, Taipei, Taiwan

Department of Life Science, Tunghai University, Taichung, Taiwan
e-mail: cac@gate.sinica.edu.tw

Keywords

Marginal reefs · Biogeography · Poleward migration · Cold dome · Upwelling

2.1 Importance of Coral Ecosystem and Threats of Climate Change

2.1.1 Scleractinian Corals and Coral Reef Ecosystems

Scleractinian corals (hereafter referred to as corals), belonging to the class Anthozoa of phylum Cnidaria, are a group of marine invertebrates that secrete a calcium carbonate skeleton over time to create a three-dimensional reef structure. In warm (shallow tropical/subtropical) waters, corals host endosymbiotic dinoflagellates of the family Symbiodiniaceae

I. Takeuchi, H. Yamashiro (eds.), *Coral Reefs of Eastern Asia under Anthropogenic Impacts*, Coral Reefs of the World 17,
https://doi.org/10.1007/978-3-031-27560-9_2

(also known as zooxanthellae) (LaJeunesse et al. 2018) that fuel coral growth and calcification through carbon fixation by photosynthesis (Goreau 1959; Muscatine 1990), and the oxygen and glycerol produced as by-products of photosynthesis can be used by coral for calcification directly and may promote maximum coral calcification rates (Colombo-Pallotta et al. 2010). With the capability of accelerating calcification, warm-water corals can grow faster, not only to overcome the rate of bioerosion and physical erosion created by strong waves induced by severe weather events (e.g., typhoons) but also to establish the largest biological entities on Earth. Although these entities occupy less than 0.1% of the ocean floor, the ecosystems built by corals provide habitat for at least a quarter of known marine species, with many coral reef species still to be discovered (Fisher et al. 2015). Overall, it has been estimated that roughly 1–9 million species live in and around warm-water coral ecosystems (Snelgrove 2010).

The services and economic values generated by the high ecosystem biodiversity built by corals—food, income, recreation, coastal protection, and cultural settings—have a direct impact on over 500 million people in at least 90 countries worldwide (Burke et al. 2011; Gattuso et al. 2014; Hoegh-Guldberg 2015; Hoegh-Guldberg et al. 2017). The global asset value of coral ecosystems is estimated to be close to $1 trillion (Hoegh-Guldberg 2015) with the economic value of goods and services from coral reefs exceeding $375 billion annually (Burke et al. 2011; Gattuso et al. 2014). The latest analysis of the coral reef economy found that a healthy coral reef is expected to deliver additional economic benefits amounting to $34.6 billion and $36.7 billion between 2017 and 2030 in the Mesoamerica reef and the Coral Triangle, respectively (UN Environment et al. 2018).

2.1.2 Threats to the Coral Reef Ecosystems

Despite their biodiversity, productivity, and economic value to human society, coral ecosystems are heavily affected by four major threats: overfishing, habitat destruction, pollution, and climate change (Jackson 2001; Hall-Spencer et al. 2002; Hughes et al. 2003, 2018; Bellwood et al. 2004; Hoegh-Guldberg et al. 2007, 2017; Burke et al. 2011; Hoegh-Guldberg 2015). While overfishing, habitat destruction, and pollution have caused a decline in coral ecosystems locally, rising ocean temperature and changes in seawater chemistry (i.e., acidification) caused by excess greenhouse gas emissions have added a synergistic effect on ecosystem degradation at the global scale (Hoegh-Guldberg et al. 2007, 2014, 2017). The latest Intergovernmental Panel for Climate Change (IPCC 2021) special report on global warming clearly pointed out that coral ecosystems would decline by 70–90% with global warming of 1.5 °C (compared to the pre-industrial level), whereas nearly all (>99%) would be lost with 2 °C warming by 2050 (IPCC 2018, 2021).

2.1.3 Research into the Identification of Potential Resilience Reefs for the Future

With the climate crisis challenging the future survival of coral ecosystems, global efforts are attempting to develop conservation plans to select and save the coral reefs from these threats (Beyer et al. 2018; Hoegh-Guldberg et al. 2018; Anthony et al. 2020). By exploring five climate and connectivity-related variables, namely, (1) historical thermal conditions, (2) predicted future thermal conditions, (3) larval connectivity and settlement, (4) cyclone wave damage, and (5) recent global bleaching events, Beyer et al. (2018) applied modern portfolio theory to identify global coral reef locations that have a better chance to survive impacts from projected climate change compared to other reefs. The results show that the regions selected for global conservation priority encompass the tropics, including portions of central and western Southeast Asia (the Philippines, Borneo, Indonesia), Australia's Great Barrier Reef, French Polynesia, East Africa, the Red Sea, and the Caribbean. Other regions, such as the East Asia coral ecosystems stretching from south China, Taiwan, and Okinawa to high-latitude Japan that have high ecological and social value were not selected because of projected higher levels of disturbances and impacts, such as intensified typhoons. Nevertheless, it is also recommended that these excluded regions be examined for local variations in projected impacts to enable a finer-scale conservation effort (Beyer et al. 2018).

2.1.4 The Aim of This Chapter

Here, we review the research updates on coral ecosystems in Taiwan, the largest continental island in East Asia. The "coral ecosystem" in this chapter refers to the shallow-water ecosystem surrounding Taiwan, including euphotic (0–30 m) and mesophotic (30–150 m) zones, where scleractinian corals are relatively dominant. Although 108 species of azooxanthellate scleractinian corals, also known as ahermatypic corals, have been recorded in the deeper waters of Taiwan (Dai and Cheng 2020), little is known about their ecology. Therefore, they were not included in this review. Two reviews have been published which focused on the biogeographic pattern of coral assemblages around Taiwan Island and adjacent islets (Chen 1999), and the role of Taiwan as the stepping-stone for coral ecosystems in East Asia, connecting the upstream Philippine corals with downstream coral communities in Okinawa and further to high-latitudinal coral communities in Shikoku within the Kuroshio Triangle (Chen and Keshavmurthy 2009). In this chapter, we synthesize the literature, including geological, environmental, biodiversity, and ecological aspects, and highlight the importance of the transitional role of Taiwan's coral ecosystems in sustaining their high-latitudinal counterparts in East Asia in the era of changing climate.

2.2 Geological and Marine Environmental Settings of Taiwan, Including Surrounding Islands and Islets

2.2.1 Geological Setting of Taiwan

Taiwan is the largest continental island (32,260 km^2) in East Asia, separated from the southeast coast of China by the Taiwan Strait (TWS), with an average width of 180 km and a depth of less than 100 m (Fig. 2.1). Taiwan lies between the East China Sea (ECS) in the north and the South China Sea (SCS) in the south, with its east coast facing the Pacific Ocean, bordered by steep slopes that reach depths of 1000 m within a few kilometers from the shore (Denis et al. 2019). In addition, the complicated geological settings and evolutionary history of the area between northeast Taiwan and the southern Ryukyu Archipelago, including the Yonaguni Depression, East Taiwan Channel, and Okinawa Trough (Figs. 2.2 and 2.3), play an important role in the routes and strength of the Kuroshio Current (reviewed in Gallagher et al. 2015) which strongly affects marine environmental settings and biodiversity, not only around Taiwan but throughout East Asia.

One hundred and twelve small islands and islets are distributed around Taiwan Island with 90 of them forming an archipelago known as Penghu Archipelago (Pescadores, 澎湖群島) in the Taiwan Strait, 50 km west of the main island (Fig. 2.2). In the north of Taiwan, Keelung Islet (基隆嶼), the so-called "northern three islets," namely, Pinnacle (Huaping yu, 花瓶嶼), Crag (Miahua yu, 棉花嶼), and Agincourt (Pengjia yu, 彭佳嶼) islets as well as Turtle Island (Guieshan dao., 龜山島) are located between 4 km and 56 km north and north-east of the main island at the edge of the continental shelf of the ECS (Figs. 2.2 and 2.3). The tiny islet of Liuqiu (Xiaoliuqiu, 小琉球) off the southwest coast, and Orchid (Lan yu, 蘭嶼) and Green (Lyu dao, 綠島) Islands to the south-east, as well as the Vele Rete Rocks (Qixing yan, 七星岩) at the southernmost border off the Philippines near the Bashi Channel, are located south of the east coast and southern Taiwan (Fig. 2.2). All islands and islets between latitudes 21.757°N and 25.631°N support both tropical coral reefs and subtropical non-reefal coral communities (Fig. 2.3) (Chen 1999; Chen and Keshavmurthy 2009; Denis et al. 2019; Keshavmurthy et al. 2019; Dai and Cheng 2020). In addition to the presence of a 27-km long biotic reef formed by crustose coralline algae (CCA) from the

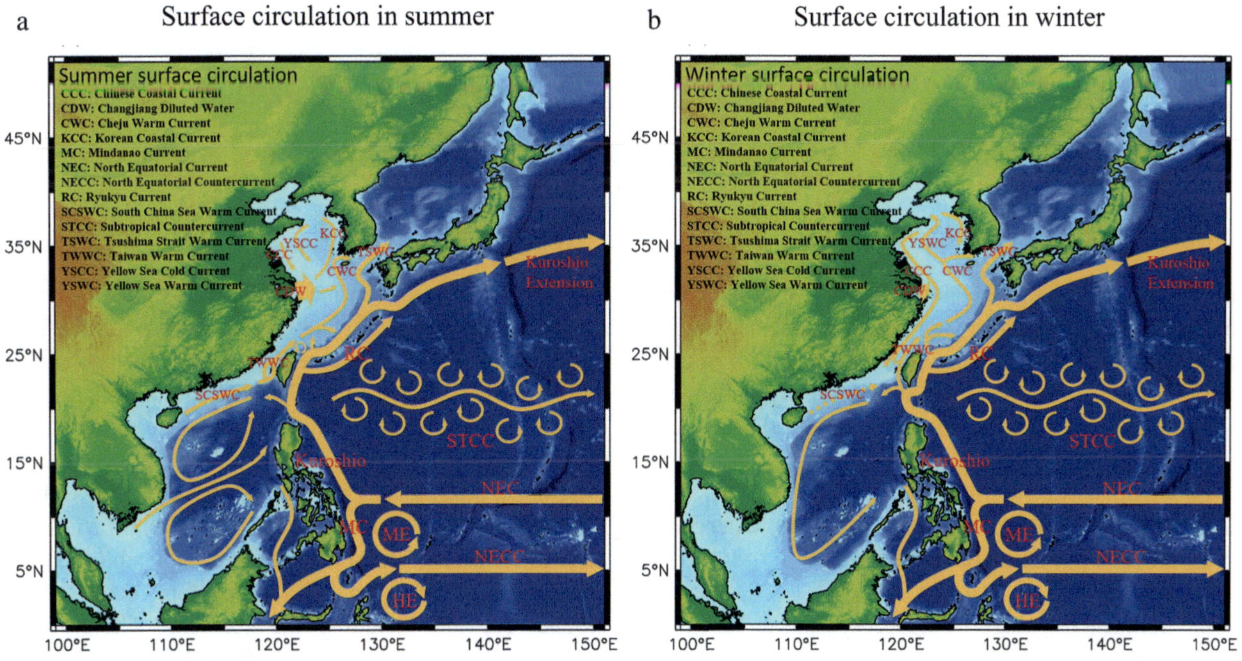

Fig. 2.1 Geological setting in East Asia along the Kuroshio Current showing all the major and minor surface current circulation paths in summer (**a**) and winter (**b**) seasons

Fig. 2.2 Types of reefs around Taiwan Island and associated small islands and islets. The three types of reefs include tropical coral reefs (orange), non-reefal coral assemblages (green), and algal reefs (pink). The horizontal black dotted line indicates the Tropic of Cancer. The red dot indicates the "*Chen's line,*" a biogeographic line that marks the boundary of the distribution of tropical coral reefs and non-reefal coral communities, as proposed in this chapter

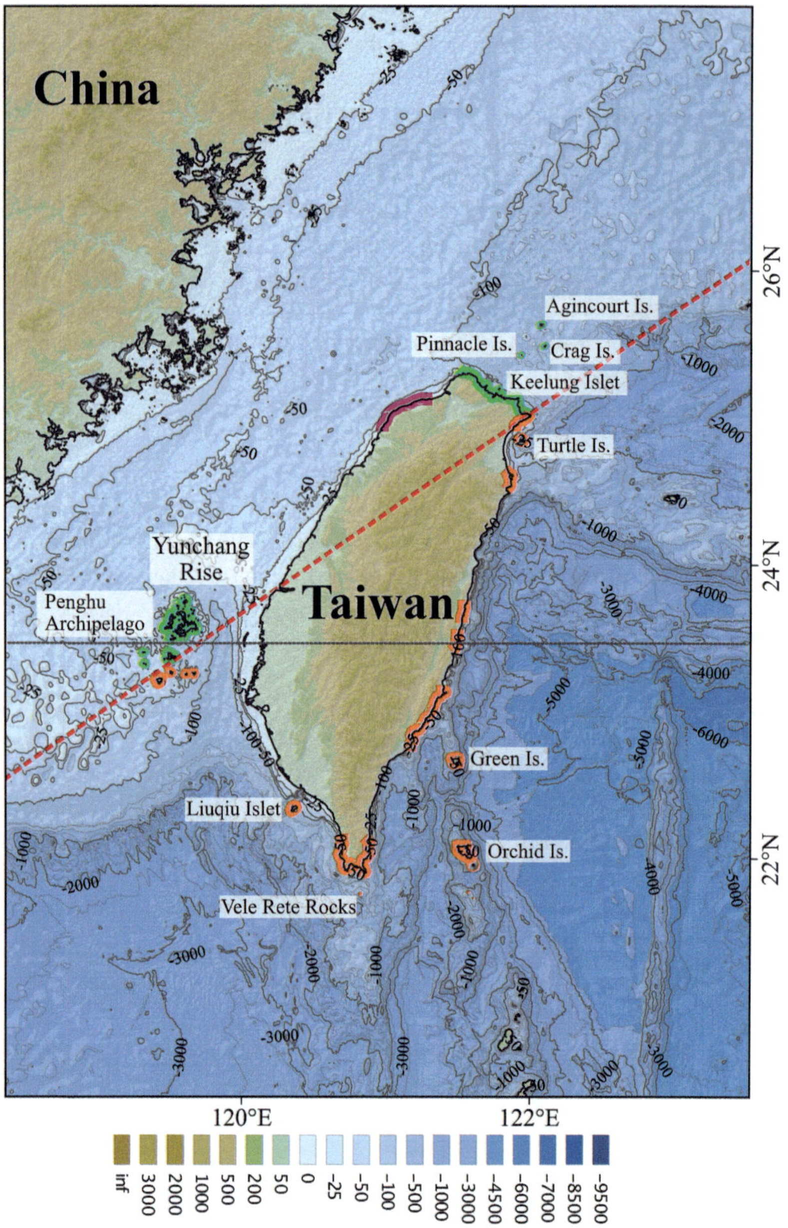

intertidal zone to 10 m deep along the northwestern coast of Taiwan (Figs. 2.3 and 2.4), it provides a unique habitat for a handful of corals, including the caryophyllid species, *Polycyathus chaishanensis,* which is critically endangered in Taiwan (Liou et al. 2017; Kuo et al. 2019, 2020).

2.2.2 Marine Environmental Settings: Currents and Upwelling/Cold Dome

The environmental settings of coral ecosystems in Taiwan are strongly influenced by complex topography, monsoons, sea surface currents and their interactions (Chen 1999; Chen and Keshavmurthy 2009; Dai and Cheng 2020). With the Tropic of Cancer passing through the Penghu Archipelago and central part of Taiwan (Fig. 2.2, black horizontal line), the climate of Taiwan is significantly influenced by the East Asian monsoon system. Winds over Taiwan are dominated by stronger northeasterly cold and dry winds during October–March and weaker southwesterly hot and humid winds from June to August, with April–May and September being the transitional periods (Jan 2018). Therefore, ocean currents around Taiwan, including the Kuroshio Current (KC), South China Sea Warm Current (SCSWC), Taiwan Warm

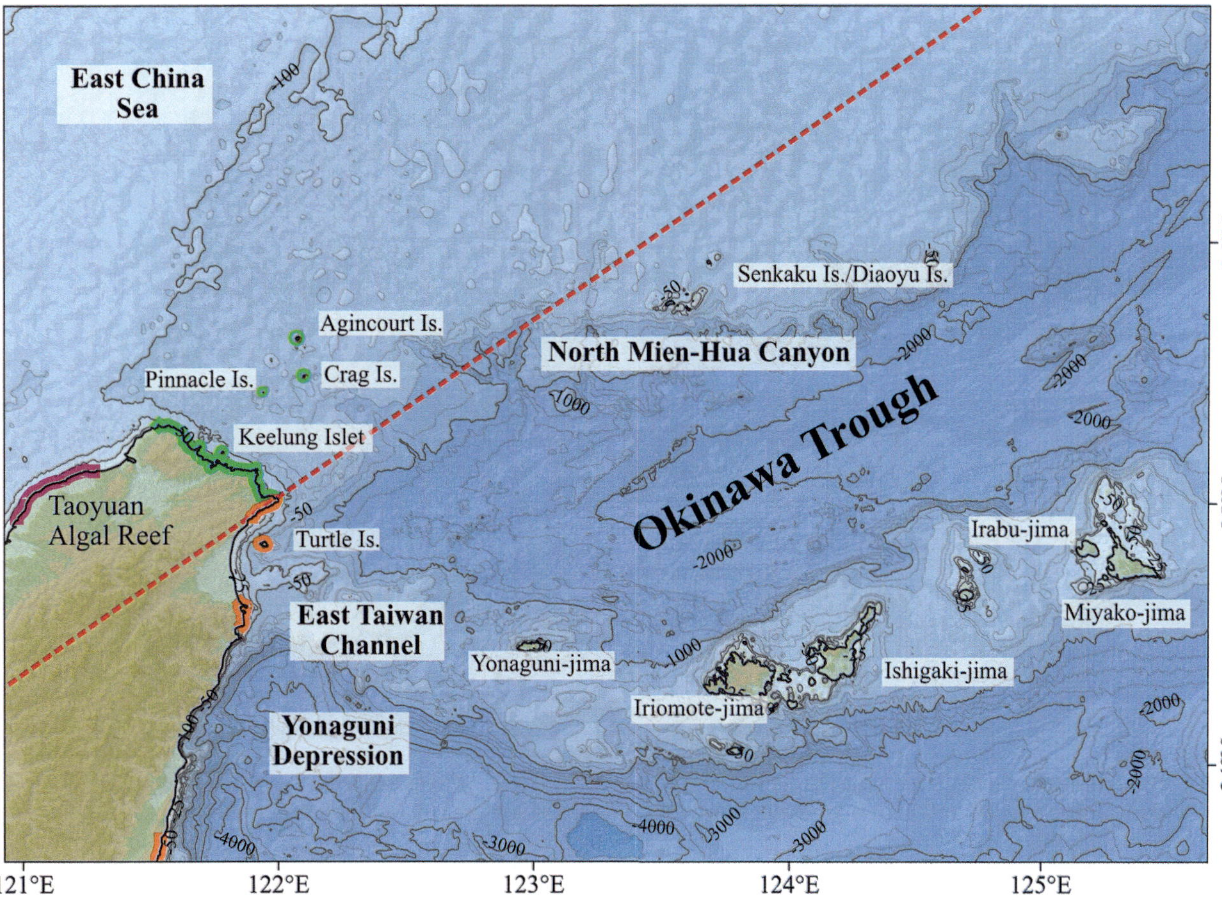

Fig. 2.3 Non-reefal (in northern Taiwan) and subtropical coral reefs (lower islands of Ryukyu Archipelago including Yonaguni Depression, East Taiwan Channel, and Okinawa Trough). Three types of reefs include tropical coral reefs (orange), non-reefal coral assemblages (green), and algal reefs (pink). The red dot indicates the "*Chen's line*," a biogeographic line marking the boundary of the distribution of tropical coral reefs and non-reefal coral communities, proposed in this chapter

Current (TWC), and China Coastal Current (CCC) (Fig. 2.1), are modulated by the topography in the vicinity of Taiwan and interactions with changing wind fields. Consequently, ocean currents show strong seasonal variations (Jan et al. 2002, 2011; Wu and Hsin 2005; Hsin et al. 2008, 2011, 2012; Jan 2018).

The KC, originating from the North Equatorial Current (NEC) in the Pacific, is the predominant western boundary current affecting the climate, hydrography, biogeochemistry, marine biodiversity, and fisheries around Taiwan (Hsin 2015; Jan 2018). As the NEC approaches the east coast of Luzon-Philippines, it splits into two branches: one turns southwards to form the Mindanao Current and the other continues northwards as the mainstream of the KC (Fig. 2.1). The KC transports significant amounts of heat, salt, and mass from the tropical ocean to high latitudes and plays an important role in the global climate and heat balance (Qu and Lukas 2003). The annual mean surface sea temperature (SST) of the KC is 24.5 °C with an estimated primary production between 150 and 300 g C m^{-2} $year^{-1}$. The KC passes through Orchid Island and Green Island off the east coast of Taiwan, across the Yonaguni Depression via the East Taiwan Channel into the East China Sea, and follows the edge of the continental shelf in the Okinawa Trough, the deep break between the ECS continental slope and the Ryukyu archipelago. It flows along the western part of the Okinawa Trough and leaves the continental slope via the Tokara Strait, reaching the southern island of Japan at ~30°N (Gallagher et al. 2015). Over the past two decades, intensive studies on the KC intrusion have shown the formation of a significant upwelling and a "cold dome" in the ECS continental shelf off the northeast coast of Taiwan (Figs. 2.5 and 2.6a, b) (Tang et al. 1999; Wu et al. 2008; Jan et al. 2011; Gopalakrishnan et al. 2013). In addition to the mainstream of the KC flowing northeastward along the ECS shelf break, a weaker northwestward branch intrudes onto the ECS shelf over the North Mien-Hua Canyon (MHC) and creates a counterclockwise circulation (i.e., cold dome), approximately 100 km in diameter, centered over the MHC. The cold dome has a long-term mean position centered at approximately 25.625°N, 122.125°E and has a lower

Fig. 2.4 Photo panel of different coral communities in Taiwan. (**a**) Tropical coral reefs; (**b**) Non-reefal coral community; (**c**) Algal reefs; and (**d**) Mesophotic corals. Photo credits: (**a**) Dr. Chia-Min Hsu; (**b** and **d**) Dr. Chao-Yang Kuo; and (**c**) Dr. Chaolun Allen Chen

temperature (<21 °C) and higher salinity (>34.5‰) than the surrounding shelf waters (Jan et al. 2011; Gopalakrishnan et al. 2013). The KC-driven upwelling off northeastern Taiwan continues to provide nutrient-rich waters for the ECS and makes the region highly biologically productive (Chen 1996).

In addition to the constant upwelling/cold dome off northeast Taiwan, year-round upwelling has also been observed at the southern Taiwan Bank off southwestern Taiwan (Figs. 2.5 and 2.6) (Hu and Wang 2016). Upwelling is mainly generated by topographical change from 100 to 200 m deep in the Penghu Channel to the shallow areas near the southern Penghu Archipelago close to the Taiwan Bank (TB-upwelling) and around the northern Penghu Archipelago (PH-upwelling) close to the Yunchang Rise (YCR) (Jan et al. 2002; Hu et al. 2003; Hong et al. 2011). In summer (April–August), the SCSWC, driven by the southwesterly monsoon, intensifies northeastward surface waters and contributes to the water in the Penghu Channel. More nutrient-rich, cool bottom waters are brought up to the surface, creating a low-temperature core (2–4 °C lower than ambient water) and relatively high chlorophyll *a* (Figs. 2.5a, b and 2.6a, b) (Fan 1979; Jan et al. 2002; Tang et al. 2002; Kuo et al. 2006; Lan et al. 2009; Jan 2018). In winter, the KC off southern Taiwan intrudes into the northern SCS through the Luzon Strait as a Kuroshio branch (KB), and a part of the KB water flows northward through the Penghu Channel to replace the SCS waters, becoming the driver of upwelling in the southern TWS (Figs. 2.5c, d and 2.6c, d). These two upwelling areas (TB-upwelling and PH-upwelling) also contribute to the important fishing grounds (for herring and squid) around the Penghu Archipelago (Lu et al. 1987; Lan et al. 2009). Although the large-scale upwellings off northeastern Taiwan and the Penghu Archipelago play a major role in local fisheries, their role in shaping Taiwanese coral ecosystems is less understood. In contrast, small-scale, tide-induced upwelling was identified in Nanwan in the Kenting National Park (KNP), southern Taiwan (Lee et al. 1999a, b). This upwelling has been proved to contribute to the local thermal resistance of coral reef ecosystems in the KNP (reviewed in Keshavmurthy et al. 2019).

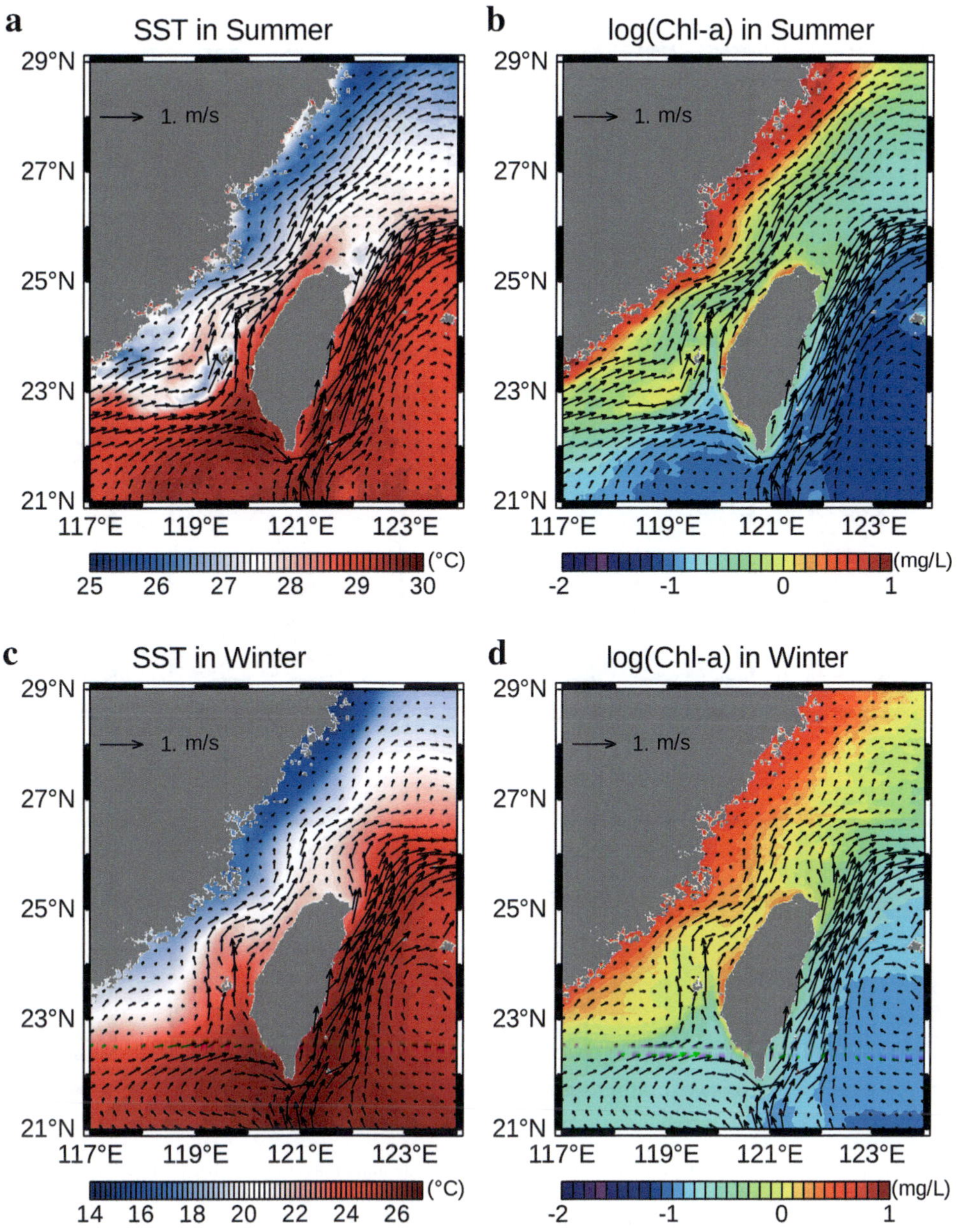

Fig. 2.5 Mean surface temperature (**a**, **c**) and chlorophyll *a* (**b**, **d**) distribution with current velocity (black arrows) in Taiwan waters in summer (**a**, **b**) and winter (**c**, **d**). Cooler temperature over northern Taiwan and Penghu Archipelago shows the presence of "cold dome/upwelling" in these regions

2.2.3 Marine Environmental Settings: Monsoons and Sea Surface Currents

Seasonal changes in monsoons and sea surface currents with different temperatures and water chemistry might also confine coral ecosystem development in the TWS, except on the west coast of the main island of Taiwan, where over 300 km of coastline is composed of estuaries, mangroves, and sandy/muddy habitats (Figs. 2.3 and 2.4). In summer, less saline SCS water intrudes northward into the TWS. On the eastern side of the TWS, the northeastward mean current is a year-round feature (Guan 1994; Huang et al. 1994; Jan and Chao 2003), connecting the SCSWC to the south and the TWC to the north of the strait (Figs. 2.1a, 2.3, and 2.5a, b). In spring, the KB water enters the Penghu Channel and escorts the edge of the YCR towards the northern TWS (Figs. 2.1b, 2.3, and 2.5c, d). This large-scale movement forces waters in the northern SCS to flow northward and enter the ECS through the TWS. The northeastward mean current splits into upper and bottom flows in summer and fall, with the upper flow

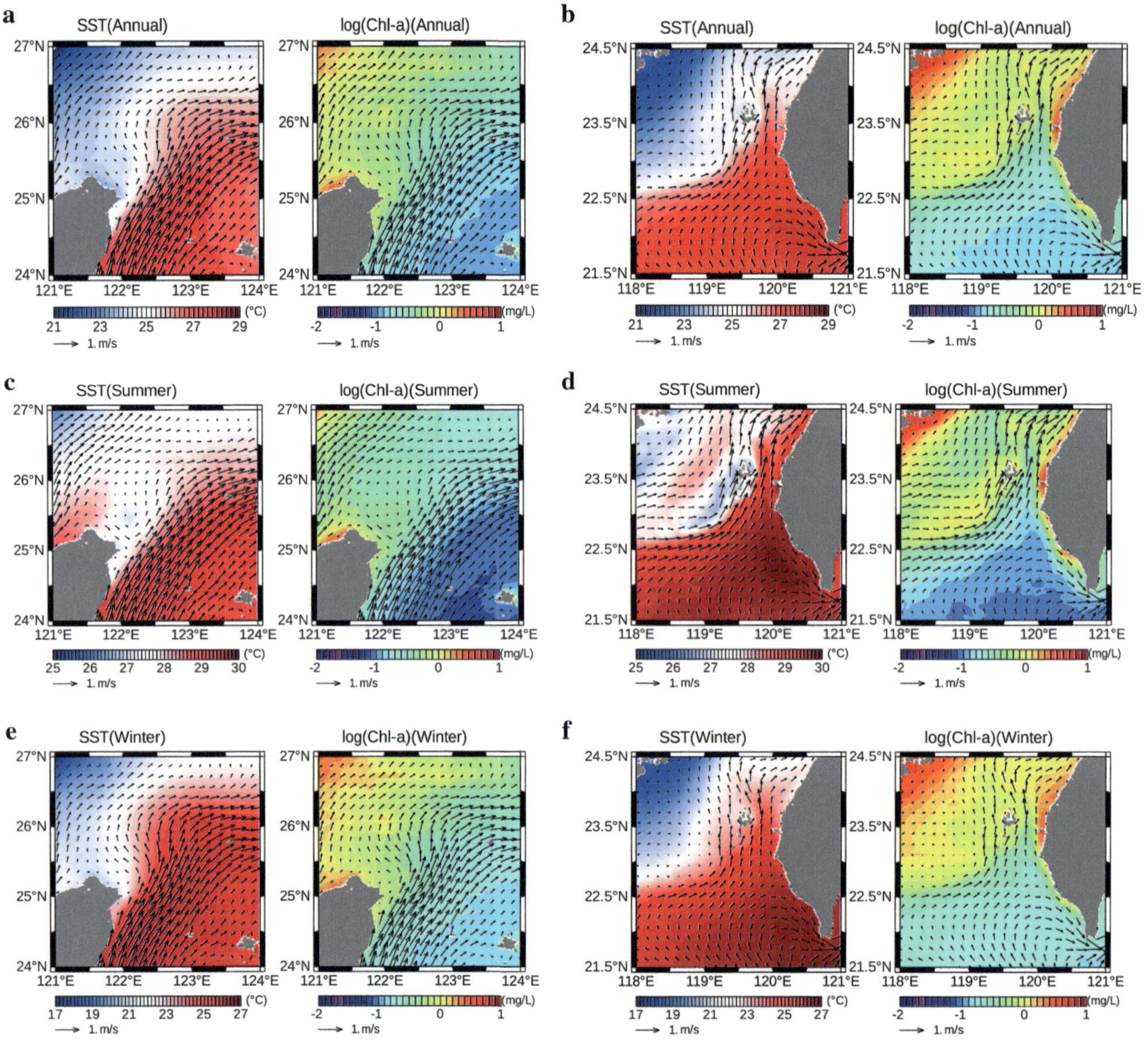

Fig. 2.6 Mean surface temperature and chlorophyll *a* distribution with current velocity (black arrows) off the northeast coast of Taiwan (**a**, **c**, **e**) and Southern Penghu (**b**, **d**, **f**). Three time periods are shown, which include annual average (**a**, **b**), summer (**c**, **d**), and winter (**e**, **f**) seasons. Cooler temperature over northern Taiwan and Penghu Archipelago shows the presence of "cold dome/upwelling" in these regions

crossing the YCR and flowing northward along the west coast of Taiwan and the bottom flow moving through the edge of the YCR at the 50 to 100 m isobaths (Fig. 2.5a, b). In late fall and winter, the cooler CCC is driven by the northeasterly monsoon and emerges in the northern TWS. The intruding CCC makes a clockwise U-turn due to the topographic blocking of the YCR and forms a cyclonic cold eddy over the deep Guanyin Depression (Fig. 2.5c, d). During this winter period, the high-salinity KB water replaces the less saline SCS water in the Penghu Channel (Jan et al. 2010). However, the current in the Penghu Channel flows consistently northward, with only rare occasions of southward flow when the winter monsoon is in full force for a prolonged period (Jan and Chao 2003; Jan 2018). Low sea surface temperatures (<18 °C) and waves caused by strong winds in the northern TWS limit coral reef development in the northern Penghu Archipelago and the northwest coast of Taiwan (Dai 2010; Dai and Cheng 2020). Interestingly, these temperatures favor CCA growth and the development of algal reefs in the sandy/shallow water on the Taoyuan coast (Dai et al. 2009; Liou et al. 2017; Kuo et al. 2020).

2.3 Biodiversity and Biogeographic Distribution of Scleractinian Corals in Taiwan: Patterns and Possible Mechanisms

2.3.1 Biodiversity of Scleractinian Corals in Taiwan

Located at the junction of the Pacific Ocean and two marginal seas, the East China Sea and the South China Sea, Taiwan and the adjacent islands and islets host a relatively high diversity of scleractinian corals and associated marine organisms (Dai 1997, 2010; Denis et al. 2019; Dai and Cheng 2020). The latest checklist of scleractinian species, based on the new systematic scheme (WoRMS Editorial Board 2021), has identified 558 species in Taiwan, with 450 recorded from the shallow-water coral ecosystems (Denis et al. 2019; Dai and Cheng 2020; Chung and Shao 2021; De Palmas et al. 2021a), including two endangered species, *Polycyathus chaishanensis* (Lin et al. 2012) and *Pseudosiderastrea formosa* (Pichon et al. 2012) (Fig. 2.7a, b).

2.3.2 Biogeographic Distribution Patterns of Scleractinian Corals in Taiwan and Japan

Although the distance between northern and southern Taiwan is less than 450 km, the biogeographic distribution of scleractinian corals in Taiwan still follows the pattern of the south-north gradient, in which species decrease towards the north latitudes (Table 2.1, Fig. 2.7). This trend is consistent with the biogeographic distribution pattern of scleractinian corals in Japan, which stretches over 2000 km from the Ryukyu Archipelago to the northern latitudes of mainland Japan (Nishihira and Veron 1995). Three hundred and thirty-nine scleractinian species were recorded in southern Taiwan (21.95°N), which is close to those of the Yaeyama Archipelago ($n = 380$) and Okinawa Islands ($n = 340$), occurring at higher latitudes in Japan between 24 and 26°N (Nishihira and Veron 1995). In Liuqiu Islet (22.34°N), Orchid Island (22.05°N), Green Island (22.66°N), and the east coast of Taiwan (22.73–23.93°N), the number of scleractinian corals ranged between 210 and 250, which corresponds to the

Fig. 2.7 Photo panel of corals including: (**a**) *Polycyathus chaishanensis* (from Datan algal reef); (**b**) *Pseudosiderastrea formosa* (from Wanlitong, Kenting National Park); (**c**) *A. japonica* (from Gupoyu, Northern Penghu Archipelago); (**d**) *A. efflorescens* (from Waimushan, Northern Taiwan); (**e**) *A. pruinosa* (from Shetoushan, central Penghu Archipelago; (**f**) *A. solitaryensis* (from Waimushan, Northern Taiwan); (**g**) *A. tumida* (from Waimushan, Northern Taiwan); (**h**) *A. hyacinthus* (Chinwan Inner Bay, central Penghu Archipelago); (**i**) *A. murciata* (Chinwan Inner Bay, central Penghu Archipelago). Photo credits: (**a**) Ms. Ai-Chi Chung; (**b**–**i**) Dr. Chaolun Allen Chen

Table 2.1 Updated number of scleractinian species recorded from tropical reefs and non-reefal coral communities around the Taiwanese juristic districts

Community type/location	Latitude (N)	No. species	Reference
Tropical reef			
Southern Taiwan	21°57′	280[a]/339[b]	[a]Dai (2007), [b]Chen (2020)
Eastern Taiwan	22°44′-23°56′	210[a]/	[a]Dai (2006)
Green island	22°30′	250[a]/	[a]Dai (2006)
Orchis Island	22°03′	248[a]/	[a]Dai (2006)
Liuqiu Islet	22°20′	239[a]/	[a]Dai (2007)
[c]Dongsha Atoll (Pratas)	20°42′	229[a]/ 257[a]	[a]Jeng et al. (2008), [a]Dai et al. (2013)
[c]Taiping Island (Itu Aba)	10°22′	197[a]/306[b]	[a]Shao et al. (2014), [b]Jeng et al. (2017)
Non-reefal community			
North + northeastern Taiwan	25°10′	136[a]/112[a]	[a]Dai et al. (2004), [a]Dai et al. (2008)
Penghu			
Northern Penghu	23°44′	107[a]/	[a]Hsieh (2008), [a]Hsieh et al. (2016)
Eastern Penghu	23°34′	84[a]/	[a]Hsieh (2008), [a]Hsieh et al. (2016)
Inner Penghu	23°30′	75[a]/	[a]Hsieh (2008), [a]Hsieh et al. (2016)
Southern Penghu	23°12′	99[a]/140[b]	[a]Hsieh (2008), [a]Hsieh et al. (2016), [b]Jeng and Dai (2020)

[a]°Survey data after 2000
[b]°Re-survey and reassemble historical data after 2017
[c]°Not included in this study

Amami Archipelago ($n = 220$, 27.70°N). On the northern and northeastern coasts of Taiwan and adjacent islets (25.16° N), the number of scleractinian corals ($n = 136$) is only about half that of the Okinawa Islands at a similar latitude, but similar to the areas between Tanegashima Island (30.62°N) and Shikoku Island (33.73°N) where 150–200 scleractinian species have been recorded (Nishihira and Veron 1995; Veron 2000). In the Penghu Archipelago (Taiwan Strait), a distinct break was observed between southern Penghu ($n = 140$) and the rest of Penghu Archipelago, including the north ($n = 107$), east ($n = 84$), and Inner Penghu ($n = 75$). Interestingly, the latter three subregions of the Penghu Archipelago have similar numbers of scleractinian species found in areas between Kyushu Island (32.60°N) and the Kii Peninsula (33.77°N) in central Japan.

2.3.3 The Forming of "Chen's Line"

Combining the data of Taiwan and Japan, we propose that a biogeographic line (hereafter named "*Chen's Line*") be drawn for scleractinian distribution in East Asia, stretching from south Penghu towards Sandiao Cape at the northeast corner of Taiwan, then extending towards Japan (see Figs. 2.2 and 2.3 for the vertical line drawn across Taiwan and extending towards Japan), encompassing the boundary between Tanegashima Island an Shikoku Island in mainland Japan (see Chap. 4 by Keshavmurthy et al. 2023 in this book for the location of Shikoku Island). This line is not only represented by the number of species but also reflects species composition. Although the proposed localities along this line are subject to further research, the line is certainly a transitional marker to separate tropical coral reefs and non-reefal coral communities in East Asia (Veron 1995, 2000; Chen and Keshavmurthy 2009). Undoubtedly, tropical coral reefs and non-reefal coral communities on both sides of Taiwan provide an ideal model for examining how coral ecosystems located in the transitional zones of East Asia respond to the impact of climate change.

Chen (1999) first examined the biogeographic patterns of scleractinian assemblages collected before 1997 for two major scleractinian groups, the genus *Acropora* and the family Merulinidae (formerly known as Faviidae in the Pacific and Indian Oceans). The results suggest two distinct provinces of scleractinian distribution following the northeast-southwest line across Taiwan (Fig. 2.3). One contains tropical reefs in southeast and southwest Taiwan, and the other contains non-reefal coral communities in north-northeast Taiwan and the Penghu Archipelago. Using hierarchical cluster analysis, reanalysis of the updated data recording the presence and absence of 69 *Acropora* and *Isopora* species surveyed in the last 20 years, showed that seven clusters best explain the species distribution (Fig. 2.8a). In addition, using multivariable statistics, non-metric multidimensional scaling (nMDS), analysis of the presence and absence of species, and association with community types, showed that over 90% of species have a tropical reef affiliation mainly found in southeast and southwest Taiwan (Fig. 2.8b). One cluster composed of non-reefal species was found in the north-northeast Taiwan and north-inner Penghu Archipelago, including *A. efflorescens, A. glauca, A. pruinosa, A. solitaryensis,* and *A. tumida* (Fig. 2.7c, g). These species, together with other cosmopolitan species, such as *A. hyacinthus* and *A. muricata* (Fig. 2.7h, i), form the coral community, which is very similar to those at higher latitudes (>30°N) in Japan (Veron 1992; Nishihira and Veron 1995; Sugihara et al. 2015; Nomura 2016a, b; also see Chap. 4 by Keshavmurthy et al. 2023 in this book). Based on this similarity, Chen and Keshavmurthy (2009) hypothesized that Taiwan could serve as a connective stepping stone for high-latitudinal coral ecosystems in East Asia.

2.3.4 The Mesophotic Coral Ecosystems (MCEs) in Taiwan

In addition to coral ecosystems in shallow water (<30 m), recent studies have begun to reveal coral diversity in mesophotic coral ecosystems (MCEs) in Taiwan (Denis et al. 2015, 2019; De Palmas et al. 2018, 2021a, 2021b; Soto et al. 2018, 2021; Ho et al. 2020). Denis et al. (2019) reviewed the environmental settings and habitat availability

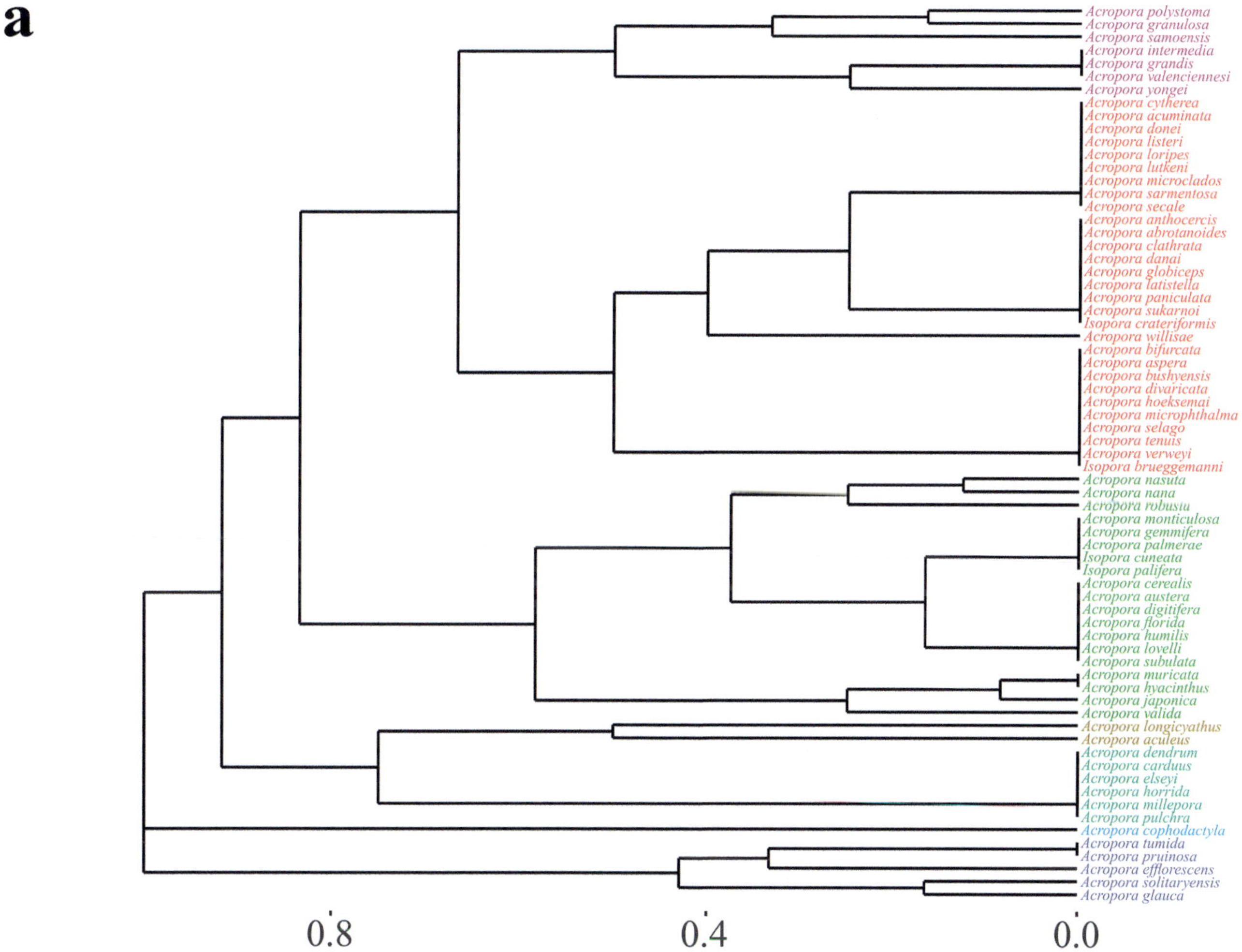

Fig. 2.8 Seven clusters of the distribution characteristics among *Acropora* and *Isopora* corals in Taiwan (**a**) and the non-metric multidimensional scaling (nMDS) (stress = 0.01) of the similarity of reefs in terms of the occurrence of *Acropora* sp. and *Isopora* sp. (**b**). The seven colors used in (**a**) and (**b**) represent the seven clusters. Hierarchical cluster analysis and the nMDS were performed using the *stats* (v. 4.0.3) (R Core Team 2020) and *vegen* (v. 2.5.7) (Oksanen et al. 2020) package in R-4.0.3 (R Core Team 2020)

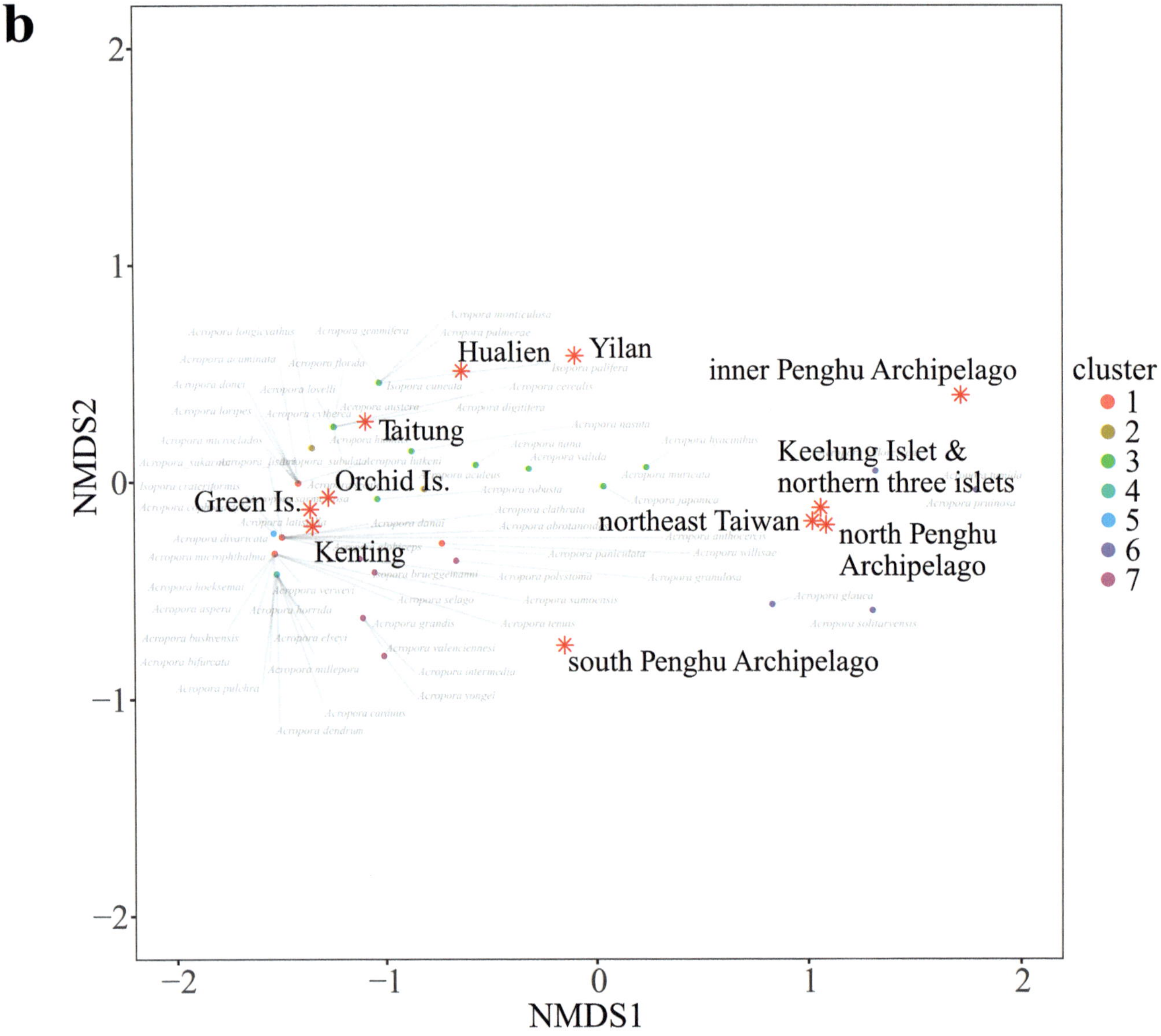

Fig. 2.8 (continued)

around Taiwan and suggested that, while the steep-sloping east coast of Taiwan can support the potential development of MCEs, low-light conditions at mesophotic depths in the north restrict the distribution of photosynthetic organisms to shallow waters. On Green Island, 102 scleractinian coral species were recorded from 30 to 60 m, with ten newly recorded species: *A. russelli, Alveopora excelsa, Anacropora pillai, Catalaphyllia jardinei, Galaxea paucisepta, Leptoria irregularis, Leptoseris amitoriensis, L. papyracea, Psammocora stellata,* and *Trachyphyllia geoffroyi* (De Palmas et al. 2021a). Of all the species recorded, 37% were similar to their shallow-water counterparts, implying that the ecological refuge hypothesis of the mesophotic zone could apply to less than half of the scleractinian coral community on Green Island (De Palmas et al. 2021a). The strong vertical gene flow of *Pocillopora verrucosa* and sharing of common recruited coral spats between euphotic and mesophotic zones also supported the ecological refuge hypothesis applied to MCEs of Green Island (De Palmas et al. 2021b; Soto et al. 2021).

2.3.5 Mechanisms Forming the Two Distinct Coral Assemblages in Taiwan

The distinct coral assemblages around Taiwan and adjacent islands and islets might have been influenced by the combined effects of oceanographic currents, monsoons, and more importantly, seafloor topology. Located at the northern border of the Coral Triangle, Taiwan may constantly receive transportation of coral larvae from upstream countries, such as the northern Philippines, through the KC. While the main KC contributes to the formation of tropical coral reefs in southeast Taiwan, the branch of the KC intruding towards southwest Taiwan, bypassing the Luzon Strait, also brings coral larvae to Liuqiu Islet, south Penghu, and further north through the channel along the west coast of Taiwan (Dai 1997; Chen 1999; Chen and Keshavmurthy 2009; Dai and Cheng 2020). In addition, the SCSWC, driven by the southwest monsoon flows northward during summer, may bring larvae from the SCS into the Taiwan Strait and contribute to marine biodiversity in the Penghu Archipelago (Ablan et al. 1999, 2002; Chen et al. 2004; Ng et al. 2019). During winter, the fresh and cold CCC, driven by the northeast monsoon, emerges into the northern TWS and encounters the YCR, which creates a cold front boundary with winter SST < 18 °C in north Penghu. At the same time, the coral communities on the northeast coast and adjacent islets are exposed to strong waves and low winter SST by the northeast monsoon. These physical ocean constraints limit the survival of tropical species throughout winter, but enable those species that can acclimate to low temperatures to proliferate on Penghu Archipelago and north-northeast Taiwan, forming non-reefal coral communities (Dai and Cheng 2020). Because of the suboptimal environmental conditions, non-reefal coral communities is also known as marginal coral community (Perry and Larcombe 2003; Dai 2018; Keshavmurthy et al. 2021).

Apart from the seasonal changes in large-scale ocean physical and climatic conditions, the Kuroshio intrusion and the resulting upwelling in northeastern Taiwan and southern Penghu may also play a role in shaping coral assemblages in these areas. Large-scale upwelling bringing up cold, more acidic (pH > 7), and nutrient-rich subsurface waters impedes the development of shallow-water reef structures in Panama and Costa Rica within the East Tropical Pacific (ETP) (Glynn and Stewart 1973; Glynn and Macintyre 1977; Glynn and Ault 2000; Chollett et al. 2010; Sánchez-Noguera et al. 2018; Wizemann et al. 2018; Randall et al. 2020). On the Pacific coast of Panama, there is a seasonally strong, wind-driven upwelling in the Gulf of Panama in the east compared to the Gulf of Chiriquí in the west, where upwelling is weaker due to the orographic blocking (Cortés 1997; Poveda et al. 2006; Glynn et al. 2017). Therefore, the thermal regime in the Gulf of Panama is much cooler and more variable, resulting in slower coral growth and less long-term reef accretion than that of the Gulf of Chiriquí (Glynn 1977; Glynn and Macintyre 1977; Toth et al. 2012, 2015). In addition, in North Pacific Costa Rica, the seasonal upwelling system of the Gulf of Papagayo also plays a major role in shaping coral communities and reef accretion along the Papagayo coast towards 338 km south of the Nicaraguan border (Jiménez et al. 2010; Rixen et al. 2012; Sánchez-Noguera et al. 2018; Wizemann et al. 2018).

2.4 Natural and Anthropogenic Disturbances Affecting Coral Ecosystems in Taiwan

2.4.1 Natural Disturbances

Taiwan is located along the major routes of typhoons originating from the tropical northwest Pacific and South China Sea regions. Therefore, strong waves and heavy precipitation caused by typhoons are the major natural disturbances in coral ecosystems in Taiwan. A total of 370 typhoons between 1911 and 2020 made landfall or passed without making a landfall in Taiwan. Either way, these typhoons had an impact, with three to four typhoons per year making landfall and causing significant destruction along the coast (Wu and Kuo 1999; Hung et al. 2020). Overall, the paths of typhoons are divided into nine typhoon path categories by the Central Weather Bureau, Taiwan (Fig. 2.9). The typhoons of Category 1 (12.76%), Category 2 (13.28%), Category 3 (12.76%), Category 7 (6.77%), and Category 9 (6.77%) passed through the non-reefal coral communities in northeast Taiwan, adjacent islets, and Penghu

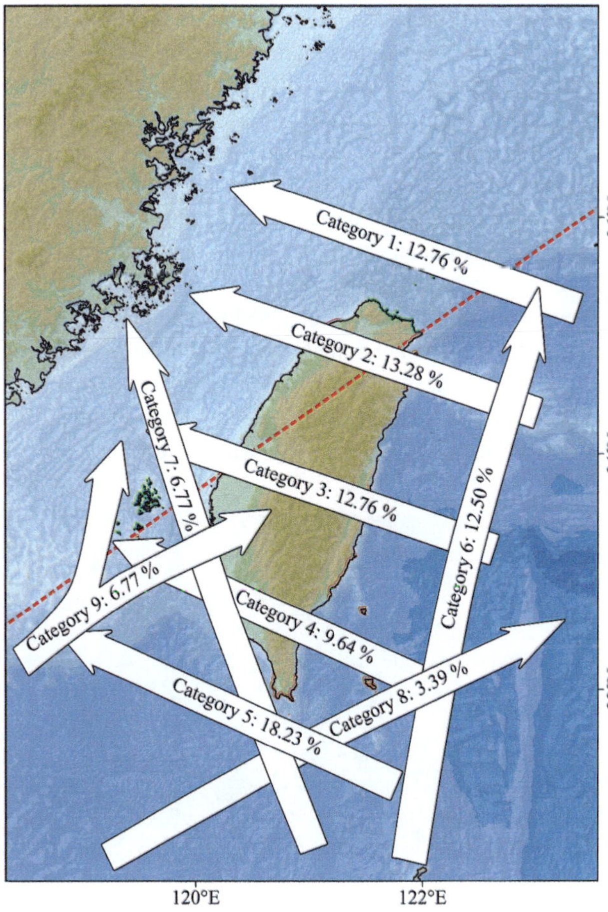

Fig. 2.9 Map of Taiwan showing the nine typhoon tracks with information on the category ranking and relative occurrence percentages

Archipelago. In addition, typhoons have significantly higher chances of passing through tropical coral reefs (53.7%) than non-reefal coral communities (38.8%), with over 35% of typhoons affecting reefs in Kenting National Park (KNP), southern Taiwan (for detail, see review in Keshavmurthy et al. 2019; Kuo et al. 2022a). Recurrent mechanical disturbances caused by typhoons and local anthropogenic disturbances have been strongly linked to the disappearance of *Acropora* and *Montipora* from the reefs of the KNP (Kuo et al. 2011, 2012). A recent study on the impact of seven typhoons in combination with heat stresses between 2015 and 2017 also showed typhoon-induced local shifts from coral to macro-algal dominance in the KNP (Ribas-Deulofeu et al. 2021). This study showed that although the effects of heat waves could be mitigated by typhoons, the mechanical damage induced by storms may decrease the structural complexity of reefs and their associated diversity. In September 2016, Super Typhoon Meranti (Cat. 4 in Fig. 2.9) passed through Green Island and triggered waves 17 m high that toppled the "big mushroom" coral *Porites* sp. bommie (Fig. 2.10). The "big mushroom," estimated to be over 1000 years old, located at about 20 m deep on the sandy bottom with a circumference of 31 m and a height of 12 m (Soong et al. 1999), was a local icon, attracting both international and local tourists to Green Island. In addition, instant freshwater runoff and sedimentation caused by extreme rainfall during typhoons are particularly harmful to fringing coral ecosystems (Devlin et al. 2001; Omija 2004; Edmunds and Gray 2014). For example, Typhoon Morakot brought record-breaking torrential rainfall, exceeding 3000 mm in 4 days over southern Taiwan in August 2009 (Chien and Kuo 2011); it caused severe damage to coral reefs in the KNP (Kuo et al. 2011).

Sea surface temperature (SST) anomalies, either prolonged and high or extremely low, sometimes overlapping with typhoons, have also caused large-scale mortality of corals in Taiwan (Hsieh et al. 2008; Chang et al. 2009, 2013; Keshavmurthy et al. 2019; Kuo et al. 2021; Ribas-Deulofeu et al. 2021). Tropical coral reefs in the southwest and southeast Taiwan have been repeatedly exposed to rising SST caused by El Niño. In 1998, mass bleaching was also observed in the KNP, Orchid Island, Green Island, and Penghu Archipelago but not in northeastern Taiwan or adjacent islets (Wilkinson 1998; reviewed in Keshavmurthy et al. 2019 and Kuo et al. 2021). Entering the new millennium, at least seven coral bleaching events caused by rising SST have been recorded in the KNP (reviewed in Keshavmurthy et al. 2019; Kuo et al. 2022a). In 2020, an unprecedented coral bleaching event was caused by the continuous warm water moving from the northern SCS towards southwest Taiwan in June, synergistic with the lack of cooling from typhoons. Heat stress of up to 18° heating weeks (DHWs) resulted in a loss of one-third of shallow-water scleractinian corals due to severe bleaching. The highest mortality rate (50%) was recorded in Liuqiu Islet because of the homogeneity of species composition (mainly *Pocillopora* and *Montipora*) in the shallow area and lack of a cooling effect at a small scale. In addition, this was the first mass bleaching event (12 DHWs) in non-reefal coral communities in Taiwan (Kuo et al. 2021). In contrast, extremely low SST occurred in the late winter of 2008, creating a minimum in-situ measured temperature of 11.73 °C resulting in extremely high marine life mortality, including corals, reef-associated fishes, and macroinvertebrates (Hsieh et al. 2008; Chang et al. 2009, 2013; Lee et al. 2014). The unusual cold shock was derived from the intrusion of the CCC from the north of the YCR into the waters surrounding the Penghu Archipelago, with wind speeds of more than 6.7 ms^{-1} lasting for 30 days (Chang et al. 2009). The satellite data showed a minimum SST of 12.6 °C on February 16, 2008, 7.8 °C lower than the 12-year average of 20.4 °C. Analysis of the long-term changes of the SST and wind speed with the El Niño/La Niña events showed that the wind variation corresponded well with El Niño/La Niña, especially, the continuous strong wind in the La Niña winter of 2008 (Chang et al. 2009).

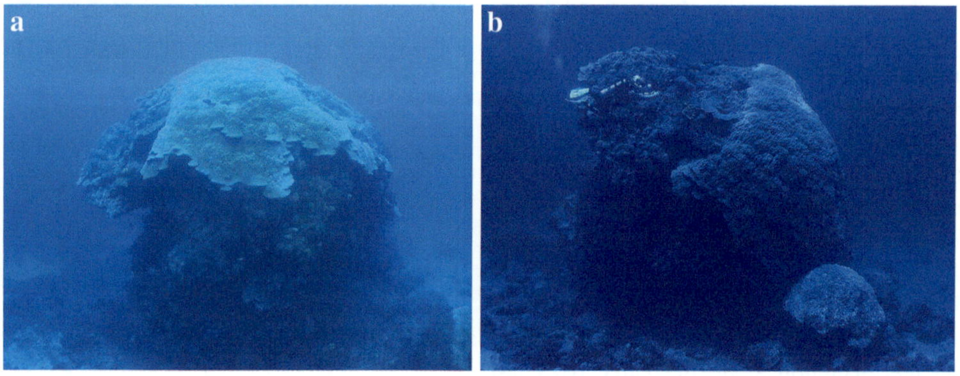

Fig. 2.10 The "big mushroom" coral *Porites* sp. bommie before (2016) (**a**); and after (2017) (**b**) Typhoon Meranti. Photo credit: Dr. Ming-Jay Ho

2.4.2 Anthropogenic Disturbances

Coral ecosystems in Taiwan, either tropical coral reefs or non-reefal coral communities, fringe the coastline (except those in Dongsha Atoll and Taiping Island in the SCS). Therefore, coral ecosystem health is strongly affected by human activities, including overfishing, habitat destruction, and pollution (Dai 1997; Chen and Dai 2004; Meng et al. 2007a, b, 2008; Liu et al. 2009, 2012; Hsieh et al. 2011; Huang et al. 2011, 2012a, b; Chen 2014; Ribas-Deulofeu et al. 2016; Kuo et al. 2022a). Reef Check surveys, conducted in the late 1990s, show that tropical coral reefs and non-reefal coral communities in Taiwan have suffered from long-lasting and severe overfishing. The relative abundance of keystone fishes and macroinvertebrates has been consistently low over the years, particularly among the reefs with several targeted taxa, such as humphead wrasse, bumphead parrotfish, lobsters, and sea urchins, which were not seen throughout entire surveys (Dai 2010; Chen 2014; Kuo et al. 2022a). The fishing effect is also evident in the mesophotic coral ecosystem down to a depth of 70 m on Green Island, with fishing lines tangled on broken coral pieces (Ho et al. 2020). The mass-balance trophic model based on fish and coral biomass showed an extremely reduced trophic level of catch, matter cycling, and trophic transfer efficiency, suggesting that the coral reefs of the KNP are overfished (Liu et al. 2009). In addition to overfishing, habitat destruction has increased since the late 1990s along the coast of the KNP; this is to accommodate housing development in response to the rapid expansion of tourism, with a record of eight million visitors in 2014 alone (Chen 2015). Consequently, sediment runoff due to heavy precipitation and eutrophication caused by sewage and other pollutant discharge has suffocated corals and induced macroalgal blooms, such as *Ulva*, *Codium*, *Laurencia*, and *Sargassum*, resulting in the continued degradation of coral reefs (Dai 1997; Dai et al. 1998; Meng et al. 2008; Liu et al. 2012).

The non-reefal coral communities in the Penghu Archipelago, particularly those in the northern and inner sections, also face similar long-term disturbances caused by human activity (Hsieh et al. 2011; Huang et al. 2011, 2012a, 2012b; Wu 2014). The communities here are composed of loose coral colonies attached to shallow flats that are largely exposed to air during low tides and are present above 10 m depth in a semi-close embayment (Hsieh et al. 2001, 2007, 2016; Huang et al. 2011, 2012a, b). For hundreds of years, residents of Penghu have collected corals as building materials and harvested marine organisms from shallow flats as protein sources, including fishery practices using stone weirs as traps (Wu 2014). This resulted in a significant modification of the seascape of coral communities in the shallow flats of the northern and inner sections. In addition, oyster aquaculture and marine cages deployed in the embayment have caused chronic nutrient enrichment in the surrounding waters in the inner section of the Penghu Archipelago, which may have resulted in the deterioration of suitable habitats for coral reef organisms (Huang et al. 2011, 2012a, b). Anthropogenic-induced degradation of the marine environment and depletion of fish and invertebrate functional groups due to overfishing and an outbreak of the corallivorous gastropod, *Drupella* spp., resulted in a fivefold decline in living coral cover from 80.9% ± 10.8% in 2001 to 16.3% ± 8.4% by 2008, with the dominant coral functional group shifting from branching species to columnar and massive species within 7 years in the first no-take area at Chinwan Inner Bay, Penghu (Hsieh et al. 2001, 2007, 2011). The latest surveys of coral communities showed no signs of ecological resilience in the Chinwan Inner Bay (Kuo et al. 2022b). Ribas-Deulofeu et al. (2016) examined the structure of benthic communities around Taiwan, and the results indicated that the existing coral diversity shows an obvious decreasing gradient from south to north, whereas macro-algal diversity is higher on the north-eastern coast. Kuo et al. (2022b) also examined the benthic community structure in both tropical and non-reefal coral communities in Taiwan, and the results indicated that the complexity of the benthic community within individual sites is more apparent than that between reef types. These results suggest that local disturbances and habitat degradation, such as the KNP and Penghu, smooth out the pattern of regional distribution of corals, as proposed by Chen (1999) and Chen and Keshavmurthy (2009).

2.5 Impact of Climate Change and Ocean Acidification on Coral Ecosystems in Taiwan

2.5.1 Rising Sea Surface Temperature

Taiwan and its adjacent islands and islets are probably among the areas most affected by climate change because they are located in the transitional zone of the tropics and subtropics in East Asia (TCCIP 2021). For instance, between 1860 and 2020, the increased temperature in Taiwan (2 °C) compared with 1.6–1.9 °C in the rest of the countries in East Asia and 1.3 °C globally (Berkeley Earth 2022). Sea surface temperature variability (both warming and extremely low temperature), changes in track, frequency and intensity of typhoons, strength and routes of ocean currents, and interaction among them have impacted on the present and will impact on the future of these transitional coral ecosystems in Taiwan.

Analysis of sea surface temperature from 1948 to 2020 in Taiwan, using the dataset of the National Oceanic and Atmospheric Administration (NOAA) (Kalnay et al. 1996), clearly

Table 2.2 The annual changing rate, representing the slope of linear regression analysis, and the actual temperature difference (ΔT) in the four regions of Taiwan, including Penghu Archipelago, northern Taiwan, Green Island, and Orchid Island in the East coast and southern Taiwan, at different time periods

	1948–1976		1976–1998		1998–2011		2011–2020		1948–2020	
	Slope	ΔT	Slope	ΔT	Slope	ΔT	Slope	ΔT	Slope	ΔT
Taiwan	−0.010	0.060	0.043	2.047	−0.047	−1.118	0.050	0.595	0.017	1.583
Penghu Archipelago	−0.016	−0.024	0.061	2.515	−0.050	−1.248	0.044	0.566	0.023	1.809
Northern Taiwan	−0.012	0.073	0.042	1.988	−0.043	−0.958	0.037	0.414	0.017	1.526
Green Is. & Orchid Is.	−0.002	0.108	0.031	1.782	−0.046	−1.043	0.078	0.735	0.015	1.582
Southern Taiwan	−0.007	0.0005	0.030	1.758	−0.048	−1.091	0.077	0.905	0.013	1.573

showed a warming trend of 1.583 °C in the last 71 years (Table 2.2, Fig. 2.11a). Linear regression analysis revealed four distinct regimes between 1948 and 2020. The first regime (1948–1976) of the stable or slightly cooling SST lasted through 1976 with the lowest slope, considered as the annual changing rate of −0.010 and 0.060 °C cooling in 28 years. The regime of 1976–1998 was the second warmest decade with a slope of 0.042 and 2 °C warming; the period of 2011–2020 had a slope of 0.049 and 0.594 °C of warming. In contrast, in the period 1999–2011, which corresponds to the global warming hiatus (Modak and Mauritsen 2021), cooling was observed with a slope of −0.047 and 1.118 °C temperature drop. The non-reefal coral communities, the Penghu Archipelago and northern Taiwan, have a higher warming speed of 0.227 °C and 0.171 °C, respectively, than the mean of the whole of Taiwan (0.166 °C) per decade, and those of the east (0.154 °C) and south (0.131 °C) (Table 2.2, Fig. 2.8b). A similar pattern was also observed when analyzing the monthly climatology of the SST in the Taiwan Strait, available from the Met Office, Hadley Centre, UK, where the magnitude and rate of the overall SST warming between 1870 and 2018 in northern Taiwan was approximately 1.5 times than those of the south (Lee et al. 2021). As a consequence, the winter SST in the non-reefal coral communities has been gradually increasing beyond 18 °C, and in 2020, was unprecedently warm (Fig. 2.12).

2.5.2 Tropical Typhoon

The latest IPCC AR6 report concluded that anthropogenic climate change may have contributed to the poleward migration of maximum tropical typhoon intensity in the western North Pacific in recent decades, related to anthropogenically forced tropical expansion (IPCC 2021). Poleward shifts have decreased the number of typhoons in the Philippine Sea and South China Sea region, including the Marianas, Philippines, Vietnam, Taiwan, and southern China, and increased typhoon numbers in the East China Sea region, including Japan (especially the Ryukyu Archipelago), the Korean Peninsula, and parts of eastern China (Kossin et al. 2016, 2020; Sobel et al. 2016; Nakamura et al. 2017; Bell et al. 2019; Cha et al. 2020; Knutson et al. 2020; Roberts et al. 2020; Yang et al. 2020). Typhoon tracks have been modified by climate change, but there is also emerging evidence of an increase in the annual global proportion of Category 4 or 5 tropical cyclones in recent decades (IPCC 2019). Under the worst-case scenario (RCP8.5), it has been estimated that while the number of typhoons affecting Taiwan will decrease by approximately 15% in the middle and by 55% towards the end of the twenty-first century, the proportion of strong typhoons (Category 4 or 5) will increase by approximately 100% in the middle and 50% towards the end of the twenty-first century, and the rate of typhoon rainfall change will increase by about 20% in the middle and 35% towards the end of this century (TCCIP 2021).

2.5.3 Ocean Acidification

Ocean acidification (OA), which has a possible negative effect on coral and coral reefs, is accelerated by increased atmospheric carbon dioxide (CO_2) concentrations (Hoegh-Guldberg et al. 2007, 2017; Matear and Lenton 2018; Mollica et al. 2018). Ocean acidification affects coral reef ecosystems at several scales, including reduced skeleton density (Mollica et al. 2018), survival rates (Cumbo et al. 2013b) and metabolic performance of coral larvae (Edmunds et al. 2013), as well as the diversity and complexity of benthic communities (Agostini et al. 2018, 2021). However, the effect of OA on coral reefs is not only complicated and species-specific (Cumbo et al. 2013a; Dufault et al. 2013; Schutter et al. 2015), but also less severe than the effect of rising temperature (Putnam et al. 2013; Hughes et al. 2017). Studies on the effect of OA on coral assemblage levels in Taiwan are limited to only one study by Bramanti et al. (2015). This study predicted the population dynamics of a brooding *Pocillopora* species in the KNP, using a size-based demographic model with life history tables. The results showed that, in 2100, a fully closed population can resist extirpation with 10% of the external larval supply under the projected scenario of RCP8.5. However, as argued by Hughes et al. (2017), the pH simulated in this study, along with most OA experiments, was too extreme to be realistic at the end of the twenty-first

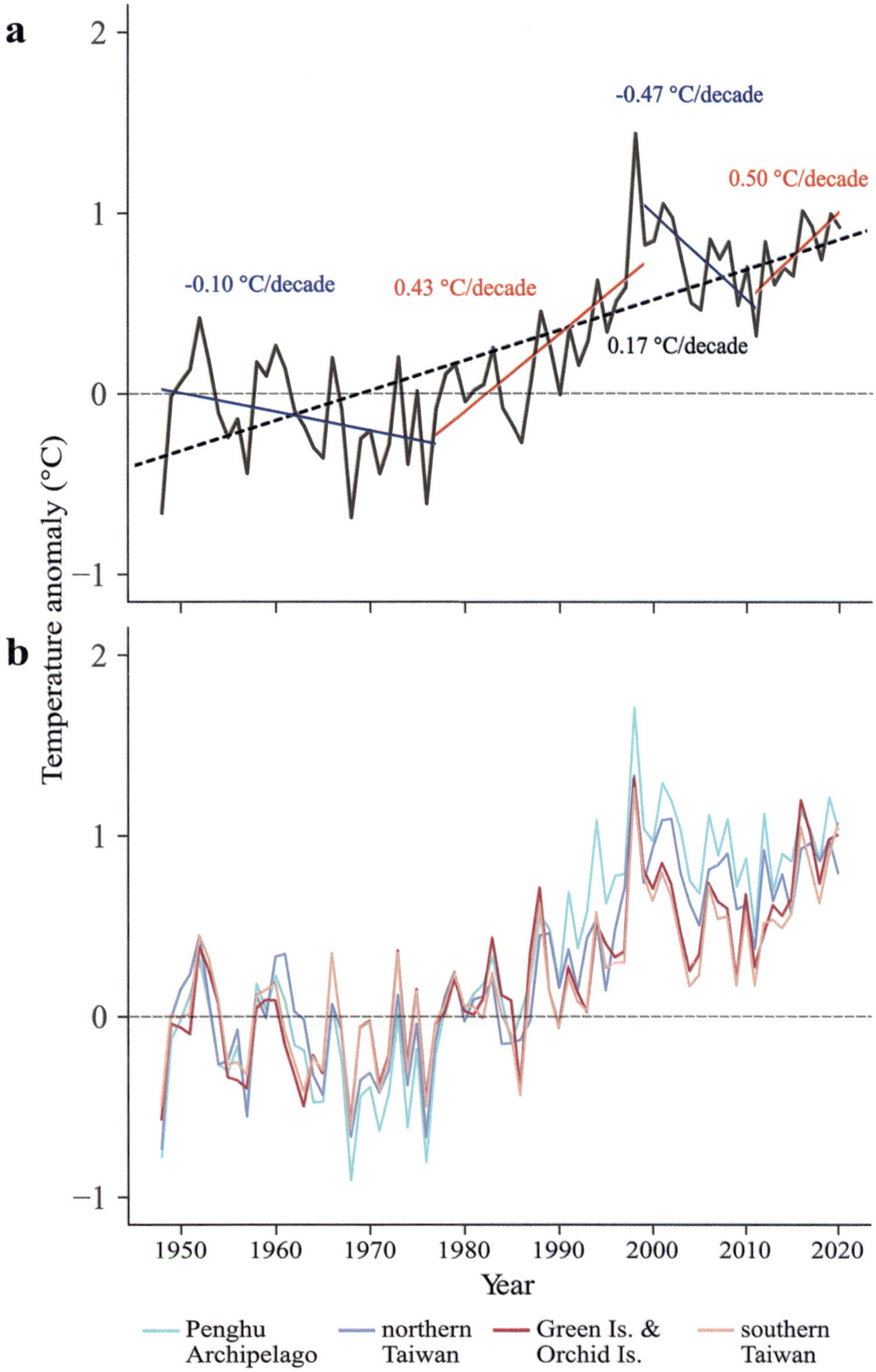

Fig. 2.11 The temperature anomaly over time in Taiwan (**a**) and four regions of Taiwan, including Penghu Archipelago, northern Taiwan, Green Island and Orchid Island in the East coast and southern Taiwan (**b**). The baseline (dashed line) is the average temperature (23.5 °C) between 1948 and 1997. The changing rates for each time period in (**a**) were calculated using linear regression analysis, as shown in Table 2.2

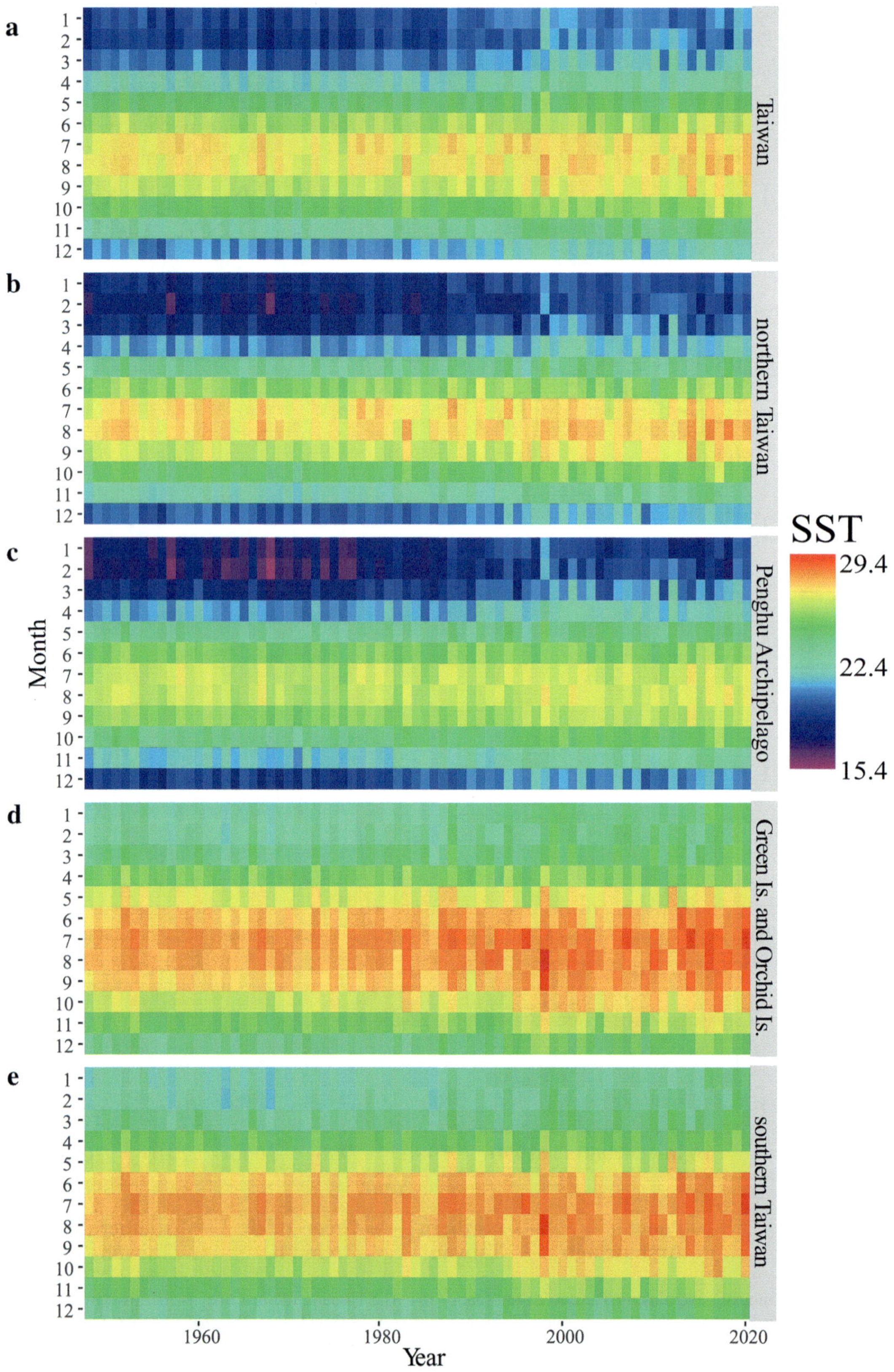

Fig. 2.12 The SST in Taiwan (**a**) and the four regions, including Penghu Archipelago (**b**), northern Taiwan (**c**), Green Island and Orchid Island in the East coast (**d**), and southern Taiwan (**e**), between 1948 and 2021. The data was extracted from the NOAA NCEP/NCAR Reanalysis I dataset

century. Nevertheless, it is important to examine in more detail the effects of seawater temperature and OA (by using realistic pH simulations) on corals in the transitional coral ecosystem in Taiwan. Such efforts will pave the way for determining the fate of corals not only in Taiwan but also in high-latitude communities.

2.5.4 Bioerosion

Bioerosion is a more profound threat to coral reef ecosystems than OA under climate change conditions because bioerosion processes are facilitated by OA (Manzello et al. 2008; Tribollet et al. 2009; Wisshak et al. 2012, 2013; Schönberg et al. 2017). OA directly facilitates the chemical bioerosion process by reducing alkalinity and pH, which then indirectly benefits the mechanical bioerosion process by softening the substrate (Tribollet et al. 2009; summary in Schönberg et al. 2017). More importantly, unlike the scenarios applied in most OA experiments, the net bioerosion rate of the sponge *Cliona orientalis*, a well-known bioeroder, increases in RCP4.5, one of the scenarios that we are likely to face in the future (Fang et al. 2013). *Cliona* spp. have been recorded in several regions of Taiwan, including Green Island, Orchid Island, and northern Taiwan (TEIA 2020; Huang et al. 2021). Although *Cliona* spp. have not caused a significant loss of coral abundance, as a consequence of the boom of *Terpios hoshinota* on Green Island and Orchid Island in the late 2000s (Liao et al. 2007; Chen et al. 2009), an increase in the cover of *Cliona* spp. has recently been observed. Future investigations are required on the spatiotemporal interaction between *Cliona* spp. and corals and the bioerosion processes under climate change in the coral ecosystems of Taiwan. Moreover, bioerosion can also occur through physical processes over a long period of time because of high wave action. Non-reefal coral communities in Taiwan are generally exposed to high wave action and average low seawater temperatures compared to their reefal counterparts. Such differences may, in the future, manifest as highly eroded communities. Such mechanisms and differences between non-reefal and reefal communities, combined with rising seawater temperature and OA (as mentioned in the previous section), need further investigation and should be a key research topic for coral reefs and coral communities in Taiwan.

2.6 Conclusion

Transitional coral ecosystems of high-latitude subtropical coral reef communities and mesophotic zones have received much attention in recent years (Beger et al. 2014; Loya et al. 2019; Soares et al. 2019, 2020; Soares 2020). First, transitional coral communities are characterized by overlapping species ranging from tropical to temperate zones, and their environmental settings, ecological dynamics, and community structure are significantly different from those of their tropical counterparts (Beger et al. 2014; Sommer et al. 2014a, b, 2017, 2018). Second, the physical environmental conditions of transitional coral ecosystems, such as temperature, nutrients, siltation, turbidity, pH, and depth, often exhibit suboptimal conditions that are unsuitable for the persistence of most tropical species. It has been shown that corals living under the suboptimal conditions in the transitional coral ecosystems could have a higher thermal plasticity (Tuckett and Wernnberg 2018; Keshavmurthy et al. 2020) that may allow them to survive in the future climate change scenarios (Pandolfi et al. 2011; Beger et al. 2014; Tuckett and Wernnberg 2018). Third, transitional coral ecosystems can provide environmental and ecological stepping stones for corals and reef-associated species from the tropics to move to higher latitudes via currents, either through larval transport or rafting during climate-driven range shifts (Harrison and Booth 2007; Chen and Keshavmurthy 2009; Budd and Pandolfi 2010; Haak et al. 2010; Beger et al. 2011, 2014; but see Kim et al. 2019; Soares 2020; Agostini et al. 2021). Therefore, the role of refuges in the range shifting of tropical species to cooler regions highlights conservation priorities for transitional coral ecosystems (Nakabayashi et al. 2019; Fifer et al. 2022).

In this chapter, we reviewed the current understanding of geological and marine environmental settings, coral diversity and biogeographic patterns, potential mechanisms shaping coral diversity, natural and anthropogenic disturbances, and the impact of climate change on coral ecosystems around Taiwan and its adjacent islands and islets. Our analyses indicate that within the total coastline of 1520 km, there are not only significant differences in abiotic settings, including underwater geo-topology, ocean currents, and monsoons, but also in the biotic formation of distinct coral ecosystems, namely tropical coral reefs and non-reefal coral communities (Fig. 2.3). These findings demonstrate that the Taiwanese coral ecosystem possesses typical features of a transitional ecosystem, including overlapping ranges of tropical and temperate species, suboptimal environmental conditions, and potential environmental and ecological stepping stones for poleward migration and range expansion to the higher latitude regions in East Asia.

2.7 Future Research Directions of Coral Ecosystems in Taiwan

Based on the results of this chapter, several future research directions, particularly in response to the impacts of climate change, are highlighted below.

2.7.1 Pattern and Process of Biodiversity Formation in the Transitional Coral Ecosystems in Taiwan

Although the inventory list of scleractinian coral species in Taiwan has recently been completed (Dai and Cheng 2020), new records have been continuously added through new surveys of less-explored habitats, such as algal reefs (Kuo et al. 2019, 2020) and mesophotic ecosystems (Ho et al. 2020; De Palmas et al. 2021a), and through classical and molecular taxonomy (Huang et al. 2020b). In addition, the distinct biogeographic boundary between tropical reefs and non-reefal coral communities, defined by coral diversity (Fig. 2.3), lays out a testable hypothesis that can be further explored at the species level or with other reef-associated organisms in Taiwan. For example, recent intensive surveys of the occurrence of two morphologically indistinguishable cauliflower corals, *Pocillopora damicornis* and *P. acuta,* clearly indicate concordance in the distribution boundary pattern in Taiwan, following the line shown in Fig. 2.3, where seawater temperature may play an important role in determining their distribution (Huang et al. 2020b).

2.7.2 Role of Kuroshio Current: A Promoter or a Barrier in Reef Formation and Connecting Coral Ecosystems in the East Asia

The KC and its branches significantly influence marine diversity and coral ecosystems, not only in Taiwan but also throughout East Asia. Upwelling caused by the KC facing a sharp rise in seabed topography on the northeast coast of Taiwan and the Penghu Archipelago in the Taiwan Strait results in seawater with significantly lower temperatures and higher nutrients. This is coupled with a slow growth rate due to low temperatures and the bioerosion effect caused by strong winds driven by the northeast monsoon during winter. Such processes result in a lower aragonite saturation environment, leading to the absence of reefs (Kleypas et al. 1999) on the northeast coast of Taiwan and the Penghu Archipelago. Further research is needed to explore the role of upwellings in shaping coral ecosystems on the northeast coast of Taiwan and the Penghu Archipelago, particularly under the projection of rising seawater temperatures, as has been documented in the eastern tropical Pacific (Randall et al. 2020).

Although the KC has been recognized as a promoter connecting marine biodiversity in the northwestern Pacific (Chen and Keshavmurthy 2009), studies have also shown that the KC might create an invisible barrier to marine fishes when they try to move across the region (Senou et al. 2006; Fujikura et al. 2010; Kuriiwa et al. 2014). On the northeast coast of Taiwan, the KC crosses the Yonaguni Depression via the East Taiwan Channel into the East China Sea. The main current flowing into the edge of the continental shelf in the Okinawa Trough and a branch pumping into the north and northeast coasts of Taiwan creating upwelling, and a cold dome surrounding the islets may also create an invisible barrier in this region (Figs. 2.2, 2.5, and 2.6a, b). This invisible barrier creates the east-west isolation of coral ecosystems between the south Ryukyu, which includes the Miyako and Yaeyama Islands, and those located on the East China Sea continental shelf, including the Senkaku Islands and islets adjacent to the northeast coast of Taiwan at the same latitude (24–25°N). While tropical coral reefs with over 350 scleractinian species are recorded in both Miyako and Yaeyama Island groups (Ministry of the Environment and Japanese Coral Reef Society 2004), the benthic communities of the islets in northeast Taiwan, which are less than 150 km west of south Ryukyu, are characterized as non-reefal coral communities with fewer scleractinian species (<120) (Table 2.1). What remains to be explored is the coral ecosystem of the Senkaku Islands, which is important not only for testing the hypothesis of west-east isolation by the Kuroshio barrier, but also for evaluating the south-north connectivity of non-reefal coral communities in Taiwan and those in high-latitude East Asia.

Recent molecular genetic studies also provide clues to the role the Kuroshio Current may have been playing as a barrier in influencing the biodiversity of the coral ecosystems in the East Asian region. Studies on *Acropora hyacinthus* (Suzuki et al. 2016; Nakabayashi et al. 2019), *Alveopora japonica* (Kang et al. 2020), and *Coelastrea aspera* (Mitsuki et al. 2021) have demonstrated the existence of distinct cryptic lineages in the Kuroshio region between Taiwan and high-latitude East Asia, including Korea and Japan. These results hint that Kuroshio Current acts as a barrier to creating isolation at evolutionary or ecological timescales in the coral ecosystem of East Asia. Continuing to reveal the mechanisms of the Kuroshio Current in driving marine biodiversity in East Asia is crucially important, as the Kuroshio is accounted for in scenarios of poleward migration or range expansion for tropical species moving up to high latitudes in response to rising seawater temperatures caused by climate change (Yamano et al. 2011, 2012; Yara et al. 2011; Nakabayashi et al. 2019). Whether modern habitats at high latitudes serve as potential short-term refuges for tropical species to migrate, or whether they are indeed long-term refugia for temperate species already isolated and adapted (Nakabayashi et al. 2019; Keshavmurthy et al. 2023-in this book) remains to be explored.

2.7.3 Long-Term Ecological Research (Including Environmental Settings) in Coral Ecosystems of Taiwan

The coral ecosystems of Taiwan are significantly influenced by spatial heterogeneity in the geological and environmental settings. Towards southwest and southeast Taiwan, fringing tropical coral reefs are actively developed from the shallow coast to the mesophotic zone. In northeastern Taiwan and the north Penghu Archipelago, corals grow on hard substrates where the mean winter seawater temperature is lower than 18 °C, the temperature limit for reef development (Kleypas et al. 1999). In facing the impacts of climate change, including high or extremely low seawater temperature, intense typhoons, and modification of the strength and route of currents, it is important to establish island-wide long-term ecological research (LTER) to reveal the adaptation, acclimatization, migration, and extinction of corals in coral ecosystems in Taiwan. The time scale for the LTER on coral reef ecosystems should be decades as a result of the time frames for coral recovery and reassembly of the coral community after disturbances from local studies in the KNP (Kuo et al. 2012, 2022a; Keshavmurthy et al. 2019), the East Asia region (Kimura et al. 2022), and other regions, such as the Great Barrier Reef (De'ath et al. 2012; Johns et al. 2014). The first LTER of coral ecosystems, including studies on both environmental monitoring and coral reef ecosystems, was established in the KNP in southern Taiwan in the early 2000s and continues until today (Meng et al. 2008; Liu et al. 2009, 2012; Kuo et al. 2012; reviewed in Keshavmurthy et al. 2019). However, the scale is currently restricted to annual monitoring of the benthic community due to funding restrictions. Future LTER should be restored not only in the KNP but also extended to other tropical coral reefs, such as the Green Island, Orchid Island, and Liuqiu Islet, and non-reefal coral communities, such as the Penghu Archipelago and northeast Taiwan.

The LTER should collect information on abiotic factors that are known to affect corals and reef organisms and establish permanent quadrats or fixed transects to document long-term trends for comparative analysis across different coral ecosystems in Taiwan. Depending on the taxon or process under investigation, biotic data can be applied to the aspects of ecosystem function (e.g., primary productivity between non-reefal and tropical coral reefs), community, population, and individual-level characteristics (e.g., herbivory function diversity and redundancy, trophic structure, demographic dynamics, and life history diversity of the same coral species between non-reefal and tropical coral reefs). All the results can be provided to the parameter settings for analytic models, particularly in conjunction with the CO_2 emission scenarios and temperature rise forecasts under the impact of climate change.

In addition, short and long-term experiments, both field and tank-based, are needed to reveal the causes and consequences of changes in the attributes of different coral ecosystems in Taiwan. For example, several lines of evidence have shown that corals from both tropical reefs in the KNP (Keshavmurthy et al. 2012, 2014, 2022; Kao et al. 2018; Carballo-Bolaños et al. 2019; Huang et al. 2020a; McRae et al. 2022a, b) and non-reefal coral communities in the Penghu Archipelago (Keshavmurthy et al. 2021) exhibit different responses to temperature anomalies. Such process-oriented studies should be extended to examine more coral species, other taxa, and those from other coral ecosystems (e.g., algal reefs and mesophotic coral ecosystems) to collectively understand the response of coral diversity in Taiwan to climate change. Other process-oriented studies, such as interactions among corals, coral reef fishes, algae, herbivores, and bioeroders (e.g., Dang et al. 2020a, b; Nozawa et al. 2020), and the causes and consequences of ecosystem resilience between tropical reefs and non-reefal communities should also be implemented.

2.7.4 Conservation and Governance of Transitional Coral Ecosystems in Taiwan in the Era of Changing Climate

Aligned with the "wicked problem" (Hughes et al. 2013) in conserving coral ecosystems in other countries of East Asia, coral ecosystems in Taiwan face threats caused by long-term anthropogenic disturbances, including overfishing, pollution, and habitat destruction. Legal frameworks, including national parks, national scenic areas, and fishery conservation zones, have been implemented to manage human activities along the coast of Taiwan since the late 1980s; coral ecosystems and other marine ecosystems, however, continue to degrade. As a result, coastal fisheries inevitably fall into a status of over-exploitation (Liu et al. 2009; Shao 2009; Dai 2010). Even with recent legal action in 2017 on the conservation status of marine animals, for example, the critically endangered status under the Wildlife Protection Act for two coral species, *Polycyathus chaishanensis* (Lin et al. 2012) and *Pseduosiderastrea formosa* (Pichon et al. 2012), and two reef fishes, humphead parrotfish, *Bolbometopon muricatum*, and humphead wrasse, *Cheilinus undulatus*, developmental projects such as the construction of hotels and industrial facilities directly threaten the survival of these legally protected species and the integrity of habitats they inhabit. For example, the Datan Algal Reef, an unprecedented biotic reef located on the northwest coast of Taiwan (Liou et al. 2017), hosts an extant population of the coral *P. chaishanensis* in Taiwan (Kuo et al. 2019, 2020). This algal reef is facing destruction due to the development and construction of the liquefied natural gas (LNG) receiving and storage terminal and port (Silver 2018). A similar confrontation occurred in northern Taiwan to save and preserve a well-developed non-reefal coral ecosystem, which is threatened by the reclamation of another LNG receiving and storage terminal and port (Wu and Chung 2020). Both confrontations are related to the national policy of the Taiwan government towards CO_2 emission reduction, air pollution, and green energy transformation, with respect to a year 2025 "non-nuclear home" policy, and, in the long-term, to a year 2050 "net-zero" carbon emission for global warming reduction.

The complexity of confrontation requires more than merely documenting the ecology or biodiversity of coral ecosystems in Taiwan. In order to resolve conservation conflicts and to meet the challenges driven by environmental catastrophes related to climate change (mass coral mortality in the Penghu Archipelago in 2008; Hsieh et al. 2008, island-wide mass coral bleaching in 2020; Kuo et al. 2021), new governance paradigms (Williams et al. 2019; Morrison et al. 2020), beyond the traditional beliefs that strong local institutions can maintain ecological and social resilience through ecosystem-based management, adaptation, and restoration, should be developed. In addition, it should examine human socioeconomic and cultural processes that might already overwhelm natural biophysical settings and become the major driving force shaping the biodiversity pattern and ecological processes of coral ecosystems in Taiwan. This will help to understand better how various practices should be adapted or changed to help conserve these ecosystems, rather than continuing to inflict additional stress. Regretfully, ecosystem-based management is still far beyond reach for any of the coral ecosystems in Taiwan, not to mention the role of local institutions in enhancing ecological and social resilience. Nonetheless, a novel approach of "socio-ecological macroecology" should be immediately adopted (Williams et al. 2019). This proposes to identify the interactive effects of biophysical (SST, currents, typhoons) and socioeconomic and cultural drivers (trade, consumer demands, human migration, fishery, tourism, CO_2 emissions) of coral ecosystems across Taiwan. Such efforts will help examine the robustness of existing coral ecosystem paradigms and whether the current paradigms can capture the dynamics of contemporary coral ecosystems. Novel social-ecological paradigms that incorporate human dynamics can be developed through in-depth interdisciplinary collaboration between the natural and social sciences to successfully manage coral ecosystems in Taiwan in this era of rapid change.

References

Ablan MCA, McManus JW, Chen CA, Shao K-T, Cabanban AS, Tuan VS, Arthana IW, Bell J (1999) Population interdependencies in the South China Sea ecosystem (submitted to the John D. Catherine T. MacArthur Foundation). ICLARM-The World Fish Center, Penang

Ablan MCA, McManus JW, Chen CA, Shao K-T, Bell J, Cabanban AS, Tuan VS, Arthana IW (2002) Meso-scale transboundary units for the management of coral reefs in the South China Sea area. Naga 25:4–9

Agostini S, Harvey BP, Wada S, Kon K, Milazzo M, Inaba K, Hall-Spencer JM (2018) Ocean acidification drives community shifts towards simplified non-calcified habitats in a subtropical−temperate transition zone. Sci Rep 8:11354. https://doi.org/10.1038/s41598-018-29251-7

Agostini S, Harvey BP, Milazzo M, Wada S, Kon K, Floc'h N, Komatsu K, Kuroyama M, Hall-Spencer JM (2021) Simplification, not "tropicalization", of temperate marine ecosystems under ocean warming and acidification. Glob Chang Biol 27:4771–4784

Anthony KRN, Helmstedt KJ, Bay LK, Fidelman P, Hussey KE, Lundgren P, Mead D, McLeod IM, Mumby PJ, Newlands M, Schaffelke B, Wilson KA, Hardisty PE (2020) Interventions to help coral reefs under global change—a complex decision challenge. PLoS One 15:e0236399. https://doi.org/10.1371/journal.pone.0236399

Beger M, Babcock R, Booth DJ, Bucher D, Condie SA, Creese B, Cvitanovic C, Dalton SJ, Harrison P, Hoey A, Jordan A, Loder J, Malcolm H, Purcell SW, Roelfsma C, Sachs P, Smith SDA, Sommer B, Stuart-Smith R, Thomson D, Wallace CC, Zann M, Pandolfi JM (2011) Research challenges to improve the management and conservation of subtropical reefs to tackle climate change threats. Ecol Manage Restor 12:e7–e10. https://doi.org/10.1111/j.1442-8903.2011.00573.x

Beger M, Sommer B, Harrison PL, Smith SDA, Pandolfi JM (2014) Conserving potential coral reef refuges at high latitudes. Divers Distrib 20:245–257

Bell SS, Chand SS, Camargo SJ, Tory KJ, Turville C, Ye H (2019) Western North Pacific tropical cyclone tracks in CMIP5 models: statistical assessment using a model-independent detection and tracking scheme. J Clim 32:7191–7208

Bellwood DR, Hughes TP, Folke C, Nyström M (2004) Confronting the coral reef crisis. Nature 429:827–833

Berkeley Earth (2022) Berkeley earth. http://berkeleyearth.org. Accessed 1 Dec 2021

Beyer HL, Kennedy EV, Beger M, Chen CA, Cinner JE, Darling ES, Eakin CM, Gates RD, Heron SF, Knowlton N, Obura DO, Palumbi SR, Possingham HP, Puotinen M, Runting RK, Skirving WJ, Spalding M, Wilson KA, Wood S, Veron JE, Hoegh-Guldberg O (2018) Risk-sensitive planning for conserving coral reefs under rapid climate change. Conserv Lett 11:e12587. https://doi.org/10.1111/conl.12587

Bramanti L, Iannelli M, Fan T-Y, Edmunds PJ (2015) Using demographic models to project the effects of climate change on scleractinian corals: *Pocillopora damicornis* as a case study. Coral Reefs 34:505–515

Budd AF, Pandolfi JM (2010) Evolutionary novelty is concentrated at the edge of coral species distributions. Science 328:1558–1561

Burke L, Reytar K, Spalding M, Perry A (2011) Reefs at risk revisited. World Resources Institute (WRI), Washington

Carballo-Bolaños R, Denis V, Huang Y-Y, Keshavmurthy S, Chen CA (2019) Temporal variation and photochemical efficiency of species in Symbiodinaceae associated with coral *Leptoria phrygia* (Scleractinia; Merulinidae) exposed to contrasting temperature regimes. PLoS One 14:e0218801. https://doi.org/10.1371/journal.pone.0218801

Cha EJ, Knutson TR, Lee T-C, Ying M, Nakaegawa T (2020) Third assessment on impacts of climate change on tropical cyclones in the typhoon committee region—part II: future projections. Trop Cyclone Res Rev 9:75–86. https://doi.org/10.1016/j.tcrr.2020.04.005

Chang Y, Lee K-T, Lee M-A, Lan K-W (2009) Satellite observation on the exceptional intrusion of cold water in the Taiwan Strait. Terr Atmos Ocean Sci 20:661. https://doi.org/10.3319/TAO.2008.08.07.01(Oc)

Chang Y, Lee M-A, Lee K-T, Shao K-T (2013) Adaptation of fisheries and mariculture management to extreme oceanic environmental changes and climate variability in Taiwan. Mar Policy 38:476–482

Chen C-TA (1996) The Kuroshio intermediate water is the major source of nutrients on the East China Sea continental shelf. Oceanol Acta 19:523–527

Chen CA (1999) Analysis of scleractinian distribution in Taiwan indicating a pattern congruent with sea surface temperatures and currents: examples from *Acropora* and Faviidae corals. Zool Stud 38:119–129

Chen CA (2014) Status of coral reefs in Taiwan. In: Kimura T, Tun K, Chou LM (eds) Status of coral reefs in east Asian seas region: 2014. Ministry of Environment, Japan, pp 69–78

Chen CA (2015) The application of coral health index (CHI) on the management of protect area: a case study on the coral reef ecosystem in Kenting National Park, Taiwan. The Construction and Planning Agency, Minister of the Interior, Taipei, Taiwan

Chen CA (2020) The long-term ecological research for coral reefs in Kenting National Park, Taiwan: 2020. Kenting National Park Headquarters, Hengchun Township, Pingtung County. in Mandarin

Chen CA, Dai C-F (2004) Local phase shift from *Acropora*-dominant to *Condylactis*-dominant community in the Tiao-Shi Reef, Kenting National Park, southern Taiwan. Coral Reefs 23:508

Chen CA, Keshavmurthy S (2009) Taiwan as a connective stepping-stone in the Kuroshio triangle and the conservation of coral ecosystems under the impacts of climate change. Kuroshio Sci 3:15–22

Chen CA, Ablan MCA, McManus JW, Bell JD, Tuan VS, Cabanban AS, Shao K-T (2004) Population structure and genetic variability of six bar wrasse (*Thallasoma hardwicke*) in northern South China Sea revealed by mitochondrial control region sequences. Mar Biotechnol 6:312–326

Chen S-L, Kuo C-Y, Chen CA (2009) Spatial distribution and temporal variation of the cyanobacteriosponge, *Terpios hoshinota*, and benthos composition on the coral reefs of Green Island. J Natl Park 19:32–45. in Mandarin with English abstract

Chien F-C, Kuo H-C (2011) On the extreme rainfall of Typhoon Morakot (2009). J Geophys Res 116:D05104. https://doi.org/10.1029/2010JD015092

Chollett I, Mumby PJ, Cortés J (2010) Upwelling areas do not guarantee refuge for coral reefs in a warming ocean. Mar Ecol Prog Ser 416:47–56

Chung K-F, Shao K-T (2021) Catalogue of life in Taiwan. http://taibnet.sinica.edu.tw. Accessed 1 Dec 2021

Colombo-Pallotta MF, Rodríguez-Román A, Iglesias-Prieto R (2010) Calcification in bleached and unbleached *Montastraea faveolata*: evaluating the role of oxygen and glycerol. Coral Reefs 29:899–907

Cortés J (1997) Biology and geology of eastern Pacific coral reefs. Coral Reefs 16:S39–S46

Cumbo VR, Edmunds PJ, Wall CB, Fan T-Y (2013a) Brooded coral larvae differ in their response to high temperature and elevated pCO_2 depending on the day of release. Mar Biol 160:2903–2917

Cumbo VR, Fan T-Y, Edmunds PJ (2013b) Effects of exposure duration on the response of *Pocillopora damicornis* larvae to elevated temperature and high pCO_2. J Exp Mar Biol Ecol 439:100–107

Dai C-F (1997) Assessment of the present health of coral reefs in Taiwan. In: Grigg RW, Birkeland C (eds) Status of coral reefs in the Pacific. Sea Grant College Program, School of Ocean and Earth Sciences and Technology, University of Hawaii, pp 123–131

Dai C-F (2006) Distribution and species diversity of reef corals in Taiwan (II). Fisheries Agency, Council of Agriculture, Executive Yuan, Taipei. in Mandarin

Dai C-F (2007) Distribution and species diversity of reef corals in Taiwan (III). Fisheries Agency, Council of Agriculture, Executive Yuan, Taipei. in Mandarin

Dai C-F (2010) Status of coral reefs in Taiwan: 2010. In: Kimura T (ed) Status of coral reefs in east Asian seas region: 2010. Ministry of Environment, Japan, pp 79–87

Dai C-F (2018) Coastal and shallower water ecosystem. In: Jan S (ed) Regional oceanography of Taiwan version II. National Taiwan University Press, Taipei, Taiwan, pp 263–299

Dai C-F, Cheng Y-R (2020) Corals of Taiwan vol 1: Scleractinia fauna. Owl Publishing House, Taipei

Dai C-F, Chen Y-T, Kuo K-M, Chuang C-H (1998) Changes of coral communities in Nanwan Bay, Kenting National Park: 1987-1997. J Natl Park 8:79–99. in Mandarin with English abstract

Dai C-F, Wang S-W, Chang J-S, Cheng A (2009) Handbook for ecological tours of Guanyin algae reef. CPC Corporation, Kaohsiung. in Mandarin

Dai C-F, Chin C-H, Cheng A (2013) Coral fauna of Dongsha Atoll in the South China Sea. Marine National Park Headquarters, Kaohsiung. in Mandarin

Dai C-F, Jan R-Q, Wang J-C, Hsu Y-F, Ho K-S, Wang S-W, Chang J-S, Chen C-S, Huang H-D (2008) The research of natural resources of the three northern Islets. Marine National Park Headquarters, Kaohsiung. in Mandarin

Dai C-F, Soong K, Chen CA, Fan T-Y, Hsieh HJ, Jeng M-S, Chen C-H, Horng S (2004) Status of coral reefs of Taiwan. In: Chou LM, Tun K, Kimura T (eds) Status of coral reefs in East Asian Seas region: 2004. Ministry of Environment, Japan, pp 153–164

Dang VDH, Cheung P-Y, Fong C-L, Mulla AJ, Shiu J-H, Lin C-H, Nozawa Y (2020a) Sea urchins play an increasingly important role for coral resilience across reefs in Taiwan. Front Mar Sci 7:581945. https://doi.org/10.3389/fmars.2020.581945

Dang VDH, Fong C-L, Shiu J-H, Nozawa Y (2020b) Grazing effects of sea urchin *Diadema savignyi* on algal abundance and coral recruitment processes. Sci Rep 10:20346. https://doi.org/10.1038/s41598-020-77494-0

De Palmas S, Soto D, Denis V, Ho M-J, Chen CA (2018) Molecular assessment of *Pocillopora verrucosa* (Scleractinia; Pocilloporidae) distribution along a depth gradient in Ludao, Taiwan. PeerJ 6:e5797. https://doi.org/10.7717/peerj.5797

De Palmas S, Denis V, Soto D, Lin YV, Ho M-J, Chen CA (2021a) Scleractinian diversity in the upper mesophotic zone of Ludao (Taiwan): a museum collection with new records from Taiwanese waters. Mar Biodivers 51:80

De Palmas S, Soto D, Ho M-J, Denis V, Chen CA (2021b) Strong horizontal and vertical connectivity in the coral *Pocillopora verrucosa* from Ludao, Taiwan, a small oceanic Island. PLoS One 16:e0258181. https://doi.org/10.1371/journal.pone.0258181

de Soares MO (2020) Marginal reef paradox: a possible refuge from environmental changes? Ocean Coast Manag 185:105063

de Soares MO, Tavares TCL, de Carneiro PBM (2019) Mesophotic ecosystems: distribution, impacts and conservation in the South Atlantic. Divers Distrib 25:255–268

de Soares MO, de Araújo JT, Ferreira SMC, Santos BA, Boavida JRH, Costantini F, Rossi S (2020) Why do mesophotic coral ecosystems have to be protected? Sci Total Environ 726:138456

De'ath G, Fabricius KE, Sweatman H, Puotinen M (2012) The 27-year decline of coral cover on the great barrier reef and its causes. Proc Natl Acad Sci U S A 109:17995–17999

Denis V, De Palmas S, Benzoni F, Chen CA (2015) Extension of the known distribution and depth range of the scleractinian coral *Psammocora stellata*: first record from a Taiwanese mesophotic reef. Mar Biodivers 45:619–620

Denis V, Soto D, De Palmas S, Lin YTV, Benayahu Y, Huang YM, Liu S-L, Chen J-W, Chen Q, Sturaro N, Ho M-J, Su Y, Dai C-F, Chen CA (2019) Mesophotic coral ecosystems: Taiwan. In: Loya Y,

Puglise KA, Bridge TCL (eds) Mesophotic coral ecosystems. Springer, Cham, pp 249–264
Devlin M, Waterhouse J, Taylor J, Brodie J (2001) Flood plumes in the great barrier reef: spatial and temporal patterns in composition and distribution. GBRMPA Research Publication 68. Great Barrier Reef Marine Park Authority, Townsville
Dufault AM, Ninokawa A, Bramanti L, Cumbo VR, Fan T-Y, Edmunds PJ (2013) The role of light in mediating the effects of ocean acidification on coral calcification. J Exp Biol 216:1570–1577
Edmunds PJ, Gray SC (2014) The effects of storms, heavy rain, and sedimentation on the shallow coral reefs of St. John, US Virgin Islands. Hydrobiologia 734:143–158
Edmunds PJ, Cumbo VR, Fan T-Y (2013) Metabolic costs of larval settlement and metamorphosis in the coral *Seriatopora caliendrum* under ambient and elevated pCO_2. J Exp Mar Biol Ecol 443:33–38
Fan K-L (1979) On upwelling off the Peng-Hu Islands. Acta Oceanogr Taiwan 9:55–57
Fang JKH, Mello-Athayde MA, Schönberg CHL, Kline DI, Hoegh-Guldberg O, Dove S (2013) Sponge biomass and bioerosion rates increase under ocean warming and acidification. Glob Chang Biol 19:3581–3591
Fifer JE, Yasuda N, Yamakita T, Bove CB, Davies SW (2022) Genetic divergence and range expansion in a western North Pacific coral. Sci Total Environ 813:152423
Fisher R, O'Leary RA, Low-Choy S, Mengersen K, Knowlton N, Brainard RE, Caley MJ (2015) Species richness on coral reefs and the pursuit of convergent global estimates. Curr Biol 25:500–505
Fujikura K, Lindsay D, Kitazato H, Nishida S, Shirayama Y (2010) Marine biodiversity in Japanese waters. PLoS One 5:e11836. https://doi.org/10.1371/journal.pone.0011836
Gallagher SJ, Kitamura A, Iryu Y, Itaki T, Koizumi I, Hoiles PW (2015) The Pliocene to recent history of the Kuroshio and Tsushima currents: a multi-proxy approach. Prog Earth Planet Sci 2:17. https://doi.org/10.1186/s40645-015-0045-6
Gattuso J-P, Hoegh-Guldberg O, Pörtner H-O (2014) Cross-chapter box on coral reefs. In: Field CB, Barros VR, Dokken DJ, Mach KJ, Mastrandrea MD, Bilir TE, Chatterjee M, Ebi KL, Estrada YO, Genova RC, Girma B, Kissel ES, Levy AN, MacCracken S, Mastrandrea PR, White LL (eds) Climate change 2014: impacts, adaptation, and vulnerability. Part A: global and sectoral aspects. Contribution of working group II to the fifth assessment report of the intergovernmental panel on climate change. Cambridge University Press, Cambridge, pp 97–100
Glynn PW (1977) Coral growth in upwelling and nonupwelling areas off the Pacific coast of Panama. J Mar Res 35:567–585
Glynn PW, Ault JS (2000) A biogeographic analysis and review of the far eastern pacific coral reef region. Coral Reefs 19:1–23
Glynn PW, Macintyre IG (1977) Growth rate and age of coral reefs on the Pacific coast of Panama. Proc 3rd Int Coral Reef Symp 2:251–259
Glynn PW, Stewart RH (1973) Distribution of coral reefs in the Pearl Islands (Gulf of Panamá) in relation to thermal conditions. Limnol Oceanogr 18:367–379
Glynn PW, Alvarado JJ, Banks S, Cortés J, Feingold JS, Jiménez C, Maragos JE, Martínez P, Maté JL, Moanga DA, Navarrete S, Reyes-Bonilla H, Riegl B, Rivera F, Vargas-Ángel B, Wieters EA, Zapata FA (2017) Eastern Pacific coral reef provinces, coral community structure and composition: an overview. In: Glynn PW, Manzello DP, Enochs IC (eds) Coral reefs of the eastern tropical pacific. Springer, Dordrecht, pp 107–176
Gopalakrishnan G, Cornuelle BD, Gawarkiewicz G, McClean JL (2013) Structure and evolution of the cold dome off northeastern Taiwan: a numerical study. Oceanography 26:66–79
Goreau TF (1959) The physiology of skeleton formation in corals. I. A method for measuring the rate of calcium deposition by corals under different conditions. Biol Bull 116:59–75
Guan B-X (1994) Patterns and structures of the currents in Bohai, Huanghai and East China seas. In: Zhou D, Liang Y-B, Zeng C-K (eds) Oceanology of China seas. Springer, Dordrecht, pp 17–26
Haak AL, Williams JE, Neville HM, Dauwalter DC, Colyer WT (2010) Conserving peripheral trout populations: the values and risks of life on the edge. Fisheries 35:530–549
Hall-Spencer J, Allain V, Fosså JH (2002) Trawling damage to Northeast Atlantic ancient coral reefs. Proc R Soc Lond B 269:507–511
Harrison PL, Booth DJ (2007) Coral reefs: naturally dynamic and increasingly disturbed ecosystems. In: Connell SD, Gillanders BM (eds) Marine ecology. Oxford University Press, London, pp 316–377
Ho M-J, Kuo C-Y, Wu H-Y, Jeng M-S, Shyue S-W, Chen CA (2020) Fishing pressure occurs down to the lower mesophotic coral ecosystem in Lyudao, Taiwan. Galaxea J Coral Reef Stud 22:29–30
Hoegh-Guldberg O (2015) Reviving the ocean economy: the case for action—2015. WWF International, Gland, Switzerland, Geneva
Hoegh-Guldberg O, Mumby PJ, Hooten AJ, Steneck RS, Greenfield P, Gomez E, Harvell CD, Sale PF, Edwards AJ, Caldeira K, Knowlton N, Eakin CM, Iglesias-Prieto R, Muthiga N, Bradbury RH, Dubi A, Hatziolos ME (2007) Coral reefs under rapid climate change and ocean acidification. Science 318:1737–1742
Hoegh-Guldberg O, Cai R, Poloczanska ES, Brewer PG, Sundby S, Hilmi K, Fabry VJ, Jung S (2014) The ocean—supplementary material. In: Barros VR, Field CB, Dokken DJ, Mastrandrea MD, Mach KJ, Bilir TE, Chatterjee M, Ebi KL, Estrada YO, Genova RC, Grima B, Kissel ES, Levy AN, MacCracken S, Mastrandrea PR, White LL (eds) Climate change 2014: impacts, adaptation, and vulnerability. Part B: regional aspects. contribution of working group II to the fifth assessment report of the intergovernmental panel on climate change. Available from www.ipcc-wg2.gov/AR5 and www.ipcc.ch
Hoegh-Guldberg O, Poloczanska ES, Skirving W, Dove S (2017) Coral reef ecosystems under climate change and ocean acidification. Front Mar Sci 4:158. https://doi.org/10.3389/fmars.2017.00158
Hoegh-Guldberg O, Kennedy EV, Beyer HL, McClennen C, Possingham HP (2018) Securing a long-term future for coral reefs. Trends Ecol Evol 33:936–944
Hong H, Chai F, Zhang C, Huang B, Jiang Y, Hu J (2011) An overview of physical and biogeochemical processes and ecosystem dynamics in the Taiwan Strait. Cont Shelf Res 31:S3–S12
Hsieh HJ (2008) Studies on subtropical scleractinian coral communities in Penghu Archipelago. PhD Thesis. National Taiwan University
Hsieh HJ, Lu Y-L, Tsai W-S, Wei N-V, Jeng M-S, Chen CA (2001) Unexpectedly high coral coverage in Chinwan Inner Bay, Pescadores: a proposed site for a marine protection area. Coral Reefs 20:316–317
Hsieh HJ, Chen CA, Dai C-F, Ou W, Tsai W-S, Su W-C (2007) From the drawing board to the field: an example for establishing an MPA in Penghu, Taiwan. Aquatic Conserv Mar Freshw Ecosyst 17:619–635
Hsieh HJ, Hsien Y-L, Jeng M-S, Tsai W-S, Su W-C, Chen CA (2008) Tropical fishes killed by the cold. Coral Reefs 27:599
Hsieh HJ, Chen K-S, Lin YI, Huang YA, Baird AH, Tsai S, Su W-C, Chen CA (2011) Establishment of a no-take area (NTA) could not guarantee the preservation of coral communities in Chinwan Inner Bay, Penghu, Taiwan. Zool Stud 50:443–453
Hsieh HJ, Wen CK-C, Huang Y-C, Chen K-S, Dai C-F, Chen CA (2016) Spatial patterns and environmental settings of non-reefal coral communities across the tropic of cancer in the Penghu archipelago (Pescadores), Taiwan. Zool Stud 55:45. https://doi.org/10.6620/ZS.2016.55-45
Hsin Y-C (2015) Multidecadal variations of the surface Kuroshio between 1950s and 2000s and its impacts on surrounding waters. J Geophys Res Oceans 120:1792–1808

Hsin Y-C, Wu C-R, Shaw P-T (2008) Spatial and temporal variations of the Kuroshio east of Taiwan, 1982–2005: a numerical study. J Geophys Res 113:C04002. https://doi.org/10.1029/2007JC004485

Hsin Y-C, Chiang T-L, Wu C-R (2011) Fluctuations of the thermal fronts off northeastern Taiwan. J Geophys Res 116:C10005. https://doi.org/10.1029/2011JC007066

Hsin Y-C, Wu C-R, Chao S-Y (2012) An updated examination of the Luzon Strait transport. J Geophys Res Oceans 117:C03022. https://doi.org/10.1029/2011JC007714

Hu J, Wang XH (2016) Progress on upwelling studies in the China seas. Rev Geophys 54:653–673

Hu J, Kawamura H, Hong H, Pan W (2003) A review of research on the upwelling in the Taiwan Strait. Bull Mar Sci 73:605–628

Huang Q-Z, Wang W-Z, Li YS, Li CW (1994) Current characteristics of the South China Sea. In: Zhou D, Liang Y-B, Zeng C-K (eds) Oceanology of China seas. Springer, Dordrecht, pp 39–47

Huang Y-CA, Hsieh HJ, Huang S-C, Meng P-J, Chen Y-S, Keshavmurthy S, Nozawa Y, Chen CA (2011) Nutrient enrichment caused by marine cage culture and its influence on subtropical coral communities in turbid waters. Mar Ecol Prog Ser 423:83–93

Huang Y-CA, Huang S-C, Hsieh HJ, Meng P-J, Chen CA (2012a) Changes in sedimentation, sediment characteristics, and benthic macrofaunal assemblages around marine cage culture under seasonal monsoon scales in a shallow-water bay in Taiwan. J Exp Mar Biol Ecol 422–423:55–63. https://doi.org/10.1016/j.jembe.2012.04.008

Huang Y-CA, Huang S-C, Meng P-J, Hsieh HJ, Chen CA (2012b) Influence of strong monsoon winds on the water quality around a marine cage-culture zone in a shallow and semi-enclosed bay in Taiwan. Mar Pollut Bull 64:851–860

Huang Y-Y, Carballo-Bolaños R, Kuo C-Y, Keshavmurthy S, Chen CA (2020a) *Leptoria phrygia* in southern Taiwan shuffles and switches symbionts to resist thermal-induced bleaching. Sci Rep 10:7808. https://doi.org/10.1038/s41598-020-64749-z

Huang Y-Y, Lai S, Wei E-C, Chang H-Y, Chen T-R, Kuo C-Y, Chung A-C, Ho M, Hsin Y-C, Hsieh HJ, Chen CA (2020b) Poleward migration or long-term misidentification: a case study of *Pocillopora acuta* and *P. damicornis*. Paper presented at the annual symposium of Taiwanese coral reef society, Taipei, 14 April 2020

Huang C-Y, Hwang J-S, Yamashiro H, Tang S-L (2021) Spatial and cross-seasonal patterns of coral diseases in reefs of Taiwan: high prevalence and regional variation. Dis Aquat Organ 146:145–156

Hughes TP, Baird AH, Bellwood DR, Card M, Connolly SR, Folke C, Grosberg R, Hoegh-Guldberg O, Jackson JBC, Kleypas J, Lough JM, Marshall P, Nyström M, Palumbi SR, Pandolfi JM, Rosen B, Roughgarden J (2003) Climate change, human impacts, and the resilience of coral reefs. Science 301:929–933

Hughes TP, Huang H, Young MAL (2013) The wicked problem of China's disappearing coral reefs. Conserv Biol 27:261–269

Hughes TP, Barnes ML, Bellwood DR, Cinner JE, Cumming GS, Jackson JBC, Kleypas J, van de Leemput IA, Lough JM, Morrison TH, Palumbi SR, Nes EHV, Scheffer M (2017) Coral reefs in the anthropocene. Nature 546:82–90

Hughes TP, Anderson KD, Connolly SR, Heron SF, Kerry JT, Lough JM, Baird AH, Baum JK, Berumen ML, Bridge TC, Claar DC, Eakin CM, Gilmour JP, Graham NAJ, Harrison H, Hobbs J-PA, Hoey AS, Hoogenboom M, Lowe RJ, McCulloch MT, Pandolfi JM, Pratchett M, Schoepf V, Torda G, Wilson SK (2018) Spatial and temporal patterns of mass bleaching of corals in the anthropocene. Science 359:80–83

Hung C-W, Shih M-F, Lin T-Y (2020) The climatological analysis of typhoon tracks, steering flow, and the pacific subtropical high in the vicinity of Taiwan and the western North Pacific. Atmos 11:543. https://doi.org/10.3390/atmos11050543

IPCC (2018) In: Masson-Delmotte V, Zhai P, Pörtner H-O, Roberts D, Skea J, Shukla PR, Pirani A, Moufouma-Okia W, Péan C, Pidcock R, Connors S, Matthews JBR, Chen Y, Zhou X, Gomis MI, Lonnoy E, Maycock T, Tignor M, Waterfield T (eds) Global warming of 1.5°C. An IPCC special report on the impacts of global warming of 1.5°C above pre-industrial levels and related global greenhouse gas emission pathways, in the context of strengthening the global response to the threat of climate change, sustainable development, and efforts to eradicate poverty. Cambridge University Press, Cambridge

IPCC (2019) In: Pörtner H-O, Roberts DC, Masson-Delmotte V, Zhai P, Tignor M, Poloczanska E, Mintenbeck K, Alegría A, Nicolai M, Okem A, Petzold J, Rama B, Weyer NM (eds) IPCC special report on the ocean and cryosphere in a changing climate. Cambridge University Press, Cambridge

IPCC (2021) In: Masson-Delmotte V, Zhai P, Pirani A, Connors SL, Péan C, Berger S, Caud N, Chen Y, Goldfarb L, Gomis MI, Huang M, Leitzell K, Lonnoy E, Matthews JBR, Maycock TK, Waterfield T, Yelekçi O, Yu R, Zhou B (eds) Climate change 2021: the physical science basis. Contribution of working group I to the sixth assessment report of the intergovernmental panel on climate change. Cambridge University Press, Cambridge

Jackson JBC (2001) Historical overfishing and the recent collapse of coastal ecosystems. Science 293:629–637

Jan S (2018) Regional oceanography of Taiwan, 2nd edn. National Taiwan University Press, Taipei

Jan S, Chao S-Y (2003) Seasonal variation of volume transport in the major inflow region of the Taiwan Strait: the Penghu Channel. Deep Sea Res Part II Top Stud Oceanogr 50:1117–1126

Jan S, Wang J, Chern C-S, Chao S-Y (2002) Seasonal variation of the circulation in the Taiwan Strait. J Mar Syst 35:249–268

Jan S, Tseng Y-H, Dietrich DE (2010) Sources of water in the Taiwan Strait. J Oceanogr 66:211–221

Jan S, Chen C-C, Tsai Y-L, Yang YJ, Wang J, Chern C-S, Gawarkiewicz G, Lien R-C, Centurioni L, Kuo J-Y (2011) Mean structure and variability of the cold dome northeast of Taiwan. Oceanography 24:100–109

Jeng M-S, Dai C-F (2020) Long-term ecological monitoring and conservation strategies in the waters of South Penghu Marine National Park. J Natl Park 30:28–45. in Mandarin with English abstract

Jeng M-S, Chen CA, Tang S-L, Liu S-L, Liu P-J, Lee C-H, Chen Y-J, Ko C-Y (2017) Aquatic animal and plant resources and assessment of ecosystem services of the coral reefs in Taiping Island (Itu Aba), Spratlys, South China Sea. Fisheries Agency, Council of Agriculture, Executive Yuan, Taipei. in Mandarin

Jeng M-S, Dai C-F, Chen J-P, Wang W-L, Meng P-J (2008) Coral reef resource surveys and monitoring of Dongsha Atoll II. Marine National Park Headquarters, Kaohsiung. in Mandarin

Jiménez C, Bassey G, Segura Á, Cortés J (2010) Characterization of the coral communities and reefs of two previously undescribed locations in the upwelling region of gulf of papagayo (Costa Rica). Rev Mar Cost 2:95–108

Johns KA, Osborne KO, Logan M (2014) Contrasting rates of coral recovery and reassembly in coral communities on the great barrier reef. Coral Reefs 33:553–563

Kang JH, Jang JE, Kim JH, Kim S, Keshavmurthy S, Agostini S, Reimer JD, Chen CA, Choi K-S, Park SR, Lee HJ (2020) The origin of the subtropical coral *Alveopora japonica* (Scleractinia: Acroporidae) in high-latitude environments. Front Ecol Evol 8:12. https://doi.org/10.3389/fevo.2020.00012

Kalnay E, Kanamitsu M, Kistler R, Collins W, Deaven D, Gandin L, Iredell M, Saha S, White G, Woollen J, Zhu Y, Chelliah M, Ebisuzaki W, Higgins W, Janowiak J, Mo KC, Ropelewski C, Wang J, Leetmaa A, Reynolds R, Jenne R, Joseph D (1996) The NCEP/NCAR 40-Year reanalysis project. Bull Am Meteorol Soc 77:437–471

Kao K-W, Keshavmurthy S, Tsao C-H, Wang J-T, Chen CA (2018) Repeated and prolonged temperature anomalies negate Symbiodiniaceae genera shuffling in the coral *Platygyra verweyi*

(Scleractinia; Merulinidae). Zool Stud 57:55. https://doi.org/10.6620/ZS.2018.57-55

Keshavmurthy S, Hsu C-M, Kuo C-Y, Meng P-J, Wang J-T, Chen CA (2012) Symbiont communities and host genetic structure of the brain coral *Platygyra verweyi*, at the outlet of a nuclear power plant and adjacent areas. Mol Ecol 21:4393–4407

Keshavmurthy S, Meng P-J, Wang J-T, Kuo C-Y, Yang S-Y, Hsu C-M, Gan C-H, Dai C-F, Chen CA (2014) Can resistant coral-Symbiodinium associations enable coral communities to survive climate change? A study of a site exposed to long-term hot water input. PeerJ 2:e327. https://doi.org/10.7717/peerj.327

Keshavmurthy S, Kuo C-Y, Huang Y-Y, Carballo-Bolaños R, Meng P-J, Wang J-T, Chen CA (2019) Coral reef resilience in Taiwan: lessons from long-term ecological research on the coral reefs of Kenting National Park (Taiwan). J Mar Sci Eng 7:388. https://doi.org/10.3390/jmse7110388

Keshavmurthy S, Tee HS, Kao K-W, Wang J-T, Chen CA (2020) Specificity trumps flexibility—location-based stable associations between Symbiodiniaceae genera and *Platygyra verweyi* (Scleractinia; Merulinidae). PeerJ 8:e8791. https://doi.org/10.7717/peerj.8791

Keshavmurthy S, Beals M, Hsieh HJ, Choi K-S, Chen CA (2021) Physiological plasticity of corals to temperature stress in marginal coral communities. Sci Total Environ 758:143628

Keshavmurthy S, Chen T-R, Liu P-J, Wang J-T, Chen CA (2022) Learning from the past is not enough to survive present and future bleaching threshold temperatures. Sci Total Environ 852:158379

Keshavmurthy S, Mezaki T, Reimer JD, Choi K-SA, Chen CA (2023) Succession and emergence of coral reefs in high-latitude (temperate) areas of eastern Asia into the future. In: Takeuchi I, Yamashiro H (eds) Coral reefs of eastern Asia under anthropogenic impacts. Coral reefs of the world, vol 17, Springer Nature Switzerland AG, Cham, pp 51–68

Kim SW, Sampayo EM, Sommer B, Sims CA, Gómez-Cabrera MDC, Dalton SJ, Beger M, Malcolm HA, Ferrari R, Fraser N, Figueira WF, Smith SDA, Heron SF, Baird AH, Byrne M, Eakin CM, Edgar R, Hughes TP, Kyriacou N, Liu G, Matis PA, Skirving WJ, Pandolfi JM (2019) Refugia under threat: mass bleaching of coral assemblages in high-latitude eastern Australia. Glob Chang Biol 25:3918–3931

Kimura T, Chou LM, Huang D, Tun K, Goh E (2022) Status and trends of east Asian coral reefs: 1983–2019. Global Coral Reef Monitoring Network, East Asia Region

Kleypas JA, McManus JW, Meñez LAB (1999) Environmental limits to coral reef development: where do we draw the line? Am Zool 39: 146–159

Knutson T, Camargo SJ, Chan JCL, Emanuel K, Ho C-H, Kossin J, Mohapatra M, Satoh M, Sugi M, Walsh K, Wu L (2020) Tropical cyclones and climate change assessment: part II: projected response to anthropogenic warming. Bull Am Meteorol Soc 101:E303–E322. https://doi.org/10.1175/BAMS-D-18-0194.1

Kossin JP, Emanuel KA, Camargo SJ (2016) Past and projected changes in western North Pacific tropical cyclone exposure. J Climate 29: 5725–5739

Kossin JP, Knapp KR, Olander TL, Velden CS (2020) Global increase in major tropical cyclone exceedance probability over the past four decades. Proc Natl Acad Sci U S A 117:11975–11980

Kuo N-J, Ho C-R, Zheng Q, Su F-C, Lo Y-T, Tsao C-C (2006) Satellite-derived chlorophyll-a concentration in the Taiwan Strait. In: Bostater CR Jr, Neyt X, Mertikas SP, Vélez-Reyes M (eds) Remote sensing of the ocean, sea ice, and large water regions 2006, vol 6360, Stockholm, p 63600R. https://doi.org/10.1117/12.689205

Kuo C-Y, Meng P-J, Ho P-H, Wang J-T, Chen J-P, Chiu Y-W, Lin H-J, Chang Y-C, Fan T-Y, Chen CA (2011) Damage to the reefs of Siangjiao Bay marine protected area of Kenting National Park, southern Taiwan during typhoon Morakot. Zool Stud 50:85

Kuo C-Y, Yuen YS, Meng P-J, Ho P-H, Wang J-T, Liu P-J, Chang Y-C, Dai C-F, Fan T-Y, Lin H-J, Baird AH, Chen CA (2012) Recurrent disturbances and the degradation of hard coral communities in Taiwan. PLoS One 7:e44364. https://doi.org/10.1371/journal.pone.0044364

Kuo C-Y, Chung A, Keshavmurthy S, Huang Y-Y, Yang S-Y, Chen CA (2019) Lonely giant on the sand: unexpected massive Taiwanese coral, *Polycyathus chaishanensis* in the Datan algal reef demands a conservation focus. Galaxea J Coral Reef Stud 21:11–12

Kuo C-Y, Keshavmurthy S, Chung A, Huang Y-Y, Yang S-Y, Chen Y-C, Chen CA (2020) Demographic census confirms a stable population of the critically-endangered caryophyllid coral *Polycyathus chaishanensis* (Scleractinia; Caryophyllidae) in the Datan Algal Reef, Taiwan. Sci Rep 10:10585. https://doi.org/10.1038/s41598-020-67653-8

Kuo C-Y, Huang Y-Y, Hsiao A-T, Ho M-J, Jeng M-S, Chung A-C, Wei C, Chen T-R, Chen C-W, Hsieh HJ, Chen CA (2021) The collapse of the transition coral ecosystem in the Northwest Pacific. Paper presented at the annual symposium of Taiwanese coral reef society, Taipei, 16 January 2021

Kuo C-Y, Ho M-J, Chen C-H, Hsieh HJ, Jeng M-S, Fan T-Y, Dai C-F, Soong K, Hwang J-S, Chen CA (2022a) Taiwan. In: Kimura T, Chou LM, Huang D, Tun K, Goh E (eds) Status and trends of east Asian coral reefs: 1983–2019. Global Coral Reef Monitoring Network, East Asia Region, pp 82–95

Kuo C-Y, Tsai C-H, Huang Y-Y, Heng WK, Hsiao A-T, Hsieh HJ, Chen CA (2022b) Fine intervals are required when using point intercept transects to assess coral reef status. Front Mar Sci 9:795512. https://doi.org/10.3389/fmars.2022.795512

Kuriiwa K, Chiba SN, Motomura H, Matsuura K (2014) Phylogeography of blacktip grouper, *Epinephelus fasciatus* (Perciformes: Serranidae), and influence of the Kuroshio current on cryptic lineages and genetic population structure. Ichthyol Res 61: 361–374

LaJeunesse TC, Parkinson JE, Gabrielson PW, Jeong HJ, Reimer JD, Voolstra CR, Santos SR (2018) Systematic revision of Symbiodiniaceae highlights the antiquity and diversity of coral endosymbionts. Curr Biol 28:2570–2580

Lan K-W, Kawamura H, Lee M-A, Chang Y, Chan J-W, Liao C-H (2009) Summertime Sea surface temperature fronts associated with upwelling around the Taiwan Bank. Cont Shelf Res 29:903–910

Lee H-J, Chao S-Y, Fan K-L (1999a) Flood–ebb disparity of tidally induced recirculation eddies in a semi-enclosed basin: Nan Wan Bay. Cont Shelf Res 19:871–890

Lee H-J, Chao S-Y, Fan K-L, Kuo T-Y (1999b) Tide-induced eddies and upwelling in a semi-enclosed basin: Nan Wan. Estuar Coast Shelf Sci 49:775–787

Lee M-A, Yang Y-C, Shen Y-L, Chang Y, Tsai W-S, Lan K-W, Kuo Y-C (2014) Effects of an unusual cold-water intrusion in 2008 on the catch of coastal fishing methods around Penghu Islands, Taiwan. Terr Atmos Ocean Sci 25:107–120

Lee M-A, Huang W-P, Shen Y-L, Weng J-S, Semedi B, Wang Y-C, Chan J-W (2021) Long-term observations of interannual and decadal variation of sea surface temperature in the Taiwan Strait. J Mar Sci Technol 29:525–536. https://doi.org/10.51400/2709-6998.1587

Liao M-H, Tang S-L, Hsu C-M, Wen K-C, Wu H, Chen W-M, Wang J-T, Meng P-J, Twan W-H, Lu C-K, Dai C-F, Soong K, Chen CA (2007) The "black disease" of reef-building corals at Green Island, Taiwan—outbreak of a cyanobacteriosponge, *Terpios hoshinota* (Suberitidae; Hadromerida). Zool Stud 46:2

Lin M-F, Kitahara MV, Tachikawa H, Keshavmurthy S, Chen CA (2012) A new shallow-water species, *Polycyathus chaishanensis* sp. nov. (Scleractinia: Caryophylliidae), from Chaishan, Kaohsiung, Taiwan. Zool Stud 51:213–221

Liou C-Y, Yang S-Y, Chen CA (2017) Unprecedented calcareous algal reefs in northern Taiwan merit high conservation priority. Coral Reefs 36:1253

Liu P-J, Shao K-T, Jan R-Q, Fan T-Y, Wong S-L, Hwang J-S, Chen J-P, Chen C-C, Lin H-J (2009) A trophic model of fringing coral reefs in Nanwan Bay, southern Taiwan suggests overfishing. Mar Environ Res 68:106–117

Liu P-J, Meng P-J, Liu L-L, Wang J-T, Leu M-Y (2012) Impacts of human activities on coral reef ecosystems of southern Taiwan: a long-term study. Mar Pollut Bull 64:1129–1135

Loya Y, Puglise KA, Bridge TCL (eds) (2019) Mesophotic coral ecosystems. Coral reefs of the world, vol 12. Springer, New York. https://doi.org/10.1007/978-3-319-92735-0

Lu T-H, Liu C-Y, Lin C-C, Chung C-S, Chen F-S, Hwang W-C (1987) Investigations of round herring and squids fishing-grounds from Pescadores to Formosa Bank in 1985, vol 6. Annual collected reports of the Penghu Branch (Penghu Fishery Laboratory) Taiwan Fisheries Research Institute, pp 1–34. in Mandarin

Manzello DP, Kleypas JA, Budd DA, Eakin CM, Glynn PW, Langdon C (2008) Poorly cemented coral reefs of the eastern tropical Pacific: possible insights into reef development in a high-CO_2 world. Proc Natl Acad Sci U S A 105:10450–10455

Matear RJ, Lenton A (2018) Carbon–climate feedbacks accelerate ocean acidification. Biogeosciences 15:1721–1732

McRae CJ, Keshavmurthy S, Chen H-K, Ye Z-M, Meng P-J, Rosset SL, Huang W-B, Chen CA, Fan T-Y, Cote IM (2022a) Seasonal dynamics of algal symbionts and host lipids in corals from reefs with contrasting thermal regimes. Mar Ecol Prog Ser. under review

McRae CJ, Keshavmurthy S, Meng P-J, Rosset SL, Huang W-B, Chen CA, Fan T-Y, Cote IM (2022b) Variable responses to chronic and acute elevated temperature of three coral species from reefs with distinct thermal regimes. Mar Biol 169:97

Meng P-J, Chung K-N, Chen J-P, Chen M-H, Liu M-C, Chang Y-C, Fan T-Y, Lin H-J, Liu B-R, Chang C-M, Fang L-S, Shao K-T (2007a) Long-term ecological monitoring on the marine ecosystem of Kenting National Park. J Natl Park 17:71–88. in Mandarin with English abstract

Meng P-J, Chung K-N, Chen J-P, Chen M-H, Liu M-C, Chang Y-C, Fan T-Y, Lin H-J, Liu B-R, Chang C-M, Fang L-S, Shao K-T (2007b) Long-term ecological monitoring and studies of human activities on the marine ecosystem of Kenting National Park. J Natl Park 17:89–111. in Mandarin with English abstract

Meng P-J, Lee H-J, Wang J-T, Chen C-C, Lin H-J, Tew KS, Hsieh W-J (2008) A long-term survey on anthropogenic impacts to the water quality of coral reefs, southern Taiwan. Environ Pollut 156:67–75

Ministry of the Environment, Japanese Coral Reef Society (2004) Coral reefs of Japan. Ministry of the Environment, Tokyo

Mitsuki Y, Isomura N, Nozawa Y, Tachikawa H, Huang D, Fukami H (2021) Distinct species hidden in the widely distributed coral *Coelastrea aspera* (Cnidaria, Anthozoa, Scleractinia). Invertebr Syst 35:876–891

Modak A, Mauritsen T (2021) The 2000–2012 global warming hiatus more likely with a low climate sensitivity. Geophys Res Lett 48: e2020GL091779. https://doi.org/10.1029/2020GL091779

Mollica NR, Guo W, Cohen AL, Huang K-F, Foster GL, Donald HK, Solow AR (2018) Ocean acidification affects coral growth by reducing skeletal density. Proc Natl Acad Sci U S A 115:1754–1759

Morrison TH, Adger N, Barnett J, Brown K, Possingham H, Hughes T (2020) Advancing coral reef governance into the anthropocene. One Earth 2:64–74. https://doi.org/10.1016/j.oneear.2019.12.014

Muscatine L (1990) The role of symbiotic algae in carbon and energy flux in reef corals. In: Dubinsky Z (ed) Ecosystems of the world, 25. coral reefs. Elsevier Science Publishing Company, Inc., Amsterdam, pp 75–87

Nakabayashi A, Yamakita T, Nakamura T, Aizawa H, Kitano YF, Iguchi A, Yamano H, Nagai S, Agostini S, Teshima KM, Yasuda N (2019) The potential role of temperate Japanese regions as refugia for the coral *Acropora hyacinthus* in the face of climate change. Sci Rep 9:1892. https://doi.org/10.1038/s41598-018-38333-5

Nakamura J, Camargo SJ, Sobel AH, Henderson N, Emanuel KA, Kumar A, LaRow TE, Murakami H, Roberts MJ, Scoccimarro E, Vidale PL, Wang H, Wehner MF, Zhao M (2017) Western North Pacific tropical cyclone model tracks in present and future climates. J Geophys Res Atmos 122:9721–9744

Ng P-K, Chiou Y-S, Liu L-C, Sun Z, Shimabukuro H, Lin S-M (2019) Phylogeography and genetic connectivity of the marine macro-alga *Sargassum ilicifolium* (Phaeophyceae, Ochrophyta) in the northwestern Pacific. J Phycol 55:7–24

Nishihira M, Veron JEN (1995) Hermatypic corals of Japan. Kaiyusha, Tokyo

Nomura K (2016a) The illustrated zooxanthellate scleractinian corals of Kushimoto II VACATINA. Marine Pavilion Suppl 6:1–69

Nomura K (2016b) The illustrated zooxanthellate scleractinian corals of Kushimoto I REFERTINA. Marine Pavilion Suppl 5:1–56

Nozawa Y, Lin C-H, Meng P-J (2020) Sea urchins (diadematids) promote coral recovery via recruitment on Taiwanese reefs. Coral Reefs 39:1199–1207

Oksanen J, Blanchet FG, Friendly M, Kindt R, Legendre P, McGlinn D, Minchin PR, O'Hara RB, Simpson GL, Solymos P, Stevens MHH, Szoecs E, Wagner H (2020) Vegan: community ecology package. Version 2.5-7

Omija T (2004) Terrestrial inflow of soils and nutrients. In: The Japanese Coral Reef Society and Ministry of the Environment (ed) Coral reefs of Japan. Ministry of the Environment, Tokyo, pp 64–68

Pandolfi JM, Connolly SR, Marshall DJ, Cohen AL (2011) Projecting coral reef futures under global warming and ocean acidification. Science 333:418–422

Perry CT, Larcombe P (2003) Marginal and non-reef-building coral environments. Coral Reefs 22:427–432

Pichon M, Chuang Y-Y, Chen CA (2012) *Pseudosiderastrea formosa* sp nov (Cnidaria: Anthozoa: Scleractinia) a new coral species endemic to Taiwan. Zool Stud 51:93–98

Poveda G, Waylen PR, Pulwarty RS (2006) Annual and inter-annual variability of the present climate in northern South America and southern Mesoamerica. Palaeogeogr Palaeoclimatol Palaeoecol 234:3–27

Putnam HM, Mayfield AB, Fan T-Y, Chen CS, Gates RD (2013) The physiological and molecular responses of larvae from the reef-building coral *Pocillopora damicornis* exposed to near-future increases in temperature and pCO_2. Mar Biol 160:2157–2173

Qu T, Lukas R (2003) The bifurcation of the north equatorial current in the Pacific. J Phys Oceanogr 33:5–18

Randall CJ, Toth LT, Leichter JJ, Maté JL, Aronson RB (2020) Upwelling buffers climate change impacts on coral reefs of the eastern tropical pacific. Ecology 101:e02918. https://doi.org/10.1002/ecy.2918

Ribas-Deulofeu L, Denis V, De Palmas S, Kuo C-Y, Hsieh HJ, Chen CA (2016) Structure of benthic communities along the Taiwan latitudinal gradient. PLoS One 11:e0160601. https://doi.org/10.1371/journal.pone.0160601

Ribas-Deulofeu L, Denis V, Château P-A, Chen CA (2021) Impacts of heat stress and storm events on the benthic communities of Kenting National Park (Taiwan). PeerJ 9:e11744. https://doi.org/10.7717/peerj.11744

Rixen T, Jiménez C, Cortés J (2012) Impact of upwelling events on the sea water carbonate chemistry and dissolved oxygen concentration in the Gulf of Papagayo (Culebra Bay), Costa Rica: implications for coral reefs. Rev Biol Trop 60:187–195

Roberts MJ, Camp J, Seddon J, Vidale PL, Hodges K, Vannière B, Mecking J, Haarsma R, Bellucci A, Scoccimarro E, Caron L-P, Chauvin F, Terray L, Valcke S, Moine M-P, Putrasahan D, Roberts CD, Senan R, Zarzycki C, Ullrich P, Yamada Y, Mizuta R,

Kodama C, Fu D, Zhang Q, Danabasoglu G, Rosenbloom N, Wang H, Wu L (2020) Projected future changes in tropical cyclones using the CMIP6 HighResMIP multimodel ensemble. Geophys Res Lett 47:e2020GL088662. https://doi.org/10.1029/2020GL088662

Sánchez-Noguera C, Stuhldreier I, Cortés J, Jiménez C, Morales Á, Wild C, Rixen T (2018) Natural ocean acidification at Papagayo upwelling system (North Pacific Costa Rica): implications for reef development. Biogeosciences 15:2349–2360

Schönberg CHL, Fang JKH, Carreiro-Silva M, Tribollet A, Wisshak M (2017) Bioerosion: the other ocean acidification problem. ICES J Mar Sci 74:895–925

Schutter M, Nozawa Y, Kurihara H (2015) The effect of elevated CO_2 and increased temperature on *in vitro* fertilization success and initial embryonic development of single male:female crosses of broad-cast spawning corals at mid- and high-latitude locations. J Mar Sci Eng 3: 216–239

Senou H, Matsuura K, Shinohara G (2006) Checklist of fishes in the Sagami Sea with zoogeographical comments on shallow water fishes occurring along the coastlines under the influence of the Kuroshio current. Mem Nat Sci Mus Tokyo 41:389–542

Shao K-T (2009) Marine biodiversity and fishery sustainability. Asia Pac J Clin Nutr 18:527–531

Shao K-T, Peng C-I, Chang H-W, Hsu Y-C, Lin H-J, Lo W-T, Tew KS, Fan T-Y, Chan T-Y, Chiu Y-W, Soong K, Cheng I-J (2014) The biodiversity of Taiping Island (Itu Aba), Spratlys, South China Sea. Construction and Planning Agency, Minister of the Interior, Taiwan, Taipei

Silver A (2018) Science doi: https://doi.org/10.1126/science.aaw1461. Accessed 1 Dec 2021

Snelgrove PVR (2010) Discoveries of the census of marine life: making ocean life count. Cambridge University Press

Sobel AH, Camargo SJ, Hall TM, Lee C-Y, Tippett MK, Wing AA (2016) Human influence on tropical cyclone intensity. Science 353: 242–246

Sommer B, Beger M, Pandolfi J (2014a) Seasonal and temporal changes in reef fauna at Julian Rocks, 2010 to 2013. Report to Cape Byron Marine Park, NSW Department of Primary Industries

Sommer B, Harrison PL, Beger M, Pandolfi JM (2014b) Trait-mediated environmental filtering drives assembly at biogeographic transition zones. Ecology 95:1000–1009

Sommer B, Sampayo EM, Beger M, Harrison PL, Babcock RC, Pandolfi JM (2017) Local and regional controls of phylogenetic structure at the high-latitude range limits of corals. Proc R Soc B 284:20170915. https://doi.org/10.1098/rspb.2017.0915

Sommer B, Beger M, Harrison PL, Babcock RC, Pandolfi JM (2018) Differential response to abiotic stress controls species distributions at biogeographic transition zones. Ecography 41:478–490

Soong K, Chen CA, Chang J-C (1999) A very large poritid colony at Green Island, Taiwan. Coral Reefs 18:42

Soto D, De Palmas S, Ho M-J, Denis V, Chen CA (2018) Spatial variation in the morphological traits of *Pocillopora verrucosa* along a depth gradient in Taiwan. PLoS One 13:e0202586. https://doi.org/10.1371/journal.pone.0202586

Soto D, De Palmas S, Ho M-J, Denis V, Chen CA (2021) A molecular census of early-life stage scleractinian corals in shallow and mesophotic zones. Ecol Evol 11:14573–14584. https://doi.org/10.1002/ece3.8122

Sugihara K, Nomura K, Yokochi H, Shimoike K, Kajiwara K, Suzuki G, Zayasu Y, Dewa H, Fukami H, Kitano YF, Matsumoto H, Mezaki T, Nagata H, Tachikawa H, Kimura T (2015) Zooxanthellate scleractinian corals of Tanegashima Island, Japan. Center for Environmental Biology and Ecosystem Studies, National Institute for Environmental Studies, Tsukuba

Suzuki G, Keshavmurthy S, Hayashibara T, Wallace CC, Shirayama Y, Chen CA, Fukami H (2016) Genetic evidence of peripheral isolation and low diversity in marginal populations of the *Acropora hyacinthus* complex. Coral Reefs 35:1419–1432

Tang TY, Hsueh Y, Yang YJ, Ma JC (1999) Continental slope flow northeast of Taiwan. J Phys Oceanogr 29:1353–1362

Tang D, Kester DR, Ni I-H, Kawamura H, Hong H (2002) Upwelling in the Taiwan Strait during the summer monsoon detected by satellite and shipboard measurements. Remote Sens Environ 83:457–471

TCCIP (2021) Taiwan climate change projection information and adaptation knowledge platform. https://tccip.ncdr.nat.gov.tw/index_eng.aspx

TEIA (2020) Reef check Taiwan annual survey report 2020. Taiwan Environmental Information Association, Taipei

Toth LT, Aronson RB, Vollmer SV, Hobbs JW, Urrego DH, Cheng H, Enochs IC, Combosch DJ, van Woesik R, Macintyre IG (2012) ENSO drove 2500-year collapse of eastern pacific coral reefs. Science 337:81–84

Toth LT, Aronson RB, Cobb KM, Cheng H, Edwards RL, Grothe PR, Sayani HR (2015) Climatic and biotic thresholds of coral-reef shutdown. Nat Clim Chang 5:369–374

Tribollet A, Godinot C, Atkinson M, Langdon C (2009) Effects of elevated pCO_2 on dissolution of coral carbonates by microbial euendoliths. Global Biogeochem Cycles 23:GB3008. https://doi.org/10.1029/2008GB003286

Tuckett CA, Wernberg T (2018) High latitude corals tolerate severe cold spell. Front Mar Sci 5:14. https://doi.org/10.3389/fmars.2018.00014

UN Environment, ISU, ICRI, Trucost (2018) The coral reef economy: the business case for investment in the protection, preservation and enhancement of coral reef health

Veron JEN (1992) Hermatypic corals of Japan. Monograph series vol 9, Australian Institute of Marine Science, Townsville

Veron JEN (1995) Corals in space and time: the biogeography and evolution of the Scleractinia. UNSW Press, Sydney

Veron JEN (2000) Corals of the world. Australian Institute of Marine Science, Townsville

Wilkinson C (1998) Status of coral reefs of the world: 1998. Australian Institute of Marine Science, Townsville

Williams GJ, Graham NAJ, Jouffray J-B, Norström AV, Nyström M, Gove JM, Heenan A, Wedding LM (2019) Coral reef ecology in the anthropocene. Funct Ecol 33:1014–1022

Wisshak M, Schönberg CHL, Form A, Freiwald A (2012) Ocean acidification accelerates reef bioerosion. PLoS One 7:e45124. https://doi.org/10.1371/journal.pone.0045124

Wisshak M, Schönberg CHL, Form A, Freiwald A (2013) Effects of ocean acidification and global warming on reef bioerosion—lessons from a clionaid sponge. Aquat Biol 19:111–127

Wizemann A, Nandini SD, Stuhldreier I, Sánchez-Noguera C, Wisshak M, Westphal H, Rixen T, Wild C, Reymond CE (2018) Rapid bioerosion in a tropical upwelling coral reef. PLoS One 13: e0202887. https://doi.org/10.1371/journal.pone.0202887

WoRMS Editorial Board (2021) World Register of Marine Species. https://www.marinespecies.org. Accessed 21 Nov 2021

Wu C-C (2014) Cross-scale and multi-level mismatch problems in marine natural resources management: case studies in the Penghu Archipelago, Taiwan. Reg Environ Change 14:2079–2087

Wu S-W, Chung J (2020) NPP calls on Taipower to stop LNG project. In: Taipei Times. https://taipeitimes.com/News/taiwan/archives/2020/12/19/2003749000. Accessed 19 Dec 2020

Wu C-R, Hsin Y-C (2005) Volume transport through the Taiwan Strait: a numerical study. Terr Atmos Ocean Sci 16:377. https://doi.org/10.3319/TAO.2005.16.2.377(Oc)

Wu C-C, Kuo Y-H (1999) Typhoons affecting Taiwan: current understanding and future challenges. Bull Am Meteorol Soc 80:67–80

Wu C-R, Lu H-F, Chao S-Y (2008) A numerical study on the formation of upwelling off northeast Taiwan. J Geophys Res 113:C08025. https://doi.org/10.1029/2007JC004697

Yamano H, Sugihara K, Nomura K (2011) Rapid poleward range expansion of tropical reef corals in response to rising sea surface temperatures. Geophys Res Lett 38:L04601. https://doi.org/10.1029/2010GL046474

Yamano H, Sugihara K, Watanabe T, Shimamura M, Hyeong K (2012) Coral reefs at 34°N, Japan: exploring the end of environmental gradients. Geology 40:835–838

Yang J-A, Kim S, Son S, Mori N, Mase H (2020) Assessment of uncertainties in projecting future changes to extreme storm surge height depending on future SST and greenhouse gas concentration scenarios. Clim Change 162:425–442

Yara Y, Oshima K, Fujii M, Yamano H, Yamanaka Y, Okada N (2011) Projection and uncertainty of the poleward range expansion of coral habitats in response to sea surface temperature warming: a multiple climate model study. Galaxea J Coral Reef Stud 13:11–20

Dynamics of Coral Reef Communities in the Sekisei Lagoon, Japan, Following the Severe Mass Bleaching Event of 2016

3

Mariyam Shidha Afzal, Kana Ikeda, Mitsuhiro Ueno, and Takashi Nakamura

Abstract

The rise in sea surface temperature due to climate change has caused severe and widespread coral bleaching phenomena in the recent decades, resulting in considerable degradation of coral reef habitats globally. The current study examined the impact of the 2016 mass coral bleaching event on Japan's largest reef system in the Sekisei Lagoon, located in the Okinawa prefecture in the southern-most part of Japan. The post-bleaching changes during 2017, 2019, and 2020 were analyzed. The results indicated that the 2016 mass bleaching was severe, with an average of 66.40 ± 16.35% of the corals bleached in the Sekisei Lagoon. The mean post-bleaching mortality rate was 12.85 ± 9.68%. The bleaching severity was the highest among the *Acropora* species, with 79.06% of the colonies bleached. Mortality rate was the highest among branching species, such as *Millepora* (28.82%), *Seriatopora* (23.52%), and *Acropora* (15.81%), whereas massive or encrusting species fared better. Between 2016 and 2020, the average coral cover in the Sekisei Lagoon decreased significantly (Kruskal-Wallis, $\chi^2 = 24.08$, $p < 0.0001$), which coincided with the increase in algal assemblages. If the current level of coral degradation continues with impeded recovery, the Sekisei Lagoon reefs could turn into algal-dominated communities with restricted coral growth and presence.

Keywords

Coral bleaching · Community structure · High temperature · Sekisei Lagoon

3.1 Introduction

Coral reefs all around the world are under threat from the rapidly changing climate and the resulting consequences of excessive thermal warming, such as coral bleaching and mass mortality (Hoegh-Guldberg et al. 2007; Frade et al. 2018). When the coral-algal symbiosis is stressed, corals expel their endosymbiotic algae (zooxanthellae) in a phenomenon known as "coral bleaching", which can result in partial or complete coral mortality if the stress factors are prolonged (Brown 1997). Several major bleaching episodes have been recorded since the 1980s by researchers on the Ryukyu Islands (Fujioka 1999). In 1983, long-term monitoring of the Japanese coral reefs was introduced in the Sekisei Lagoon that is located between the Ishigaki and Iriomote Islands of Okinawa (Muko et al. 2019). With an area of 300 km^2, the Sekisei Lagoon is known to be the largest coral reef in Japan (Muko et al. 2019).

During the large-scale bleaching event of 1998, several Indo-Pacific areas experienced high mortality of reef-building corals. Similarly, coral communities in the Ryukyu Archipelago were also severely affected (Loya et al. 2001; Wilkinson et al. 1999). The first detailed study on the impacts of bleaching and its influence on coral communities was carried out on the Ishigaki Island during the 1998 global mass bleaching event (Fujioka, 1999). According to Fujioka (1999), the bleaching of 1998 was unlike other previously recorded events, as it was observed to be widespread among

M. S. Afzal
Faculty of Science, Graduate School of Engineering and Science, University of the Ryukyus, Nishihara, Okinawa, Japan

K. Ikeda
Faculty of Science, Graduate School of Engineering and Science, University of the Ryukyus, Nishihara, Okinawa, Japan

Regeneration Network of the Sea in Yoron Island (NPO), Yoron, Kagoshima, Japan

M. Ueno
Coral Reef Research in Sekisei Lagoon, Ishigaki, Okinawa, Japan

T. Nakamura (✉)
Faculty of Science, Graduate School of Engineering and Science, University of the Ryukyus, Nishihara, Okinawa, Japan

Sesoko Station, Tropical Biosphere Research Center, University of the Ryukyus, Nishihara, Okinawa, Japan
e-mail: takasuke@sci.u-ryukyu.ac.jp

I. Takeuchi, H. Yamashiro (eds.), *Coral Reefs of Eastern Asia under Anthropogenic Impacts*, Coral Reefs of the World 17,
https://doi.org/10.1007/978-3-031-27560-9_3

almost all the coral species and within varying depths from reef flat to 23 m deep.

Following that of 1998, the bleaching event of 2001 induced further deterioration of coral communities in the Sekisei Lagoon (Okamoto et al. 2007). In 2007, the Sekisei Lagoon reefs were again subjected to severe bleaching; consequently, 60% mortality rate of *Acropora* corals was observed (Okamoto et al. 2007; Nojima and Okamoto 2008; Harii et al. 2014). In addition to the Sekisei Lagoon reefs being subjected to chronic bleaching disturbances, the area is prone to highly frequent typhoons, coral diseases, and crown-of-thorns starfish outbreaks (Muko et al. 2019). Recurrent outbreaks of the crown-of-thorns starfish which feeds on coral polyps has been observed and recorded in the Sekisei Lagoon since the beginning of the 1980s (Yaeyama Marine Park Research Station 1983; Muko et al. 2019). Coinciding with the 2007 bleaching, the reefs were also decimated by five consecutive typhoons between 2004 and 2007 that passed through the Ryukyu archipelago, resulting in loss of *Acropora* and *Montipora* assemblages (Harii et al. 2014; Muko et al. 2019). Prior to the 2007 bleaching event, the average coral cover of the Sekisei Lagoon remained above 40%, according to the Coral Reef Monitoring data of Japan. However, between 2007 and 2010, Kimura et al. (2014) observed a periodic decline in coral cover with only a slight recovery that started in 2011. Furthermore, between 2014 and 2017, there were several consecutive coral bleaching events with adverse effects globally. In 2016, the Sekisei Lagoon reefs were subjected to one of the most severe bleaching events ever recorded, in which a relatively high mortality was observed among massive coral species that are otherwise known to be "winners" when it comes to resilience to thermal stress (Nakamura, 2017). More than 91% of reefs are expected to experience yearly bleaching-level thermal stress by 2050, according to future projections (Heron et al. 2016). Systematic monitoring and long-term research data have enabled an understanding of the changes in the coral community structure following acute and cumulative disturbances. Moreover, this enables researchers to determine the key factors that influence changes among coral communities.

Bleaching susceptibility and rates of mortality following the disturbances vary according to coral species (McClanahan 2004). For accurate prediction of future reef community structure and subsequent species, understanding the patterns of bleaching susceptibility among key coral taxa (Marshall and Baird, 2000) and the spatiotemporal variations of degraded reef recovery (Graham et al. 2015) is crucial. Awareness of the changes among the coral communities under environmental perturbations, such as large-scale bleaching events, will facilitate a better understanding and prediction of their responses to future climate change events (Muko et al. 2019).

The coral communities in the Sekisei Lagoon and its surrounding Islands are vital for the future of Japanese coral reefs as they are considered a principal seeding population for the northern Japanese coral reefs that are connected by the Kuroshio Current (Furushima et al. 2002). This study examined the reef substrate composition along with coral cover percentages and species taxa in the Sekisei Lagoon during (September 2016) and the following 3 years (September of 2017, 2019, and 2020) after the bleaching to quantify the impact and the subsequent changes that occurred after the bleaching. Additionally, we compared the distribution patterns and the difference in bleaching susceptibility among coral genera levels.

3.2 Methods

3.2.1 Study Area

This study was carried out in the semi barrier reefs, fringing reefs, and small patch reefs of the Sekisei Lagoon, which is located between the Iriomote and Ishigaki Islands in the Ryukyu Archipelago of Japan (Fig. 3.1). This reef system is 20 km wide in the east-west direction and 15 km long in north-south direction, where large parts of the inner reef system are relatively shallow (2–5 m) but can extend to a maximum depth of >20 m (Muko et al. 2019). The outer reefs of the northern and southern barrier reefs extend down to a depth of 25 m and are mainly rocky (Okamoto et al. 2007). According to Shimoike (2004), the coral reef communities in the Sekisei Lagoon were dominated by fast growing *Acropora* and *Montipora* species and were highly diverse, accounting for more than 363 hermatypic coral species.

Thirty reefs in the Sekisei Lagoon were initially surveyed in September of 2016 during the peak of a mass scale bleaching event. These included outer reef sites (Anadomari farside, Urabishi east farside, St. M, Kuroshima Nakamoto, St. 20, St.18, and Katagua farside). The rest were more sheltered patch reef sites inside the lagoon system (Fig 3.1b). All 30 survey sites were revisited in September 2017, to record the changes in community composition and coral cover. The same surveys were repeated in September 2019 and 2020, for all reef sites (except St. F, Milky way, and Kuroshima Route South in 2019, Katagua farside in 2020) to assess the post-bleaching changes in substrate cover as well as potential recovery of coral communities.

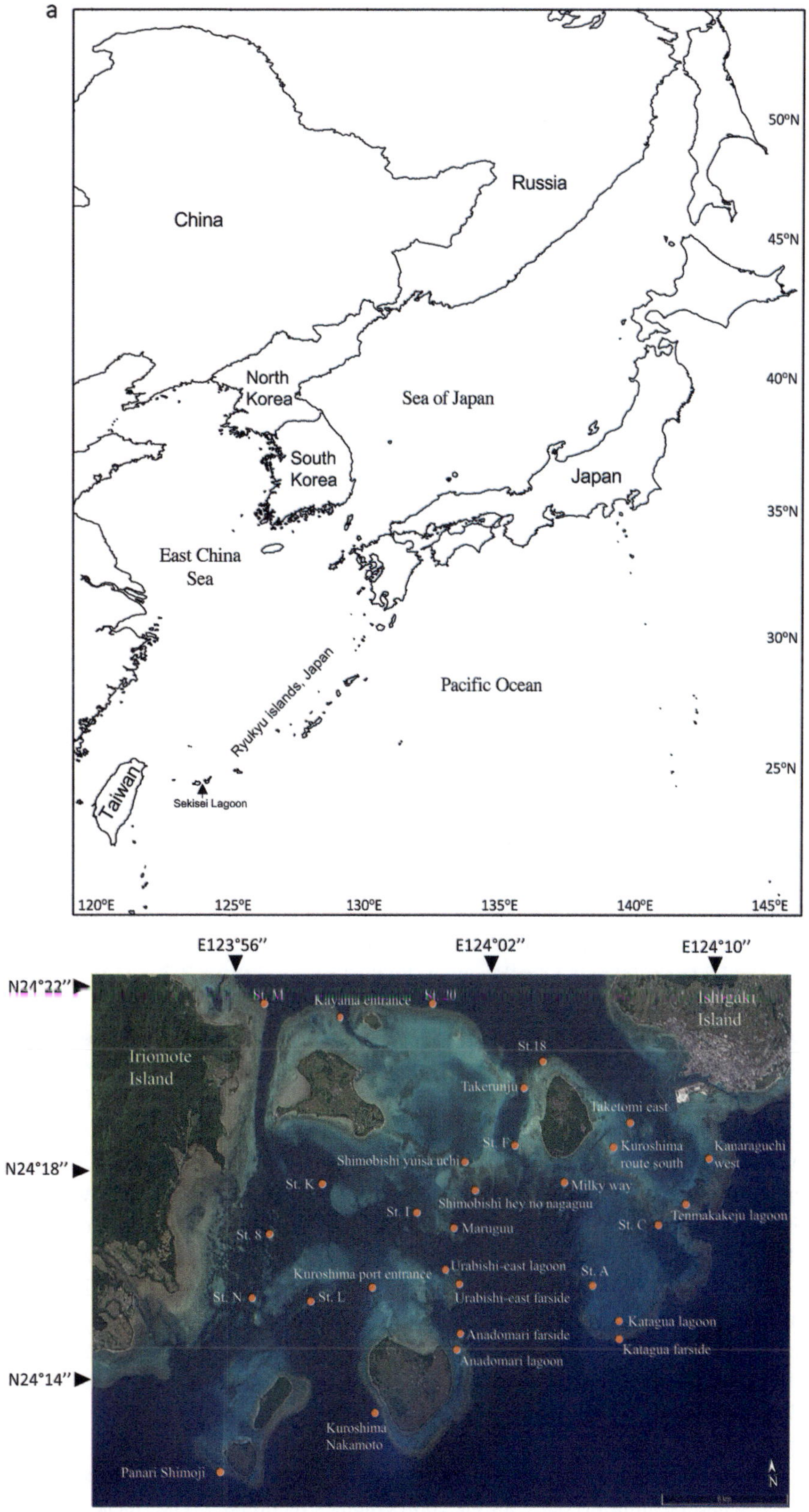

Fig. 3.1 (**a**) Map of Japan with the Sekisei Lagoon located in the Ryukyu Island chain, southern Japan. Map generated from ArcGIS 10.3 software (Environment Systems Research Institute (ESRI), 2015). (**b**) Magnified map (Base map from Google Earth Pro V 7.3.4.8248 (2015)) showing overview of the survey locations. A total of 30 survey sites were assessed for this study ranging from patch reefs inside the lagoon system to outer fringing reefs. The depth of the survey sites ranged from 1.5 to 8 m

3.2.2 Sea Surface Temperature (SST) and Degree Heating Week (DHW)

Raw data of daily maximum SST and DHW from 1998, 2007, and 2016 (three worst bleaching events in the history of the Sekisei Lagoon) were downloaded from the United States National Oceanic and Atmospheric Administration's (NOAA) Coral Reef Watch (CRW) website (https://coralreefwatch.noaa.gov/products/vs/data.php). NOAA CRW uses data from satellite sensors to track global SST anomalies and create near real-time coral bleaching "hotspot" maps with a 0.5° (50 km) resolution (Kayanne 2017). To anticipate bleaching, the hotspot anomalies were added together to calculate DHW, which is calculated by the accumulating temperature anomalies exceeding the maximum of the monthly mean SST for a given location over a 12-week period (Kayanne 2017; Liu et al. 2017; Skirving et al. 2020). DHW values of >4 °C-weeks are known to trigger coral bleaching, whereas DHW values of >8 °C-weeks are thought to cause widespread bleaching and bleaching-induced coral death (Kayanne, 2017; NOAA Coral Reef Watch, 2019). The downloaded raw temperature data were pooled together and analyzed into time series graphs to observe the temperature changes and DHW during severe bleaching years.

3.2.3 Field Methods and Statistical Analysis

Benthic cover, bleaching intensity, and coral genera composition were recorded from five replicates of a 20 m line transect laid parallel to the depth contour by the line intercept transect method (data recorded every 0.5 m). Benthic cover was quantified within 11 different groups, including live hard coral, soft coral, sponge, bare rock, rubble, sand, macro algae, turf algae, crustose coralline algae, silt, and "others" group for substratum such as zoanthids, anemones, tunicates etc. During the bleaching event of 2016, each coral colony or individual was recorded into one of the four categories set by Frade et al. (2018): healthy/non-bleached (no signs of bleaching observed in the colony), minor bleaching (paling or 1–50% bleached), severe bleaching (51–100% bleached), and recently dead (bare skeletal features visible with filamentous algae starting to settle). Based on field observations, the bleaching intensity in the Sekisei Lagoon was severe, as almost all the corals that we encountered recorded as "severe" appeared fully bleached. Following coral taxonomic guide by Veron and Stafford-Smith (2000), Nishihira and Veron (1995), and Nishihira (2004), all the coral colonies recorded in transect were identified up to their genera. In addition to coral genus identification, Simpson's diversity index was calculated for each site to determine the temporal changes and the overall live percentage cover of each coral genera over the four sampling years (Fig. 3.8). The coral genera cover for the year 2016 also included coral colonies undergoing bleaching, as these colonies could regain their lost zooxanthellae and return to normal conditions. However, dead coral colonies due to bleaching were not included in this count.

Overall bleaching intensity, and bleaching-induced death categorized as "mortality" among each coral genera were recorded during the 2016 surveys, and the results were analyzed into a scatterplot showing bleaching percentage vs. mortality (Fig. 3.4) among the most commonly found genera in the Sekisei Lagoon.

The mean and standard deviation of substratum composition percentage cover were averaged across five transects for each reef and compared among the survey years. Since raw and log transformed data were not distributed normally, comparisons were made using the non-parametric Kruskal-Wallis H test. The Wilcoxon signed-rank test was also used to compare the differences in substrate covers among the sampling years. Similarities among the survey years and sites were visualized using non-Metric MultiDimensional Scaling (nMDS) ordinations using the Bray–Curtis similarity index performed by the metaMDS function of the "vegan" package in R Studio software (Oksanen et al. 2013; Kruskal 1964). In addition, community-level differences between the years were tested with PERMANOVA of Bray–Curtis dissimilarities using the Adonis function. All statistical calculations were carried out using R Studio (R Studio Team 2020).

3.3 Results

3.3.1 Sea Surface Temperature (SST) and Degree Heating Week (DHW) of the Sekisei Lagoon

Maximum sea surface temperatures of the Sekisei Lagoon ranged between an average of 29–30 °C in summer months (July–September) during the severe bleaching events. The maximum temperatures recorded from NOAA SST data were 31.51 °C in August and 31.12 °C in September of 1998. For 2007, the maximum temperature recorded were 31.36 °C in July and 31.49 °C in August. The maximum SST data of 2016 revealed maximum recorded temperatures of 30.86 °C in July, 31.27 °C in August, and 30.63 °C in September (Fig. 3.2a). These temperature spikes corresponded with bleaching episodes in the Sekisei Lagoon. The bleaching threshold level for the southern Ryukyu Islands where the Sekisei Lagoon is situated was 30 °C according to NOAA Coral Reef Watch (NOAA Coral Reef Watch, 2019) satellite data. In comparison with the three major bleaching events, the highest recorded DHW value

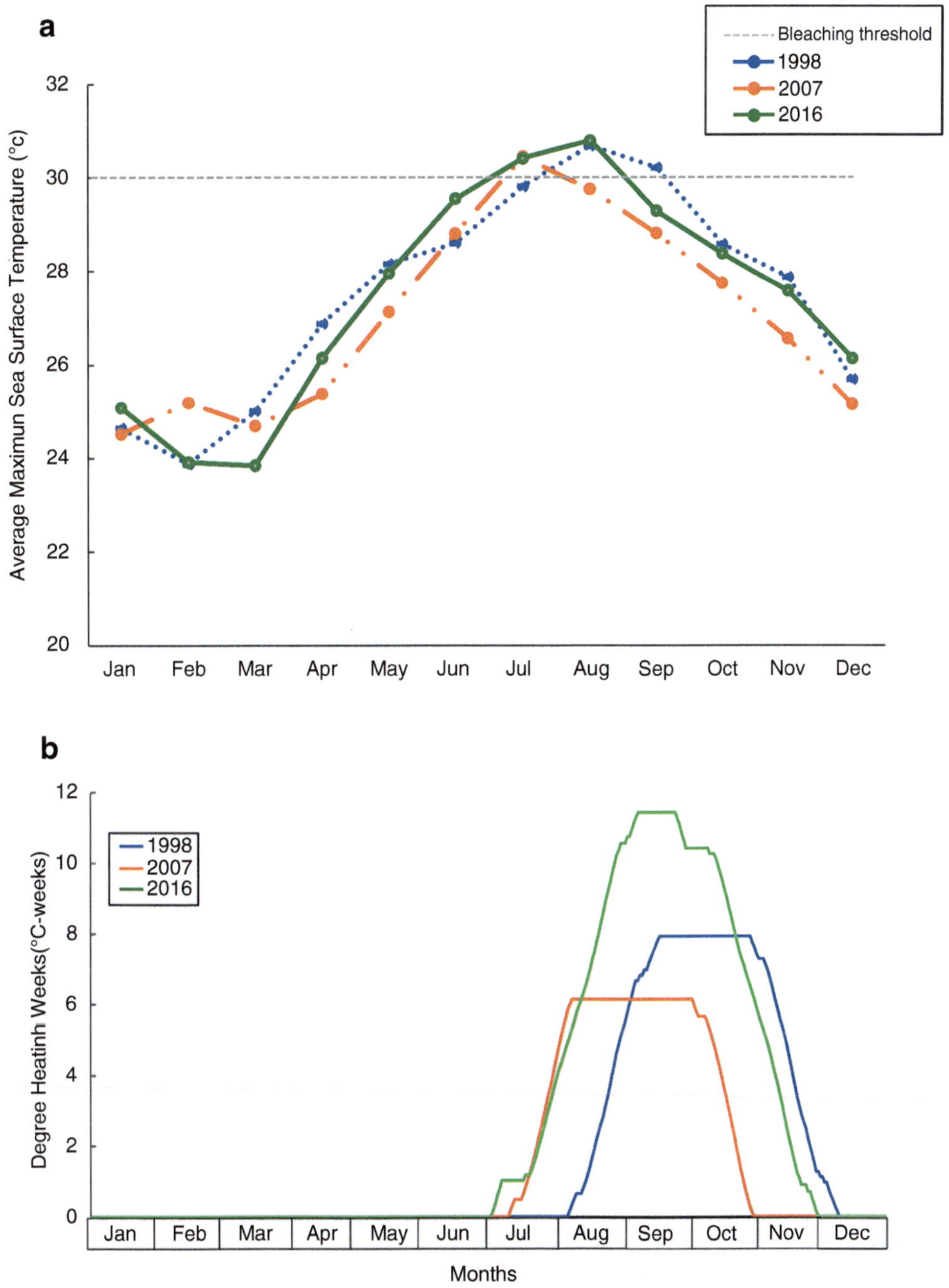

Fig. 3.2 (**a**) Monthly means of maximum sea surface temperature (SST) during major coral bleaching years of 1998, 2007, and 2016 for the Sekisei Lagoon, estimated from satellite data (NOAA Coral Reef Watch, 2019). (**b**) Degree heating week (DHW) estimated from satellite data off the Sekisei Lagoon during major bleaching years of 1998, 2007, and 2016 (NOAA Coral Reef Watch, 2019)

was 11.42 °C-weeks in September of 2016, whereas in 1998 bleaching event it was 7.91 °C-weeks from September to October (Fig. 3.2b). The maximum DHW recorded for 2007 bleaching event was 6.13 °C-weeks from August to September of 2007. There is significant difference between the 1998 and 2016 bleaching events (Wilcoxon signed-rank test, $p < 0.00001$). In contrast to 1998 and 2007 bleaching events, the severity of 2016 bleaching was predictably higher. According to NOAA CRW, DHW values >8 °C-weeks is known to cause widespread bleaching and coral mortality. Similar level of widespread bleaching and coral mortality was quite evident for the Sekisei Lagoon reefs in September of 2016 during the survey period. In addition, DHW did not exceed 8 °C-weeks in 1998 and 2007, but it did in 2016 with a maximum of 11.42 °C-weeks.

3.3.2 Coral Bleaching Patterns and Susceptibility Among Coral Genera in the Sekisei Lagoon

The recorded coral bleaching in 2016 of all the coral genera in the Sekisei Lagoon was relatively high, with an average of 66.40 ± 16.35% (Mean ± SD). The mean mortality rate recorded in the survey (during the bleaching event) was 12.85 ± 9.68%. Overall, the majority of the bleached coral colonies (73.9%) observed in the Sekisei Lagoon in 2016 experienced severe bleaching. Only 11% of the coral colonies were recorded to be in healthy condition, and 2.5% of coral colonies were undergoing minor bleaching.

A site-based comparison between six sites (Urabishi east lagoon, St. M, Milky way, St. C, Katagua Lagoon, and Taketomi east) revealed that all the recorded coral colonies were either severely bleached or recently dead (Fig. 3.3). Only the following sites, Panari Shimoji South Lagoon, Kanaraguchi west, and Tenmakakejyu Lagoon had healthy coral colonies amounting to above 40%. These sites also exhibited the lowest mortality rate (1–2.5%) during the bleaching event.

The low mortality and high number of healthy/non-bleached coral cover during 2016 in these three sites could be attributed to the fact that these sites were dominated by massive coral species such as *Porites* and *Montipora* which are more resistant to thermal stress compared to the bleaching-sensitive *Acropora* species. In terms of bleaching severity in the context of reef exposure, there was no significant difference ($F = 0.205$, $p < 0.65$) between the outer reef sites (68.14 ± 17.82%) and sheltered inner reef sites (65.88 ± 15.90%).

In September 2016, a total of 34 coral genera were recorded with a mean diversity index of 0.56 ± 0.15. Different levels of bleaching were observed for almost all the recorded coral genera; however, there were distinct differences observed among the genera in their susceptibility to bleaching and mortality.

Overall, the most abundant genera were *Acropora* (49.83 ± 22.98%), *Montipora* (13.45 ± 18.54%), *Millepora* (9.62 ± 7.92%), and *Porites* (8.22 ± 8.11%). The rest of the genera represented less than 3% of overall coral cover in the surveyed sites of the Sekisei Lagoon. Of all the *Acropora* corals recorded, 79.06% of the observed colonies were undergoing "severe bleaching" and 15.81% of them

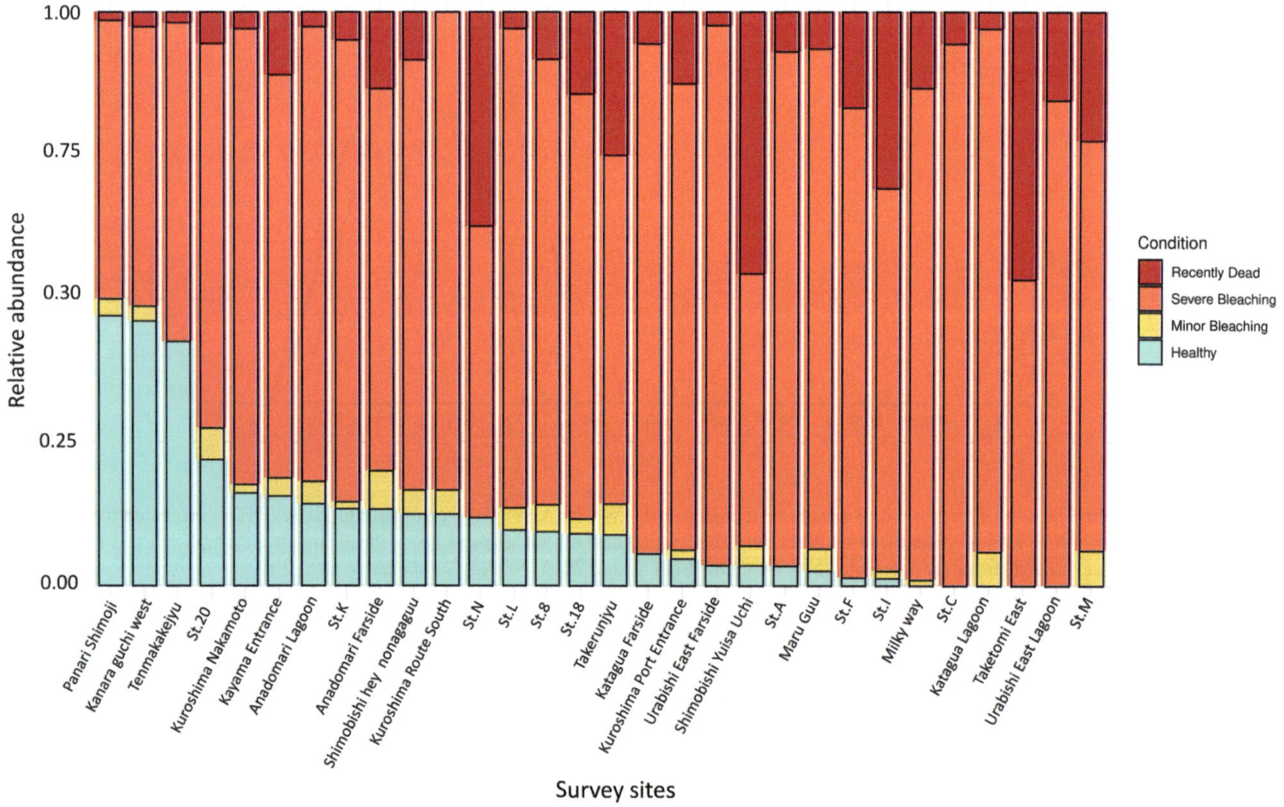

Fig. 3.3 Stacked bar graphs provide the relative abundance of different bleaching categories in each site. The data was collected between the depths of 2 m and 8 m in the Sekisei Lagoon during the peak of the bleaching event from the 3rd to 12th of September, 2016. The stacked bars are arranged by site according to relative abundance of healthy coral cover from high to low

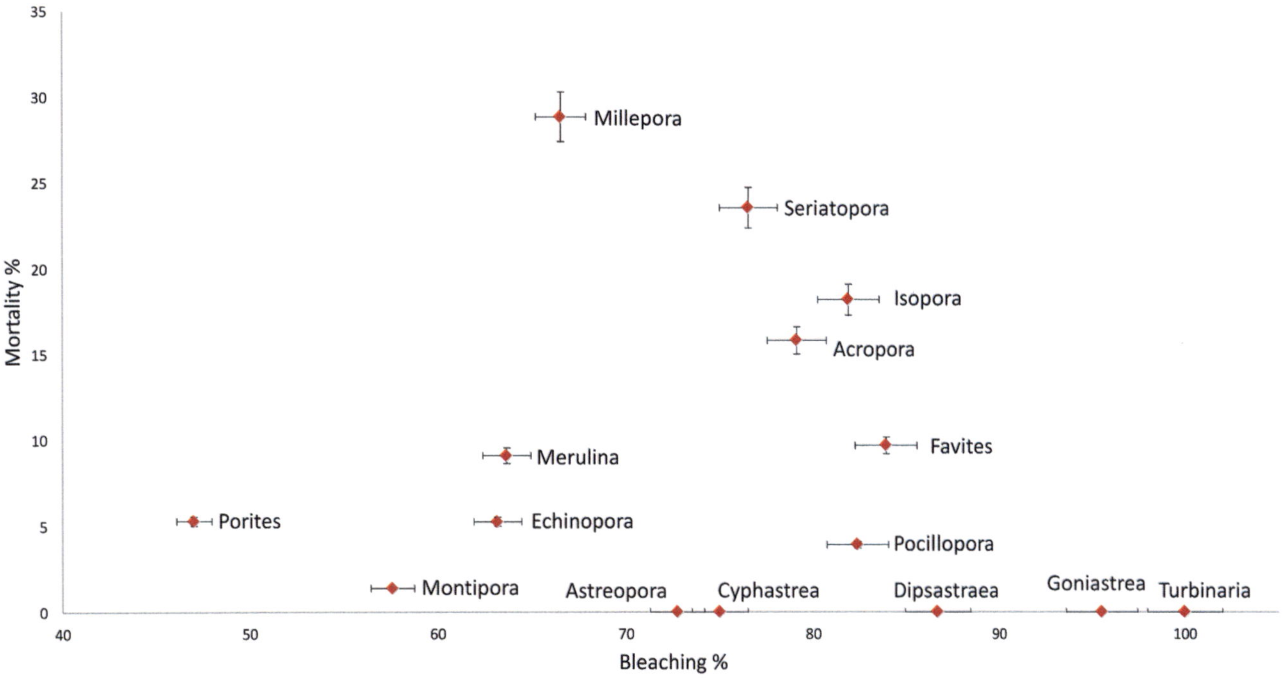

Fig. 3.4 Scatterplot showing 15 of the most abundant coral genera in the Sekisei Lagoon with their bleaching susceptibility and mortality rates (bleaching-induced death) during the mass bleaching event in September of 2016

experienced mortality (bleaching-induced death, Fig. 3.4). Only 3.70% of *Acropora* corals were recorded as "healthy" during the survey. Among the predominant genera coverage, tabular and branching *Acropora* species were the most severely impacted taxa in the majority of the surveyed locations. Among the *Montipora* species, 57.60% was recorded as "severely bleached," while only 1.38% was observed to have experienced bleaching-induced death and 37.32% were in "healthy" condition. Among the *Millepora* species, "severe bleaching" was recorded to be at 66.47% with 28.82% mortality, which was the highest among all coral genera, while only 3.52% of them were in "healthy" condition. For the *Porites* genera, "severe bleaching" was observed at 47.01% and the mortality rate was 5.28%. Among all the genera, *Porites* had the highest percentage of healthy individuals, with 44.37%.

Mortality was high among branching species such as *Millepora* (28.82%), *Seriatopora* (23.52%), and *Acropora* (15.81%), whereas massive or encrusting species fared better. It is known that a coral colony's growth morphology and species type influences the susceptibility to bleaching, where finer branched species are more sensitive to bleaching stresses than massive or encrusting species (Loya et al. 2001, Hongo and Yamano 2013). However, the coral genera *Turbinaria, Goniastrea, Dipsastraea,* and *Favites* were generally found as the most sensitive in the severely bleached group (average of 80–100% of the population bleached) during the 2016 survey in the Sekisei Lagoon.

3.3.3 Temporal Changes Observed in Reef Community Assemblages Following Bleaching

Relative benthic cover on the surveyed sites in the Sekisei Lagoon was dominated by "bare rock" (46.26 ± 19.48) and "hard coral" (28.91 ± 13.74%) in 2016 (Fig. 3.5). However, 1 year after the bleaching event in 2017, dominant benthic assemblage had shifted to "bare rock" (26.85 ± 14.13%) followed by "turf algae" (26.42 ± 17.38%).

A significant decrease (Kruskal-Wallis, $\chi^2 = 25.3$, $p < 0.00001$) in overall hard coral cover was observed in the entire lagoon system during the post-bleaching surveys, where the coral cover decreased from 28.91 ± 13.74% in 2016 to 13.68 ± 6.03% in 2020. Wilcoxon signed-rank test further supported that there was a significant change ($p < 0.001$) in hard coral cover in the years of 2016, 2017, 2019, and 2020 (Table 3.1). The recovery rate of hard coral cover has been relatively slow and non-uniform across the sites over the 4 year post-bleaching period. Outer reef sites (Urabishi east farside, St. 20, and St. 18) and those located at the large channel system (St. M and Kayama entrance) exhibited relatively better recovery than those at the sheltered patch reef sites inside the lagoon system. This could be because the survey sites near St. M have a strong tidal flow velocity at high tide, which exceeds 2 kns. Moreover, offshore water flows from the southern side of the lagoon system and passes through the channel on the north side, reducing the water temperature in the process (Furushima et al. 2002).

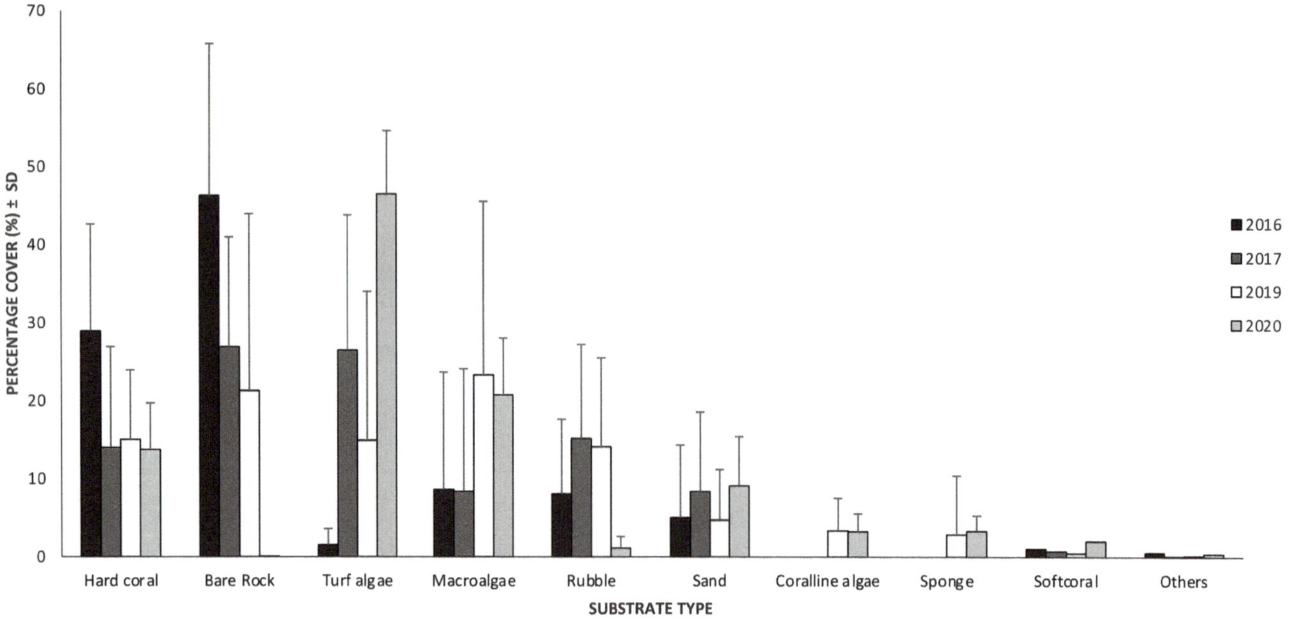

Fig. 3.5 Changes in averages of the substrate categories of reefs surveyed in the Sekisei Lagoon in 2016, 2017, 2019, and 2020. The bars represent the average percentage cover for each year with standard deviation

The decline in hard coral cover following the bleaching event coincides with the increase of algal assemblages in both categories of macroalgae and turf algal cover. The mean turf algal cover recorded in the Sekisei Lagoon in 2016 was 1.51 ± 2.08%, followed by an unprecedented increase to 24.42 ± 17.38% in 2017. A significant increase (Kruskal-Wallis, $\chi^2 = 80.4$, $p < 0.00001$) up to 46.44 ± 8.18% in turf algal cover was recorded in 2020. Wilcoxon signed-rank test showed significant difference ($p < 0.001$) between the years 2016, 2017, 2019, and 2020. The overall macroalgal cover in 2016 was 8.57 ± 15.08% and the cover remained almost identical (8.34 ± 15.74%) in 2017. However, 3 years post-bleaching in 2019, macroalgae became most dominant overall substrate category in Sekisei lagoon, with coverage up to 23.28 ± 22.25%. Wilcoxon signed-rank test showed significant difference ($p < 0.001$) in macroalgal cover from 2016 to

Table 3.1 Comparison of average substrate cover between the years in the Sekisei Lagoon, and the result of the Kruskal-Wallis test on the significance of the substrate categories in relation to the sampling years. Wilcoxon signed-rank test showed the significance between each separate year in relation to the substrate categories. Significance codes for p values are as follows: <0.0001 "***," <0.001 "**," <0.01 "*," and >0.05 Not significant (NS)

	Hard coral		Turf algae		Macro algae		Bare rock		Rubble		Sand	
Month/year	Mean	SD	Mean	SD	Mean	SD	Mean	SD	Mean	SD	Mean	SD
September 2016	28.91	13.74	1.51	2.08	8.57	15.08	46.26	19.48	8.02	9.63	4.98	9.31
September 2017	13.97	12.95	26.42	17.38	8.34	15.74	26.85	14.13	15.14	12.06	8.31	10.25
September 2019	14.99	8.94	14.85	19.12	23.28	22.25	21.30	22.65	14.07	11.42	4.71	6.46
September 2020	13.68	6.03	46.44	8.18	20.74	7.27	0.03	0.05	1.14	1.48	9.08	6.29
Kruskall–Wallis test												
Chi- squared	25.3		80.4		30.8		79.5		32		4.95	
df	3		3		3		3		3		3	
P value	***		***		***		***		***		0.17	
Wilcoxson signed rank test	*p* value		*p* value		*p* value		*p* value		*p* value		*p* value	
Cover 2016–2017	***		***		NS		***		NS		NS	
Cover 2016–2019	***		*		***		***		NS		NS	
Cover 2016–2020	***		***		***		***		***		NS	
Cover 2017–2019	NS		***		***		NS		NS		NS	
Cover 2017–2020	NS		***		***		***		***		NS	
Cover 2019–2020	NS		***		NS		***		***		NS	

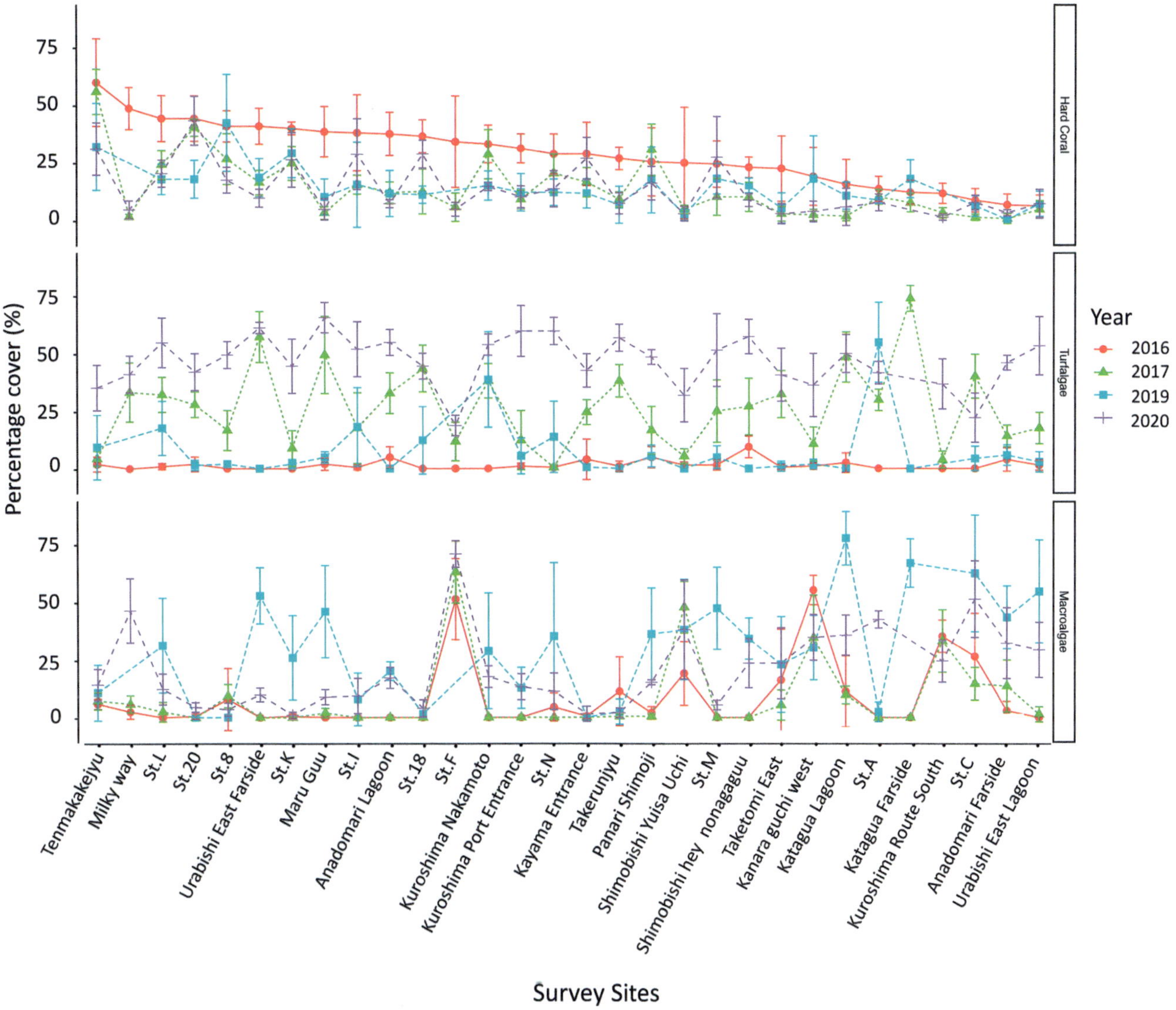

Fig. 3.6 Site-based comparison of changes in the average hard coral, turf algae, and macroalgae cover of all the reefs surveyed in the Sekisei Lagoon in 2016, 2017, 2019, and 2020. The line plots are plotted site-wise according to percentage cover of hard coral in 2016 from high to low

2017, 2016 to 2019, 2016 to 2020, 2017 to 2019, and 2017 to 2020.

Individual site comparison showed sharp decline in hard coral cover (20% or more decrease in initial coral cover) 1 year post-bleaching in the sites of the Kuroshima port entrance, Anadomari lagoon, Urabishi east farside, St. F, Shimobishi yuisa uchi, Milky way, St. L, Maruguu, St. I, St. 18, and the Taketomi island (Fig. 3.6). One of the most severely impacted sites was the Milky way, where the overall coral cover in 2016 was 48.74 ± 9.12%, but has declined to 1.95 ± 1.09% in 2017 due to thermal stress. In 2020, the recovery of hard corals in Milky way was still impeded as the coral cover remained below 5%. Prior to bleaching, no turf algae were recorded in this site, and macro-algal cover was 2.43 ± 2.98% in 2016. However, with the loss of hard coral cover, the site became dominated by turf algae (33.17 ± 12.84%) just 1 year post bleaching (Fig. 3.7). This was the case in 2017 for most of the sites in the Sekisei Lagoon. This pattern of steep increase in turf algae and macroalgal cover following mass coral mortality was evident in almost all of Sekisei Lagoon reefs (Fig. 3.6).

Even though overall hard coral recovery is slow within the whole lagoon system, some of the survey areas have exhibited optimistic resilience. Sites St. M, St. I, Kayama entrance, St. 20, and St.18 demonstrated exceptional recovery with more than 10–15% increase in overall hard coral cover from 2017 to 2020. In that same timeframe, these sites also showed an increase in the total number of hard corals

Fig. 3.7 Temporal changes in substrate assemblage observed for S13 ("Milky way") in the Sekisei Lagoon during a six-year period. Picture (**a**) healthy *Acropora* assemblages in September 2015, (**b**) September 2016 during the peak of bleaching event, (**c**) May 2017, 8 months post-bleaching, (**d**) April 2019, 31 months post-bleaching. (**e**) September 2020, 48 months post-bleaching. Note the high macroalgal cover replacing the hard coral cover on the reef after 4 years of the mass bleaching event

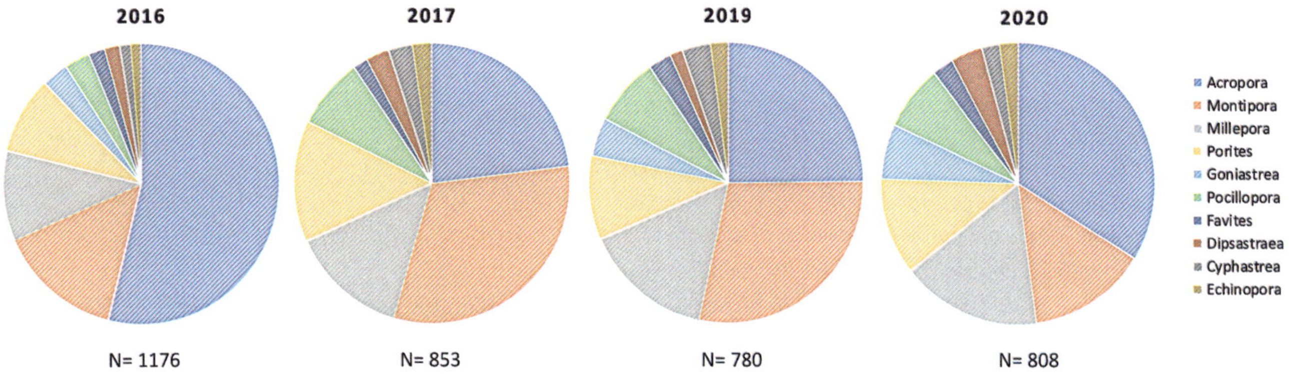

Fig. 3.8 Temporal changes in live coral cover percentage of the coral genera found in the Sekisei Lagoon in 2016, 2017, 2019, and 2020. "N" represents total number of observations

(Table 3.3). *Acropora* populations at the sites St. M, Kayama entrance, and St. 20 exhibited recovery following the bleaching event, although site St. I and St. 18 had shifts from *Acropora* dominance in 2016 to *Millepora* dominance in 2020. These sites were identified as "resilient," which is important for further understanding and analysis of bleaching implications and phase shifts in the lagoon.

3.3.4 Temporal Changes in Percentage Cover of Coral Genera

The onset of thermal stress led to changes in the coral genera composition of the Sekisei Lagoon in the following years. With 34 coral genera recorded, the most dominant genera during the 2016 bleaching event were *Acropora* (49.83 ± 22.98%), *Montipora* (13.45 ± 18.94%), *Millepora* (9.62 ± 7.92%), and *Porites* (8.22 ± 8.11%) (Fig. 3.8). However, 1 year post-bleaching, the dominant coral genus in the Sekisei Lagoon shifted to *Montipora* (27.43 ± 20.44%).

A 29% decrease was observed in the *Acropora* species from its 2017 cover rate of 20.04 ± 12.13%. A slight increase in overall cover of *Porites* and *Millepora* species to 12.19 ± 9.47% and 11.95 ± 5.13%, respectively, was observed in the same year.

In 2017, a total of 32 genera of corals were recorded. However, some of the less abundant genera in 2016 were not present in 2017 surveys, including *Goniastrea*, *Caulastraea*, and *Seriatopora*. In 2019, 3 years post-bleaching, *Montipora* still remained the most dominant genus as it did in 2017, though its overall cover slightly fell to 24.61 ± 14%. The second most abundant genus as per the 2019 survey was *Acropora* (21.41 ± 9.07%).

Four years post-bleaching, there was a shift of dominance back to branching and tabular *Acropora* (26.36 ± 13.77%) and *Millepora* (12.87 ± 8.41%) in the Sekisei Lagoon. However, a notable decrease to 10.89 ± 9.01% was observed among the *Montipora* species in 2020. Even though there was an overall decline recorded for most of the genera during the 2020 survey, the overall diversity index was the highest (Simpson's diversity index: 0.73 ± 0.17) that year with a total of 46 genera recorded.

3.4 Discussion

Between 2015 and 2016, prolonged periods of excessively high SSTs caused severe bleaching in most of the coral reef regions globally, including in the Pacific Ocean (Hughes et al. 2017). SSTs recorded for the Sekisei region were substantially above the recognized bleaching threshold of 30 °C (NOAA Coral Reef Watch, 2019; Liu et al. 2017). Hence, severe coral bleaching was recorded throughout southern Japan in the summer of 2016 (Kayanne et al. 2017). Our results indicate that the 2016 mass bleaching event caused major loss of coral cover in most of the surveyed sites in the Sekisei Lagoon, which is also the largest coral reef system in Japan with key areas for marine conservation of Japanese coral reef ecosystems. Within 1 year of the onset of the bleaching event, a total loss of 14.94% of the hard coral cover was observed. By 2017, the bleaching damage was quite evident in the Sekisei Lagoon, as dead colonies of *Acropora* increased on the shallow reef flat. This was similar to the observation by Fujioka (1999) following the 1998 bleaching. With the decrease in the hard coral cover following bleaching, the density of corallivores and herbivorous fishes also diminished tremendously in the Sekisei Lagoon (Sato et al. 2020). This further encouraged the rapid increase of algal assemblages in most of the sites, which potentially limits coral regeneration and larval settlements. The mass scale coral mortality following bleaching not only increased the available space for algae, but also enhanced the uptake of nutrients and the carbon production, both of which coincided with the intensified algal competition (Davey et al. 2008). A shift in algal cover type was also observed in the

Table 3.2 Result of PERMANOVA based on Bray–Curtis dissimilarities of the relative abundance of coral genera and substrate categories over the years. Fourth root transformations were done on the data before the analysis. Significance codes for *p* values are as follows: <0.0001 "***," <0.001 "**," <0.01 "*" and >0.05 Not significant (NS)

	df	Sum of Sqs	*MeanSqs*	*F.Model*	R^2	Pr (>F)
Substrate: year	1	0.574	0.57405	2.6322	0.02257	0.007 **
Residuals	114	24.862	0.21809		0.97743	
Total	115	25.436			1.00	

DF degrees of freedom, *SS* sum of squares, *MS* mean sum of squares

Sekisei Lagoon where there was rapid increase in turf algae assemblages, from 1.5% to 26.42% just 1 year following the bleaching event. We then observed a takeover of macroalgae as the most dominant substrate assemblage in 2019 (Fig. 3.4). However, by 2020, there was again a shift back to turf algal dominance with a cover percentage of 46.44%, the highest ever recorded in the Sekisei Lagoon. Macroalgal abundance is known to fluctuate seasonally with peak abundance in cooler months of November to April, followed by a seasonal die-off in the warmer months of June to September (Tsai et al. 2004; Kuo et al. 2012). This trend could be the reason for the slight decrease in macroalgal cover.

Our results also showed a dramatic decrease in the cover of bare rock to 0.03% in 2020, which further supports the fact that these available benthic surfaces were replaced by turf algae assemblages. nMDS based on Bray–Curtis dissimilarity plots revealed a clear separation (PERMANOVA, $R^2 = 0.022$, $p = 0.007$) (Table 3.2) between hard coral and bare rock cover, with turf algae and macroalgal cover as arrow vectors that demonstrate the strength and direction of the correlation between these substrates (Fig. 3.9).

Variation in coral diversity and functional composition of coral assemblages play a major role in spatial recovery of coral assemblages among different sites in the Sekisei Lagoon. In contrast to 2016 surveys, the total abundance of *Acropora* species have decreased tremendously over the years in Sekisei, implying that recruitment failure could be a factor that has contributed to the slow recovery of this taxon, potentially even within the conditions prior to bleaching. These corals were formerly the most prominent on the Sekisei reefs. The sites with the highest decrement in the coral cover following 2016 were generally dominated by branching and tabular *Acropora* species (Table 3.3). Sites such as St. F, Shimobishi yuisa uchi, Milky way, Katagua Lagoon, and Maruguu had approximately 80–95% of *Acropora* coral dominance prior to the 2016 bleaching event. However, 1 year post bleaching in 2017, the cover rate of *Acropora* species dropped to below 5%. In addition, no colonies of *Acropora* corals were observed at the sites of Shimobishi yuisa uchi, Milky way, Maruguu, and St. I in the 2017 surveys, and they showed no signs of recovery even by 2020. The number of *Acropora* colonies observed in 2020 still remained 0 at Shimobishi yuisa uchi, 1 at Milky way and Maruguu, and 2 at St. I from our transect surveys.

In the years following the bleaching event, multi-species coral assemblages were more common in the Sekisei Lagoon compared to the high proportion of *Acropora* species that used to dominate the reefs in 2016 (Table 3.3). Transition from *Acropora* dominance to other coral genera, such as *Dipsastraea* and *Porites,* has been commonly observed after periods of disturbances such as coral bleaching or typhoon damage (Muko et al. 2019; Burt et al. 2008). Apart from the *Acropora* species, the Hydrozoan coral *Millepora* species was the most susceptible to mortality during bleaching events, where 28.82% of the population died after the 2016 event. A similar vulnerability of the *Millepora* corals to bleaching and mortality has been reported from bleaching episodes across the globe by several studies (Glynn and Weerdt 1991; Marshall and Baird 2000).

The remnant coral populations in the Sekisei Lagoon were mostly encrusting and massive species such as *Montipora* and *Porites,* which were also the least susceptible genera to mortality. According to Nakamura (2017), massive corals were considered less susceptible to bleaching stress, however during the 2016 bleaching event in the Sekisei Lagoon, the mortality rate of these massive species was higher compared to that in the bleaching surveys conducted in the lagoon in 2007. Although massive *Porites* such as *P. lobata* and *P. lutea* are categorized as "winners" when it comes to thermal stress bleaching resistance, they are not immune to repeated bleaching events and prolonged bleaching (Grottoli et al. 2014). Larger *Porites* corals may survive severe bleaching events, as evidenced by stress bands in their skeletons; although high rates of partial mortality may result in a reduction in the colony's effective size, impeding its reproductive ability (Barkley et al. 2018; Pisapia et al. 2019). Hence, with the current species assemblages, the coral community's recovery rates and potential processes will be strongly influenced by the recurrence of additional bleaching events. In addition to thermal stress episodes, it is known that after periods of heavy rains, there is a significant red soil influx into the Sekisei Lagoon from the mouths of the Todoroki and Miyara Rivers located in the south-eastern Ishigaki Island, and might have significant nutrient loads that is known to be a threat to coral reefs (Suzuki et al. 2016). Additional disturbances due to anthropogenic activities in the region make full recovery of these reefs questionable in the near future.

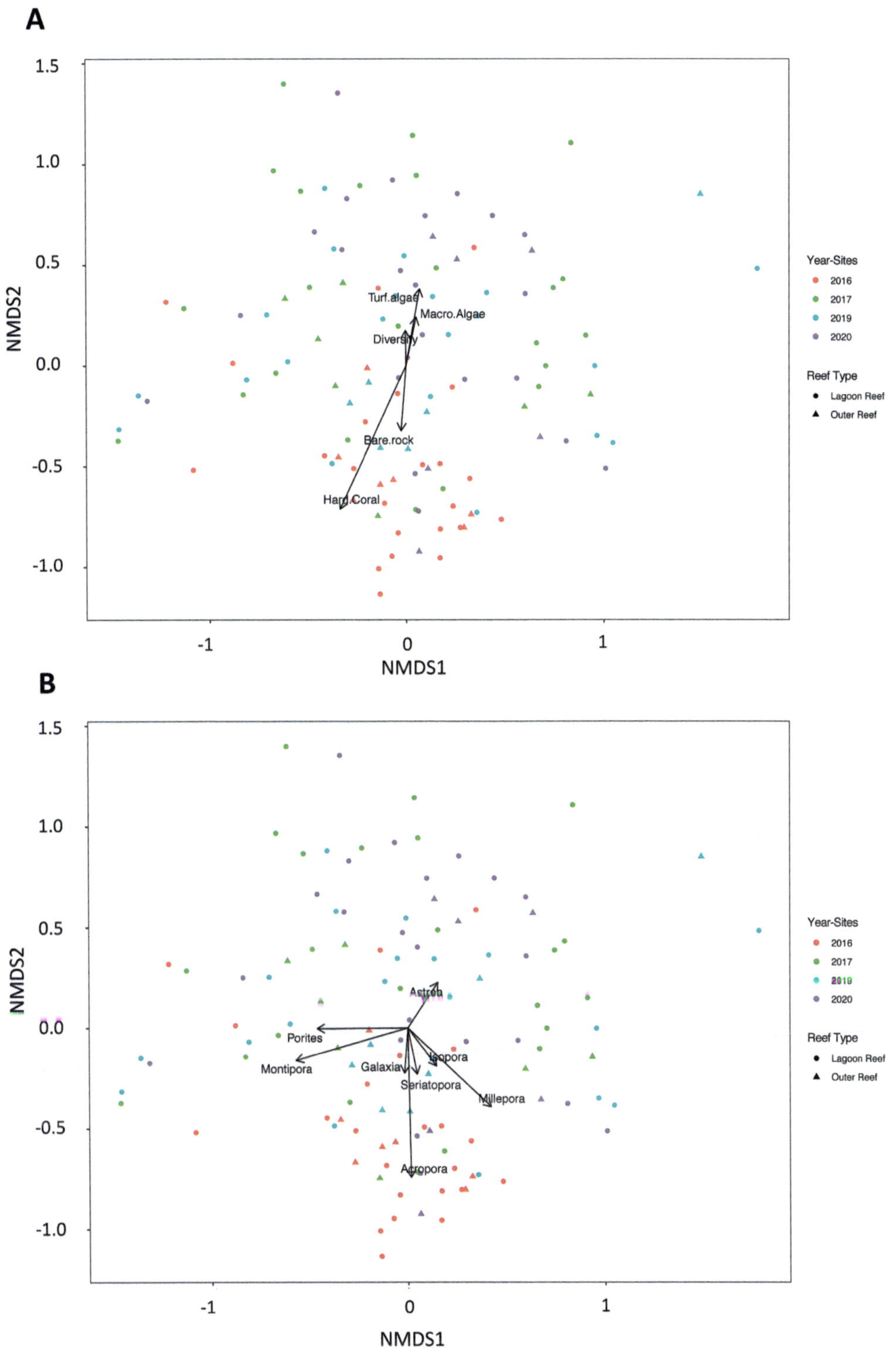

Fig. 3.9 Non-metric multidimensional scaling (nMDS) ordination plot representing the coral genera from 2016, 2017, 2019, and 2020 surveys (Dimension = 2, Stress = 0.2). (**a**) Arrows are the vectors that represent the strength and direction of the correlation of each significant substrate abundance with the NMDS space. Only the substrate with significant levels ($p < 0.05$) are plotted. (**b**) Arrows are the vectors that represent the strength and direction of the correlation of each significant coral genera abundance with the nMDS space. Only the species with significant levels ($p < 0.05$) are plotted as arrow vectors

Table 3.3 Site-wise changes in percentage cover of the four most abundant coral genera (*Acropora, Millepora, Montipora,* and *Porites*) in the Sekisei Lagoon in 2016, 2017, 2019, and 2020. The column N represents the total count of all hard corals recorded for that particular site and column D_s represents the Simpsons diversity index for that particular site

	2016						2017						2019						2020					
Sites	Acropora	Millepora	Montipora	Porites	N	Ds	Acropora	Millepora	Montipora	Porites	N	Ds	Acropora	Millepora	Montipora	Porites	N	Ds	Acropora	Millepora	Montipora	Porites	N	Ds
Panari shimoji lagoon	0	0	0	80.77	52	0.34	0	0	3.17	79.37	63	0.99	2.86	0	14.29	51.43	35	0.70	0	0	15.15	48.48	33	0.72
Kuroshima port entrance	40.63	31.25	7.81	1.56	64	0.73	0	63.16	10.53	5.26	19	0.40	0	64.00	0	4.00	25	0.56	14.29	47.62	0	4.76	21	0.68
Anadomari farside	21.43	0	0	14.29	14	0.85	0	0	50.00	0	2	0.50	0	0	0	0	1	0.00	42.86	14.29	0	0	7	0.73
Anadomari lagoon	29.87	1.30	12.99	12.99	77	0.81	0	0	33.33	16.67	24	0.71	0	0	0	0	3	0.44	0	6.67	6.67	20.00	15	0.85
Urabishi East Farside	53.57	1.19	17.86	0	84	0.68	20.59	2.94	35.29	0	34	0.72	26.32	0	15.79	0	38	0.84	5.26	5.26	0	0	19	0.91
Urabishi East Lagoon	30.77	0.00	15.38	23.08	13	0.77	40.00	0	10.00	0	10	0.28	13.33	13.33	20.00	0	15	0.86	13.33	0	6.67	6.67	15	0.91
St.M	73.47	22.45	0	4.08	49	0.41	4.76	80.95	9.52	0	21	0.14	64.86	5.41	16.22	10.81	37	0.54	78.57	5.36	0	7.14	56	0.37
St.F	81.43	10.00	0	1.43	70	0.34	41.67	25.00	0	0	12	0.25	NA	NA	NA	NA	NA	NA	30.00	30.00	10.00	0	10	0.78
Takerunjyu	41.82	12.73	32.73	10.91	55	0.69	0	26.32	0	26.32	19	0.44	14.29	0	14.29	28.57	14	0.79	20.00	0	26.67	0	15	0.76
Shimobishi hey no nakagu	21.28	6.38	4.26	19.15	47	0.86	0	14.29	23.81	23.81	21	0.86	6.90	6.90	13.79	13.79	29	0.90	5.56	0	5.56	22.22	18	0.88
Shimobishi yuisa uchi	82.35	9.80	0	3.92	51	0.31	0	50.00	20.00	0	10	0.20	0	20.00	60.00	20.00	5	0.56	0	16.67	16.67	50.00	6	0.67
Tenmakakejyu Lagoon	7.32	0	81.30	11.38	123	0.35	0	0	93.91	6.09	115	0.09	0	0	98.48	1.52	66	0.03	0	0	76.56	21.88	64	0.37
Milky way	95.00	1.00	2.00	0.00	100	0.08	0	0	75.00	0	4	0.00	NA	NA	NA	NA	NA	NA	10.00	0	30.00	0	10	0.84
Kuroshima Route South	70.83	4.17	0	8.33	24	0.34	12.50	12.50	25.00	0	8	0.83	NA	NA	NA	NA	NA	NA	0	0	33.33	0	3	0.67
St.C	77.78	0	0	0.00	18	0.43	33.33	0	33.33	0	3	0.67	23.08	0	7.69	23.08	13	0.83	50.00	0	12.50	6.25	16	0.70
Katagua Lagoon	84.38	0	3.13	0.00	32	0.28	25.00	0	0	0	4	0.75	13.64	4.55	4.55	31.82	22	0.83	8.33	0	8.33	8.33	12	0.89
St.A	28.57	10.71	3.57	7.14	28	0.81	19.05	9.52	14.29	19.05	21	0.71	29.41	11.76	5.88	5.88	17	0.84	12.50	0	12.50	6.25	16	0.90
Katagua Farside	32.00	0	32.00	8.00	25	0.76	11.76	0	17.65	5.88	17	0.94	24.32	0	16.22	5.41	37	0.84	NA	NA	NA	NA	NA	NA
St.L	23.08	0	16.48	5.49	91	0.86	4.08	0	30.61	2.04	49	0.80	2.70	0	48.65	0	37	0.69	11.90	0	0	7.14	42	0.87
St.N	72.88	1.69	3.39	11.86	59	0.45	26.19	7.14	0	40.48	42	0.70	44.00	4.00	4.00	8.00	25	0.77	17.86	32.14	0	0	28	0.82
St.K	65.85	21.95	1.22	1.22	82	0.53	60.78	13.73	3.92	0	51	0.52	46.67	1.67	28.33	15.00	60	0.68	57.41	11.11	9.26	1.85	54	0.64
Maru Guu	92.41	2.53	5.06	0	79	0.16	0	0	28.57	0	7	0.90	0	15.79	5.26	5.26	19	0.80	10.00	0	0	10.00	10	0.88
St.I	55.13	33.33	3.85	1.28	78	0.59	0	53.13	6.25	3.13	32	0.61	0	75.00	0	0	32	0.42	3.39	74.58	0	0	59	0.42
St.8	73.81	0	0	4.76	84	0.45	75.47	7.55	3.77	1.89	53	0.41	34.48	39.08	3.45	1.15	87	0.71	25.00	8.33	13.89	11.11	36	0.88
Kayama entrance	62.07	0	3.45	10.34	58	0.57	9.09	0	21.21	3.03	33	0.77	41.67	0	0	8.33	24	0.79	43.64	3.64	0	3.64	55	0.78
St.20	43.96	7.69	8.79	3.30	91	0.78	59.04	1.20	4.82	0	83	0.63	56.76	0	5.41	0	37	0.65	65.88	4.71	0	0	85	0.58
St.18	44.00	22.67	0	1.33	75	0.73	19.23	57.69	3.85	0	26	0.47	13.04	4.35	0	8.70	23	0.90	6.78	20.34	0	0	59	0.91
Kanara guchi west	5.13	2.56	51.28	41.03	39	0.57	20.00	20.00	60.00	0	5	0.10	0	0	97.30	0	37	0.05	0	12.50	62.50	12.50	8	0.56
Taketomi east	47.83	50.00	0	0	46	0.52	33.33	50.00	0	0	6	0.09	18.18	72.73	0	0	11	0.43	33.33	0	0	66.67	6	0.44
Kurosima Nakamoto	13.24	23.53	29.41	5.88	68	0.82	1.69	3.39	69.49	10.17	59	0.51	0	12.90	35.48	6.45	31	0.72	6.67	10.00	3.33	16.67	30	0.90

In addition to the current results of coral community shift, the nMDS biplot of coral genera sampled within 4 years showed significant species as arrow vectors that represent the strength and direction of the correlation of these species. From the plot, it can be depicted that *Acropora* assemblages strongly correlates with the high coral cover sites in 2016, whereas *Montipora* and *Porites* correlates with the high coral cover sites in 2017 and 2019 (Fig. 3.9). A multi-species combination with low coral coverage implies that there are only a few colonies in the community, indicating degradation of the coral community after severe disturbance or adverse conditions (Muko et al. 2019; Burt et al. 2008).

Disturbances over the last few decades in the Sekisei Lagoon particularly, the outbreaks of *Acanthaster planci* starfish, typhoons, and the 6 bleaching events in 1984, 1998, 2001, 2003, 2007, and 2016 (Kamezaki and Ui 1984, Furushima et al. 2002; Ministry of the Environment 2002, 2004; Okamoto et al. 2007; Nakamura, 2017) have caused long-term alteration of the reef ecosystem. The likelihood of widespread bleaching related to long-term increases in sea water temperature, as indicated in the current study, must be considered.

As seen from the results of this study, a maximum DHW of 11.42 °C-weeks was observed in the Sekisei Lagoon, representing an increasing trend compared to that of the 1998 and 2007 thermal anomalies. A DHW value of >8 °C-weeks corresponds to a mean monthly anomaly of greater than 2 °C, a thermal environment that will easily arise under the anticipated 2 °C global warming conditions where one third of the world's coral reefs are expected to face long-term degradation as described by Frieler et al. (2013) and Kayanne (2017). Furthermore, local stressors such as fishing, coastal development, red soil inflow/sedimentation, and agricultural runoffs that may lead to nutrient rich environments should not be overlooked, as they also contribute to reef degradation. If the current level of coral reef degradation continues with impeded recovery potential, the Sekisei Lagoon reefs could turn into macro and turf algae dominated communities with restricted coral presence and growth. Without major interventions, such as allocating rehabilitation areas or implementing effective marine protected areas that are crucial for the seeding populations, the Sekisei Lagoon reefs are less likely to recover as functional reefs in the near future.

Acknowledgements This research was supported by the Environment Research and Technology Development Fund (grant no. JPMEERF20194007) of the Environmental Restoration and Conservation Agency of Japan, Grants-in-Aids from the Japan Society for the Promotion of Science (KAKENHI grant nos. 16H02490, 18K06424 and 19H00953). MSA has been supported by Japanese government (MEXT) scholarship (Research student).

References

Barkley HC, Cohen AL, Mollica NR, Brainard RE, Rivera HE, Decarlo TM, Lohmann GP, Drenkard EJ, Alpert AE, Young CW, Vargas-Ángel B, Lino KC, Oliver TA, Pietro KR, Luu VH (2018) Repeat bleaching of a central pacific coral reef over the past six decades (1960–2016). Commun Biol 1:177. https://doi.org/10.1038/s42003-018-0183-7

Brown BE (1997) Coral bleaching: causes and consequences. Coral Reefs 16:S129–S138

Burt J, Bartholomew A, Usseglio P (2008) Recovery of corals a decade after a bleaching event in Dubai, United Arab Emirates. Mar Biol 154:27–36

Davey M, Holmes G, Johnstone R (2008) High rates of nitrogen fixation (acetylene reduction) on coral skeletons following bleaching mortality. Coral Reefs 27:227–236

Environmental Systems Research Institute (ESRI) (2015) ArcGIS release 10.3. Redlands, CA

Frade PR, Bongaerts P, Englebert N, Rogers A, Gonzalez-Rivero M, Hoegh-Guldberg O (2018) Deep reefs of the great barrier reef offer limited thermal refuge during mass coral bleaching. Nat Commun 9: 3447. https://doi.org/10.1038/s41467-018-05741-0

Frieler K, Meinshausen M, Golly A, Mengel M, Lebek K, Donner SD, Hoegh-Guldberg O (2013) Limiting global warming to 2 °C is unlikely to save most coral reefs. Nat Clim Chang 3:165–170. https://doi.org/10.1038/nclimate1674

Fujioka Y (1999) Mass destruction of the hermatypic corals during a bleaching event in Ishigaki Island, southwestern Japan. J Japan Coral Reef Soc 1999:41–50

Furushima Y, Okamoto M, Nijima S, Kanno S (2002) Characteristics of physical environment of Japanese coral reef region in Sekisei lagoon (Okinawa prefecture, Japan). Oceans 02 MTS/IEEE. https://doi.org/10.1109/oceans.2002.1192048

Glynn PW, Weerdt WHD (1991) Elimination of two reef-building hydrocorals following the 1982-83 El Nino warming event. Science 253:69–71

Google Earth Pro V 7.3.4.8248 (December 14, 2015) Sekisei Lagoon, Japan 24°18′2131″N, 124°01′2585″ E Maxar Technologies 2022 http://wwwearthgooglecom Accessed 14 Jan 2022

Graham NAJ, Jennings S, Macneil MA, Mouillot D, Wilson SK (2015) Predicting climate-driven regime shifts versus rebound potential in coral reefs. Nature 518:94–97

Grottoli AG, Warner ME, Levas SJ, Aschaffenburg MD, Schoepf V, Mcginley M, Baumann J, Matsui Y (2014) The cumulative impact of annual coral bleaching can turn some coral species winners into losers. Glob Chang Biol 20:3823–3833. https://doi.org/10.1111/gcb.12658

Harii S, Hongo C, Ishihara M, Ide Y, Kayanne H (2014) Impacts of multiple disturbances on coral communities at Ishigaki Island, Okinawa, Japan, during a 15 year survey. Mar Ecol Prog Ser 509: 171–180

Heron SF, Maynard JA, Hooidonk RV, Eakin CM (2016) Warming trends and bleaching stress of the World's coral reefs 1985–2012. Sci Rep 6:38402. https://doi.org/10.1038/srep38402

Hoegh-Guldberg O, Mumby P.J, Hooten AJ, Steneck RS, Greenfield P, Gomez E, Harvell CD, Sale PF, Edwards AJ, Caldeira K, Knowlton N, Eakin CM, Iglesias-Prieto R, Muthiga N, Bradbury RH, Dubi A, Hatziolos ME (2007) Coral reefs under rapid climate change and ocean acidification. Science 318: 1737–1742

Hongo C, Yamano H (2013) Species-specific responses of corals to bleaching events on anthropogenically turbid reefs on Okinawa Island, Japan, over a 15-year period (1995–2009). PLoS One 8: e60952. https://doi.org/10.1371/journal.pone.0060952

Hughes TP, Kerry JT, Álvarez-Noriega M, Álvarez-Romero JG, Anderson KD, Baird AH, Babcock RC, Beger M, Bellwood DR,

Berkelmans R, Bridge TC (2017) Global warming and recurrent mass bleaching of corals. Nature 543(7645):373–377
Kamezaki N, Ui S (1984) Bleaching of hermatypic corals in Yaeyama Islands. Marine Park J 61:10–13. in Japanese
Kayanne H (2017) Validation of degree heating weeks as a coral bleaching index in the northwestern Pacific. Coral Reefs 36:63–70
Kayanne H, Suzuki R, Liu G (2017) Bleaching in the Ryukyu Islands in 2016 and associated degree heating week threshold. Galaxea J Coral Reef Stud 19:17–18
Kimura T, Tun K, Chou LM (2014) Status of coral reefs in east Asian seas region: 2014. Global Coral Reef Monitoring Network. Ministry of the Environment, Japan
Kruskal JB (1964) Multidimensional scaling by optimizing goodness of fit to a nonmetric hypothesis. Psychometrika 29:1–27
Kuo CY, Yuen YS, Meng PJ, Ho PH, Wang JT, Liu PJ, Chang YC, Dai CF, Fan TY, Lin HJ, Baird AH, Chen CA (2012) Recurrent disturbances and the degradation of hard coral communities in Taiwan. PLoS One 7:e44364. https://doi.org/10.1371/journal.pone.0044364
Liu G, Skirving WJ, Geiger EF, De La Cour JL, Marsh BL, Heron SF, Tirak KV, Strong AE, Eakin CM (2017) NOAA coral reef Watch's 5km satellite coral bleaching heat stress monitoring product suite version 3 and four-month outlook version 4. Reef Encounter 32:39–45
Loya Y, Sakai K, Yamazato K, Nakano Y, Sambali H, Woesik RV (2001) Coral bleaching: the winners and the losers. Ecol Lett 4:122–131
Marshall PA, Baird AH (2000) Bleaching of corals on the great barrier reef: differential susceptibilities among taxa. Coral Reefs 19:155–116
McClanahan TR (2004) The relationship between bleaching and mortality of common corals. Mar Biol 144:1239–1245
Ministry of the Environment (2002) Report of the coral reef monitoring survey of the Sekisei Lagoon and neighbouring areas, Iriomote National Park, 2001, 68. (in Japanese)
Ministry of the Environment (2004) In: Ministry of the Environment and JCRS (ed) Coral reefs of Japan, Tokyo
Muko S, Suzuki G, Saito M, Nakamura T, Nadaoka K (2019) Transitions in coral communities over 17 years in the Sekisei Lagoon and adjacent reef areas in Okinawa, Japan. Ecol Res 34:524–534
Nakamura T (2017) Mass coral bleaching event in Sekisei Lagoon observed in the summer of 2016. J Japan Coral Reef Soc 19:29–40
Nishihira M (2004). Hermatypic corals of Japan. The Japanese coral reef society. 1st edn. Japanese Ministry of the Environment, pp 10–13
Nishihira M, Veron JEN (1995) Hermatypic corals of Japan (in Japanese). Kaiyusha Publishers, Tokyo, pp 29–30
NOAA Coral Reef Watch (2019) Updated daily. NOAA Coral Reef Watch Version 3.1 Daily 5km satellite regional virtual station time series data for Southeast Florida, Mar. 12, 2013–Mar. 11, 2014. College Park, Maryland, USA: NOAA Coral Reef Watch. Data set https://coralreefwatch.noaa.gov/product/vs/data.php. Accessed 5 Feb 2020
Nojima S, Okamoto M (2008) Enlargement of habitats of scleractinian corals to north and coral bleaching events. Nippon Suisan Gakkaishi 74:884–888
Okamoto M, Nojima S, Furushima Y (2007) Temperature environments during coral bleaching events in Sekisei Lagoon. Bull Japan Soc Fish Oceanogr 71:112–121
Oksanen J, Blanchet FG, Kindt R, Legendre P, Minchin PR, O'hara RB, Simpson GL, Solymos P, Stevens MHH, Wagner H, Oksanen MJ (2013) Package 'vegan'. Commun Ecol Package Version 2(9):1–295
Pisapia C, Burn D, Pratchett MS (2019) Changes in the population and community structure of corals during recent disturbances (February 2016-October 2017) on Maldivian coral reefs. Sci Rep 9:8402. https://doi.org/10.1038/s41598-019-44809-9
R Studio Team (2020) RStudio: integrated development for R. RStudio, PBC, Boston, MA. http://www.rstudio.com/
Sato M, Nanami A, Bayne CJ, Makino M, Hori M (2020) Changes in the potential stocks of coral reef ecosystem services following coral bleaching in Sekisei Lagoon, southern Japan: implications for the future under global warming. Sustain Sci 15:863–883
Shimoike K (2004). Sekisei Lagoon. Coral Reefs of Japan. The Japanese Coral Reef Society.1st edn. Japanese Ministry of the Environment, pp 219–224
Skirving W, Marsh B, De La Cour J, Liu G, Harris A, Maturi E, Geiger E, Eakin CM (2020) CoralTemp and the coral reef watch coral bleaching heat stress product suite version 3.1. Remote Sens (Basel) 12:3856. https://doi.org/10.3390/rs12233856
Suzuki G, Yasuda N, Ikehara K, Fukuoka K, Kameda T, Kai S, Nagai S, Watanabe A, Nakamura T, Kitazawa S, Bernardo LPC (2016) Detection of a high-density brachiolaria-stage larval population of crown-of-thorns sea star (*Acanthaster planci*) in Sekisei Lagoon (Okinawa, Japan). Diversity 8:9. https://doi.org/10.3390/d8020009
Tsai CC, Wong SL, Chang JS, Hwang RL, Dai CF, Yu YC, Shyu YT, Sheu F, Lee TM (2004) Macroalgal assemblage structure on a coral reef in Nanwan Bay in southern Taiwan. Bot Mar 47:439–453
Veron JEN, Stafford-Smith M (2000) Corals of the world. Australian Institute of Marine Science, Townsville
Wilkinson C, Lindén O, Cesar H, Hodgson G, Rubens J, Strong AE (1999) Ecological and socioeconomic impacts of 1998 coral mortality in the Indian Ocean: an ENSO impact and a warning of future change? Ambio 28:188–196
Yaeyama Marine Park Research Station (1983) Report of the monitoring survey of crown-of-thorns starfish and corals in FY1983. Okinawa, Japan

Succession and Emergence of Corals in High-Latitude (Temperate) Areas of Eastern Asia into the Future 4

Shashank Keshavmurthy, Takuma Mezaki, James Davis Reimer, Kwang-Sik Choi, and Chaolun Allen Chen

Abstract

High-latitude areas have been hypothesized as potential refugia in the future for those corals which can range-shift across the latitudes (from tropical to high or low latitudes). However, whether high latitude will be the future hope for corals either through succession (proliferation of regionally or locally endemic species) or emergence (range-shifts) needs more research. In this chapter, we argue that the future corals in high latitudes will be more due to succession than emergence. Recent data from molecular studies indicate that rather than poleward range shifts, increasing abundances of previously less abundant local coral species may be responsible for these changes, thereby hinting towards succession of species. The proliferation of rare or cryptic species (e.g., acroporid corals) adapted to the environmental features of high latitudes could form majority of the future benthos in these areas. Considering that high-latitude locations are 'oases' for native coral species, it is important to designate these areas for conservation to protect endemic species and lineages. Also, to better facilitate future conservation, it is necessary to conduct more research on high-latitude coral communities, particularly on those endemic species and lineages, by including eco-physiological, molecular ecological, and taxonomic (molecular and morphological) approaches to understand whether future coral communities will be dominated by the succession of these local species or the poleward range-shifting of lower latitude species or a mixture of both scenarios.

Keywords

Range-shift · Tropicalization · Marginalization · High-latitude Oases · Endemic species · Coral communities · Climate change

S. Keshavmurthy (✉)
Biodiversity Research Centre, Academia Sinica, Taipei, Taiwan

T. Mezaki
Kuroshio Biological Research Foundation, Otsuki, Kochi, Japan
e-mail: mezaki@kuroshio.or.jp

J. D. Reimer
Molecular Invertebrate Systematics and Ecology Lab, Faculty of Science, University of the Ryukyus, Okinawa, Japan

Tropical Biosphere Research Center, University of the Ryukyus, Okinawa, Japan
e-mail: jreimer@sci.u-ryukyu.ac.jp

K.-S. Choi
Department of Marine Life Science (BK 21 FOUR), Jeju National University, Jeju, Republic of Korea
e-mail: skchoi@jejunu.ac.kr

C. A. Chen
Biodiversity Research Centre, Academia Sinica, Taipei, Taiwan

Department of Life Science, National Taiwan Normal University, Taipei, Taiwan

Department of Life Science, Tunghai University, Taichung, Taiwan
e-mail: cac@gate.sinica.edu.tw

4.1 Introduction

In the Anthropocene, because of increasing average seawater temperatures and carbon dioxide emissions due to climate change, both the atmosphere and oceans are facing unprecedented warming. Climate change leads to a grim future for marine biodiversity as has been reiterated in the latest IPCC report (IPCC 2021). This has specifically resulted in increased incidences of thermal anomalies that drive mass shallow water coral bleaching and mortality in recent years (Hoegh-Guldberg et al. 2007; Hughes et al. 2017a, 2018b; Zande et al. 2020). Zooxanthellate scleractinian corals (Anthozoa: Hexacorallia: Scleractinia) are holobionts, having a symbiotic relationship with a multitude of microorganisms including their most important partner, photosymbiotic dinoflagellates of the family Symbiodiniaceae. This complex relationship has maintained itself via long-term adaptation and short-term acclimatization to persist under local environmental conditions. Whether some species of zooxanthellate

I. Takeuchi, H. Yamashiro (eds.), *Coral Reefs of Eastern Asia under Anthropogenic Impacts*, Coral Reefs of the World 17,
https://doi.org/10.1007/978-3-031-27560-9_4

corals go extinct and/or experience dramatic drops in abundances by the end of this century will depend on the severity of climate change, the extent of other environmental disturbances, and the ability of corals to adapt. Due to repeated and prolonged extreme seawater temperature anomaly events, it will become more challenging for many coral species to continue to effectively reproduce and survive, resulting in declines of populations, large-scale reef degradation, and local extinctions of many coral species (Carpenter et al. 2008; also, Hughes et al. 2015, 2017b). Such changes happening on a large scale to these important ecosystems will have irreversible impacts on coral reef biodiversity as well as on humans and their livelihoods in coral reef coastal regions (Carpenter et al. 2008).

The effects of changing climate on coral and coral reefs in tropical and sub-tropical locations are well known (Mydlarz et al. 2010; Schoepf et al. 2015; Hoey et al. 2016), and many corals in these areas have been shown to already be at the limits of their temperature thresholds (Schoepf et al. 2015). Frequent and prolonged incidences of higher-than-average seawater temperatures have had devastating effects on corals, for example on the Great Barrier Reef (Hughes et al. 2015, 2017b, 2018a). While corals in tropical and subtropical locations are undergoing drastic changes, more 'marginal' reefs may also provide important information in predicting the future of coral reefs (Tuckett and Wernberg 2018). Such so-called "marginal reefs" include zooxanthellate coral communities in more extreme environments, such as high pCO_2 seeps (Inoue et al. 2013; Agostini et al. 2021; Reimer et al. 2021) or in eutrophied or urban waters (Heery et al. 2018; Cook et al. 2022), and non-accreting communities in high latitude (Harriot and Banks 1995; Denis et al. 2013) and temperate environments (Yamano et al. 2001; Beger et al. 2014; Vieira et al. 2016).

In this chapter we will examine some of the high latitude and marginal coral communities in Japan, Korea, and Taiwan in the context of what future holds in terms of emergence and succession of corals. We discuss aspects of range-shifts due to tropical-like environments being created in high latitude locations, and how this will impact potential future coral ecosystems in high latitude locations in East Asia.

4.1.1 High-Latitude Coral Communities

Shallow water zooxanthellate coral reefs are generally found between 25°N and 25°S latitudes, and are most abundant in tropical areas, with diversity especially high in the Coral Triangle in the equatorial west Pacific (Pontasch 2014). Diversity and densities decrease towards the upper and lower latitudes (sub-tropical and temperate coral reefs and communities; Veron 2000; Roberts et al. 2002). In the past few decades, ocean warming (increases in both winter and summer average seawater temperatures) has resulted in increases in the abundance as well as coverage of corals in various locations in East Asian high latitude regions; this chapter focuses on such locations in Japan, Korea, and Taiwan. This trend is similar as seen at many other high-latitude locations such as along the eastern coast of the USA (Perry and Larcombe 2003), in the Bermuda Islands in the North Atlantic Ocean (Cook et al. 1990), on the Natal coast of South Africa (Schleyer et al. 2018), and on the low-latitude reefs of Australia's Lord Howe (Veron and Done 1979; Harriott et al. 1995) and Solitary islands (Sommer et al. 2017). At high latitudes, sea surface temperatures show considerable seasonal variability each year, and can be very low during winter months. For example, at the Iki Islands of Japan, corals experience temperatures as low as 14 °C during winter and as high as 27.5 °C during summer (Brown and Cossins 2011). In this way, zooxanthellate corals at high latitudes are known for their tolerances to wide ranges of seawater temperatures (Wicks 2009; Wicks et al. 2010; Tuckett and Wernberg 2018; Keshavmurthy et al. 2021).

4.1.2 Overview of High-Latitude Coral Communities in Japan, Korea, and Taiwan

Japan

The occurrence of zooxanthellate corals in Japanese waters is facilitated by the Kuroshio Current, which brings warm tropical water northwards along the Pacific coast of Japan all the way to Tateyama, Chiba (Yamano et al. 2001). On the western side of Japan, the Tsushima Current branches from the Kuroshio Current south of the islands of Yakushima and Tanegashima (Fig. 4.1). Yakushima and Tanegashima islands are also the current northern limit of coral reef formation in Japan (Reimer et al. 2008; Okano 2013). Zooxanthellate corals have been reported from temperate Japan widely (Yabe and Sugiyama 1931), including from Kyushu (roughly 32°N) (Veron and Minchin 1992; Nojima 2004), Shikoku (Veron and Minchin 1992; Iwase 2004; Denis et al. 2013; Reimer et al. 2020), and the Pacific coast of mainland Japan (Veron and Minchin 1992; Harii et al. 2001) northwards to waters near Amatsukominato in Chiba Prefecture on the Pacific coast (34°N) (Veron and Minchin 1992; Yamano et al. 2012), and even into the Sea of Japan, in the Amakusa, Goto, Tsushima, and Iki islands (Yamano et al. 2001, 2004) (Fig. 4.1). In general, it is thought as one moves poleward, the species diversity of zooxanthellate corals drops in this region (see Chen and Keshavmurthy 2009). Many high-latitude coral communities of Japan are often dominated by various *Acropora* spp., often tabular (e.g., *Acropora pruinosa*, *A. solitaryensis*, *A. hyacinthus*) (Veron and Minchin 1992; Denis et al. 2013). Although the minimum

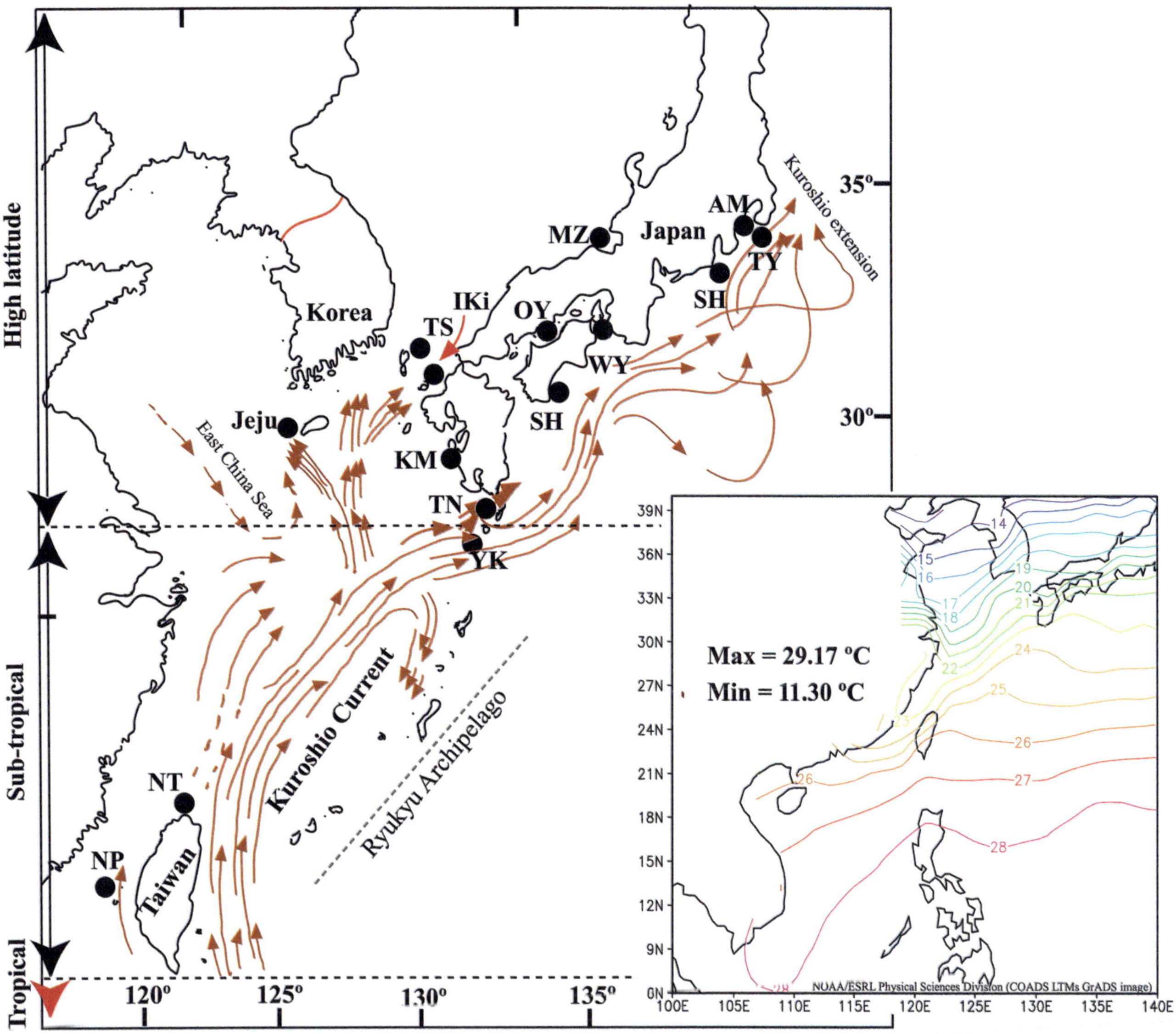

Fig. 4.1 Map of East Asia region showing locations of non-reefal (Taiwan) and high-latitude (Korea and Japan) coral communities in East Asia. The inset figure shows yearly average seawater temperature variation along the latitudinal gradient from the tropics to high latitudes in the western Pacific region. Influence of warm water currents along the central and northern west Pacific, including southern Korea, plays an important role in defining the distribution of coral communities in East Asia. From top, abbreviations are AM – Amatsukominato, TY – Tateyama, SH – Shizuoka, MZ -, OY – Okayama, WY – Wakayama, SH – Shikoku, TS – Tsushima, KM – Kushimoto, TN – Tanegashima, YK – Yakushima (all in Japan), NT – Northern Taiwan, NP – Northern Penghu. Location map with currents was re-drawn following map in the Fig. 4.1 of Veron and Minchin (1992). Seawater temperature map was drawn using data from NOAA/ESRL, Physical Sciences Division (CADS LTMs GrADS image)

temperature required for coral reef formation and development is 18 °C (Kleypas et al. 1999), 48% of coral species in the high-latitude locations of Japan tolerate a minimum seawater temperature of 14.1 °C, and approximately 25% tolerate minimum temperatures between 10.4 °C and 13.2 °C (Veron and Minchin 1992). Thus, approximately half of all zooxanthellate coral species in Japan tolerate temperatures of at least 4 °C below the 18 °C minimum required for reef development. As well, much of this region is predicted to be among marine regions experiencing the largest temperature increases from global warming (e.g., Yara et al. 2009), making continued studies in this region of particular importance.

Korea

In Korea, zooxanthellate coral species are mainly found at Jeju Island (Fig. 4.1). Located off the southern coast of the mainland Korea peninsula, Jeju Island is influenced by the Tsushima Current, a branch of the warm Kuroshio Current (Fig. 4.1; Sugihara et al. 2014). Seawater temperatures at Jeju are 13–18 °C during winter and 24.0–25.0 °C during summer (Sugihara et al. 2014; Vieira et al. 2016; Park et al. 2019). The seawater temperature differences in summer between the colder northern and warmer southern coasts of Jeju Island are approximately 0.5–1.0 °C. Seawater temperature trends have shown an increase of 1.2–1.3 °C over the last 100 years around the Korean coast, including Jeju (Sugihara et al. 2014). Such changes in environmental conditions have led to alterations in the Jeju marine community (Denis et al. 2013, 2014; Sugihara et al. 2014; Hong et al. 2015; Palmas et al. 2015; Vieira et al. 2016).

For example, the coral *Alveopora japonica* has increased in number over the past two decades around Jeju Island, and more so within the past 15 years, due to its fast turnover rates (Vieira et al. 2016) and high recruitment rates (Denis et al. 2015; Hong et al. 2015). At the same time, macroalgal cover (e.g., kelp forests) is also present around Jeju (see Lee et al. 2022). The reason for the dominance of this coral over macroalgae in many locations around Jeju Island could be due to the aggressive colonization characteristics of this species, combined with environmental changes, as suggested in Denis et al. (2015).

Taiwan

Taiwan, a continental island with several offshore islets, is located at the centre junction of the Philippine-Japan Island Arc, and the Tropic of Cancer runs through the middle of Taiwan. Northern Taiwan is influenced by seawater currents originating from the Penghu Archipelago as well as branches of the Kuroshio Current, resulting in internal waves and eddies (Fig. 4.1, see Chap. 2 in this book by Kuo et al. 2023). Although the environment in this region is not conducive to coral reef formation, the coral species here resemble to those of Shikoku and Kyushu, Japan (see Chap. 2 in this book by Kuo et al. 2023). High-latitude-like coral environments can also be seen in the northern Penghu Archipelago and the northern, northeastern, and rocky eastern coasts of Taiwan (Chen 1999; Chen and Keshavmurthy 2009; Ho and Dai 2014). The northern part of the Penghu Islands and the northeastern coast of Taiwan are mainly made up of coral communities and algal reefs due to low sea temperatures (<18 °C). Chen and Keshavmurthy (2009) have proposed that Taiwan might be a stepping-stone connecting upstream reefs with the downstream outlying coral communities along Kuroshio Current coastlines (also see for more detailed information on Taiwan in Chap. 2 in this book by Kuo et al. 2023).

4.1.3 Climate Change and Future High-Latitude Coral Communities: Possible Responses of Species

Increases in average seawater temperatures have led to the hypothesis that higher latitude locations may become possible refugia for zooxanthellate corals escaping thermal stress in tropical and subtropical locations. It has been speculated that many coral species may be able to migrate and/or expand their ranges to counter the negative effects of climate change at lower latitudes (Precht and Aronson 2004; Greenstein and Pandolfi 2008; Yamano et al. 2011; Nakamura et al. 2021), and numerous studies have examined the possibility for high-latitude locations to provide refuge (Riegl 2003; Riegl and Piller 2003; Yamano et al. 2011; Dalton & Roff 2013; Beger et al. 2014; Table 4.1). If such populations persist, the result will be the modification of marine assemblages in high-latitude locations (Precht and Aronson 2004). The future shape of such modified assemblages remains uncertain and may be site or region-specific. Moreover, corals in high-latitude locations may have certain limitations in their resilience and resistance to the ongoing effects of climate change, and their resilience may be lost in a few decades if proper

Table 4.1 Some studies reporting range shifts or poleward migration of coral species

Year	Country	Location	Species mentioned	Main conclusion	Study
2004	USA	Caribbean	*Acropora cervicornis, A. palmata*	Northward expansion along the Florida Peninsula and into the northern gulf	Precht and Aronson (2004)
2011	Japan	High-latitude Japan	*A. muricata, A. solitaryensis, A. hyacinthus, Pavona decussata*	Evidence for rapid poleward expansion since 1930s	Yamano et al. (2011)
2012	Australia	Solitary Island	4 *Acropora* species		[a]Baird et al. (2012)
2018		High-latitude Japan	*A. muricata*	Evidence for species migration	Kumagai et al. (2018)
2019	Japan	Sub-tropical and high-latitude Japan	*A. hyacinthus*	Cryptic lineages found and one specific to temperate region	Nakabayashi et al. (2019)
2021	Japan	Kushimoto	*A. hyacinthus, A. muricata*		Nakamura et al. (2021)

Note = [a]This previous report on possible range-shift of 4 *Acropora* species observed at Solitary Island has been refuted in a new paper published by Mizerek et al. (2021)

conservation and management strategies are not implemented sooner rather than later (Cruz et al. 2018; Soares 2020).

In temperate environments, zooxanthellate corals live very close to their survival thresholds regarding temperature, light, and aragonite availability (Kleypas et al. 1999). Moreover, high-latitude marine ecosystems are often dominated by macroalgae, and increases in zooxanthellate corals may indicate increased competition for space between corals and macroalgae (Hoey et al. 2011).

Indeed, in recent years, marine communities in high-latitude locations in East Asia have been shown to be experiencing such effects from increasing seawater temperatures (Yara et al. 2011; Tanaka et al. 2012; Kumagai et al. 2018; Kim et al. 2020; Agostini et al. 2021) and ocean acidification levels (Yara et al. 2012; Agostini et al. 2021). Numerous studies have documented changes in native communities via the arrival of non-native species, especially certain tropical fishes (Nakamura et al. 2013; Kim et al. 2015), macroalgae (Tanaka et al. 2012; Kumagai et al. 2018), and corals (Yamano et al. 2011; Nakamura et al. 2021). In regions affected by poleward warm currents, such as Korea (Tsushima Current) and Japan (Kuroshio Current; Fig. 4.1), corals are distributed at higher latitudes than in other regions. For example, the northern limit for zooxanthellate coral distribution off Korea and Japan is the interface of the warm northward flowing Kuroshio with the cold southward flowing Oyashio Current off eastern Honshu (Veron and Minchin 1992).

How various species of corals, reef-associated organisms, and their interactions respond to climate change can be summarized into following broad categories: (1) *extinction*—species that cannot cope with these changes; (2) *acclimatization and/or adaptation*—species that can acclimate via physiological or ecological plasticity at cellular and molecular levels; (3) *migration or range shifting*—species that can escape from stressors and migrate to regions more suitable for survival (see Bates et al. 2014). Whether these four potential responses occur individually or jointly on species remains unclear. In the section below, we explain these responses and what they would entail for zooxanthellate corals at high latitudes in East Asia.

Extinction

The possibility of extinction of species of zooxanthellate corals under increasing anthropogenic stress is an issue that has hotly debated (Dietzel et al. 2021; Muir et al. 2022). Although there are no clear answers yet, this issue must be considered. Recent studies have shown presence of endemic, cryptic species and high-latitude species haplotypes of several coral species and undergoing local expansion (see Nakabayashi et al. 2019; Fifer et al. 2021. However, persistence of species into the future will depend on successful reproduction and recruitment. If the presence of unique haplotypes and endemic coals species in high latitudes cannot cope up with future changes, there is the possibility of widespread local extinctions of multiple coral species.

Acclimatization and Adaptation

Unlike tropical and sub-tropical locations, corals in high latitudes do not form reefs, and instead coral colonies predominantly settle on hard substrates such as rock to make up the coral community. In past decade, in some locations, a very high density of coral colonies can result in the area strongly resembling a reef (Mezaki and Kubota 2012; Mezaki 2014, Fig. 4.2). Such increasing densities (e.g., Fig. 4.2) can be attributed to increasing seawater temperatures over the last hundred years (up to +1.0–1.25 °C) in Japan (Sugihara et al. 2014) and Jeju, Korea (Yamano et al. 2012) and winter in these locations has been getting progressively warmer (Denis et al. 2013, 2014).

Acclimatization is particularly relevant at high-latitude locations where sea surface temperatures and light availability vary considerably every year. It has been hypothesized that corals at high-latitude sites are better equipped to withstand environmental variability (Cook et al. 1990; Wicks et al. 2010) due to the natural fluctuations of their environment over the annual course of the seasons. However, few studies have analyzed the cellular mechanisms that high-latitude corals use to cope with these variable conditions. At the same time, there is very little detailed information about coral recruitment in high-latitude non-reefal areas (Nakamura et al. 2021). For the continued maintenance and persistence of corals in high latitudes, the recruitment of new individuals is critical (Underwood and Fairweather 1989). Hence, it is necessary to conduct long-term studies on the processes and efficiency of recruitment, which are influenced by complex biological and environmental factors (Adjeroud et al. 2017).

While most tropical and subtropical corals reefs experience limited thermal flexibility (differences between upper and lower limits being 4–5 °C) (Mydlarz et al. 2010; Schoepf et al. 2015; Hoey et al. 2016), high-latitude locations have temperatures near and on the edges of acceptable conditions for zooxanthellate corals (see Kleypas et al. 1999). Even if high-latitude habitats might act as future refugia for tropical and subtropical coral species (Yamano et al. 2011), they remain at best an uncertain prospect. This is because of increases in frequency and occurrences of heatwaves (Fordyce et al. 2019), large variations in seasonal seawater temperatures (Keshavmurthy et al. 2014; Higuchi et al. 2015; Shiu et al. 2017; Park et al. 2019), and extreme lows and highs of seawater temperatures (Riegl and Piller 2003; Abdo et al. 2012; Foster et al. 2012; Camp et al. 2018; Tuckett and Wernberg 2018; Schoepf et al. 2019; Duarte et al. 2020).

Ongoing changes in abundances of various coral species (usually increases) have been occurring in the high latitudes

Fig. 4.2 Increasing coral cover and densities in Shikoku, Japan in past decade in various locations. The photos show coral cover in Tatsukushi, Kochi (32.7885° N, 132.8674° E) (**a**–**d**), *Pavona decussata* micro-atoll at Tatsukushi, Kochi (**d**), dominant presence of *Acropora muricata* at Sunokawa, Ehime (**e**) and Nahari, Kochi (33.4110819, 134.0306909) (**f**, **g**) in past decade

of East Asia during the past few decades, already altering some coral communities (Yamano et al. 2011; Yara et al. 2011). Such observed changes may not necessarily only relate to increases in seawater temperatures (Yara et al. 2011), and there have been other factors implicated in such range shifts. For example, in many locations such as Kochi, Japan, and Jeju, Korea, coastlines have been historically dominated by seaweeds and kelp forests (Serisawa et al. 2004; Tanaka et al. 2012), which have played important roles in the thriving coastal shellfish and other fisheries

Fig. 4.3 Stronghold of *Alveopora japonica* colonies (**a**) in the waters of Jeju (especially in the northern part of the island). Macro image of *A. japonica* polyps (**b**). Ever since the seawater has become warm, the competition between macroalgae (such as kelp) and coral has become the norm, generally with kelp losing the competition for space (**c**)

(Peteiro 2017; Kang 2010; Sato et al. 2010; Steneck and Johnson 2014; Bennett et al. 2016). Under past conditions, due to several ecological interactions (shading, competition for space, abrasion, etc.), zooxanthellate coral populations were suppressed (Coyer et al. 1993; Miller and Hay 1996; Leong et al. 2018; Thompson et al. 2018).

However, in the past three decades, many macroalgal communities in high latitudes in Korea and Japan have been decimated by the effects of rising temperatures (Tanaka et al. 2012; Haraguchi et al. 2009 in Kochi, Japan; Kim 2006 in Jeju, South Korea) and other factors including increased herbivory from urchins (Fujita 2010; Jeon et al. 2015) and increased dominance of tropical fish (Nakamura et al. 2013). The results of such increased stressors are barren grounds known as 'getnoguem' in Korean and 'isoyake' in Japan (Serisawa et al. 2004; Vieira et al. 2016), with coralline algae and little else covering the rocky substrates. These conditions have led to several coral species increasing in recruitment and abundance at many locations such as Jeju Island (Denis et al. 2014; Hong et al. 2015) and Kochi, Japan (Mezaki 2012; Mezaki and Kubota 2012).

One very dynamic example of succession and population growth in high latitudes is of the coral *Alveopora japonica* (Fig. 4.3a). Previously categorized as threatened (IUCN Red List), this species has been experiencing population growth and expansion at several locations in East Asia, including at Jeju Island and further north at high latitudes in Japan (the southern limit of this coral is Taiwan; see Kang et al. (2020); Vieira et al. (2016); Keshavmurthy et al. (2021). These increases are theorized to be mainly due to decreasing macroalgal (*Sargassum, Laminaria*, and *Ecklonia*) cover (Vieira et al. 2016). Because of its high recruitment rates and weedy character, *A. japonica* thrives alongside kelp (Denis et al. 2014; Hong et al. 2015; Lee et al. 2022) (also see Fig. 4.3b, c). Interestingly, a recent study on the population genetics of *A. japonica* in East Asia showed cryptic species diversity in this region (Kang et al. 2020). Thus, while research supports the hypothesis of recent population growth around Jeju Island, it appears that this coral species has been present for a long evolutionary time at Jeju and does not represent a case of poleward migration from subtropical environments following recent climate changes (Kang et al. 2020). This work serves as an example of the need for confirmation of proposed range expansions via molecular analyses, as an absence of past records from northern regions for a certain species may be not due to its true absence but instead due to species in question being present in small numbers or in uninvestigated areas.

Migration and Range Shifts

Across the globe, there are many examples of recent zooxanthellate coral range shifts, including recent expansion of acroporids towards higher latitudes in the western Atlantic (Precht and Aronson 2004) and the western and lower latitudes of Australia (Solitary Islands) (Greenstein and Pandolfi 2008; Baird et al. 2012). Specifically, Baird et al. (2012) observed four tropical *Acropora* species for the first time in the Solitary Islands (but see Mizerek et al. 2021).

In East Asia, in recent years there have been new records of tropical corals, fish, seaweeds, and invertebrate species appearing at high-latitude environments, indicating possible northern range shifts (Haraguchi and Sekida 2008, Haraguchi et al. 2009; Mezaki 2012; Mezaki and Kubota 2012; Terazono et al. 2012; Nakamura et al. 2013; Richards et al. 2016; Vergés et al. 2014; Wernberg et al. 2011, 2016; Reimer et al. 2018).

Regarding zooxanthellate corals, during the last 30 years, there have been documented changes in the marine communities around Shikoku, with approximately 139 species of scleractinian corals recorded (Veron and Minchin 1992; also T. Mezaki unpublished). The dominance of kelp and *Sargassum* populations is now being slowly succeeded by *Acropora*-dominated coral communities (Yamano et al. 2011; Yara et al. 2011; Denis et al. 2013). Species such as *Acropora cytherea, Acropora nana, Millepora exaesa* (Nomura and Mezaki, 2005), *Acropora* sp. (Mezaki, 2012), *A. digitifera, A. robusta, A. subulata,* and *Millepora platyphylla* (T. Mezaki, unpublished data) are some species that have undergone potential range shifts. Compared to 1993, a survey conducted in 2011 showed higher coral densities and diversity in Tei-Konan City in Kochi, Shikoku, Japan (Mezaki and Kubota 2012; Mezaki 2014, see Fig. 4.4, Table 4.2). This is particularly obvious in case of *Acropora nana*, which was first recorded in Shikoku in 2004 (Nomura and Mezaki 2005) and was subsequently spotted multiple times in different locations in Kochi (Mezaki and Kubota 2012). *A. nana* in Kochi is reproductively active and this may potentially be contributing to the proliferation of local populations (see Mezaki 2014).

Another case is the coral *Acropora pruinosa*, which has been increasingly found in temperate *Sargassum* and kelp beds in Ooshima (33°22’ N), in southwestern Ehime, Shikoku (Fig. 4.5), and is believed to be comprised of two genetic lineages (Pipithkul et al. 2021). Like many taxonomically problematic *Acropora* species, it is very difficult to differentiate the two lineages morphologically. However, a potential gradual range-shift from northwestern Kyushu and the northeastern Pacific towards Shikoku may be ongoing (Pipithkul et al. 2021).

One species that has been well examined in East Asia is the coral *Acropora hyacinthus*, which has previously been thought to be extending its range into high-latitude Japan (Yamano et al. 2011). Although *A. hyacinthus* is ubiquitous in the tropical and subtropical Indo-Pacific, it is also common in high-latitude non-reefal regions such as Kochi, Shikoku (see previous section in this chapter) and Kyushu (Veron 2000; Yamano et al. 2011; Suzuki et al. 2016; Nakabayashi et al. 2019; Nakamura et al. 2021). After 2011, *A. hyacinthus* has become the dominant species at Kochi (Denis et al. 2013), and Yamano et al. (2011) speculated that this coral species may have undergone a range shift, migrating into high-latitude Japan. However, Suzuki et al. (2016) demonstrated that the *A. hyacinthus* present in high-latitude Japan is in fact a cryptic lineage different from subtropical and tropical *A. hyacinthus*.

Subsequently, Nakabayashi et al. (2019) showed that among three cryptic lineages of *A. hyacinthus*, only one was distributed in temperate regions (also Fifer et al. 2021; Rivera et al. 2021). These results imply that high-latitude non-reefal populations of *A. hyacinthus* are distinct and not recent arrivals from the southern subtropical areas of Japan (Nakabayashi et al. 2019) as had been previously hypothesized. Moreover, low genetic diversity in high-latitude *A. hyacinthus* populations (Nakabayashi et al. 2019) and low reproductive success due to a lack of reef and currents has led to low recruitment (Misaki 2017). Thus, the present unique high-latitude *A. hyacinthus* populations are likely maintained by the survival and growth of existing colonies and fragmentation due to disturbances from typhoons (Nakamura et al. 2021).

For these reasons, Fifer et al. (2021) argued that high-latitude regions should be considered as refugia and given high conservation priority due to their unique diversity, and both local adaptation and slow expansion could facilitate further range expansion of these populations in high-latitude regions. In other words, range expansions of zooxanthellate

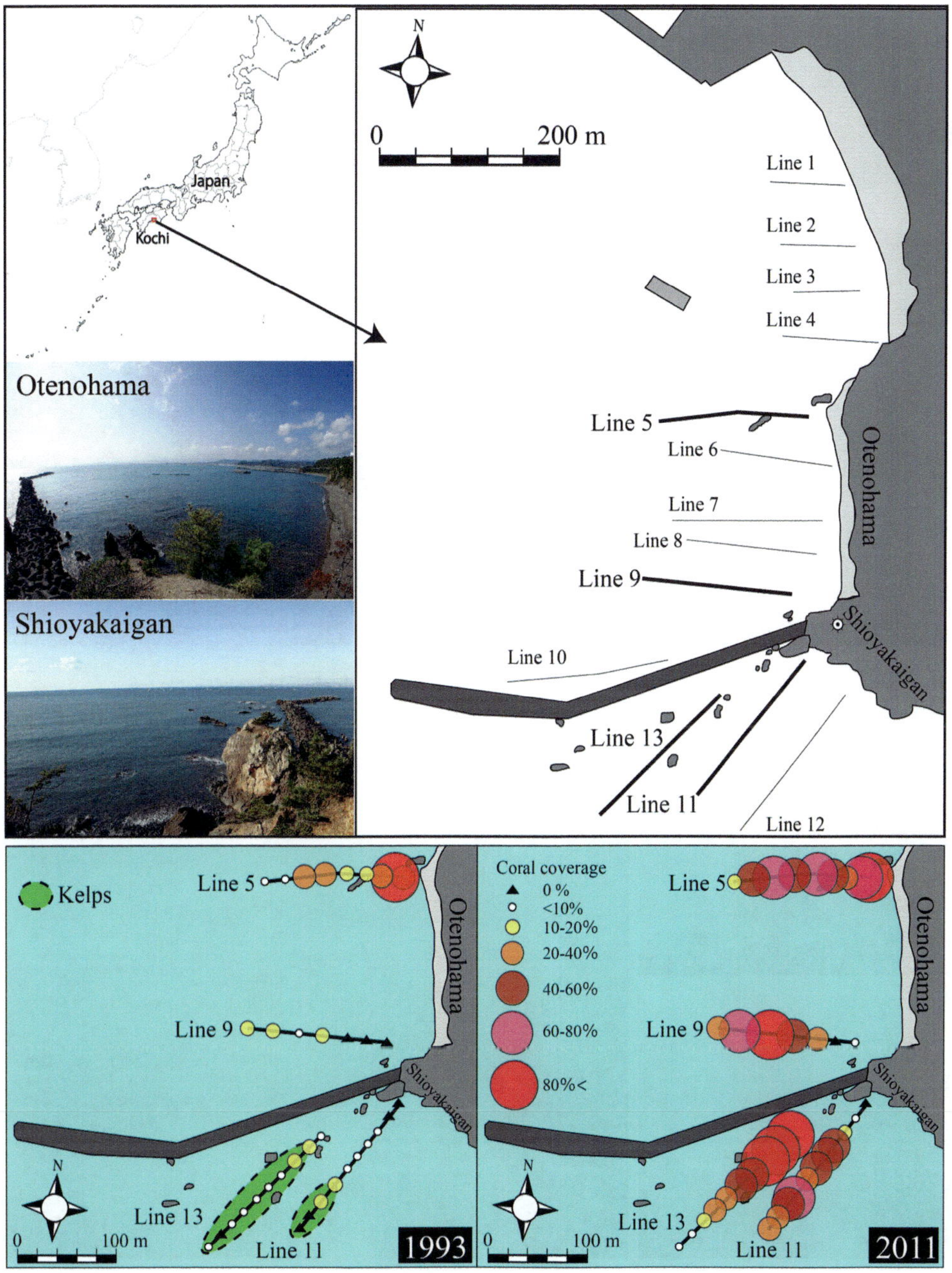

Fig. 4.4 Comparison of benthos at a location in Kochi between 1993 and 2011. Over time, observations and surveys have shown at two locations—Otenohama and Shioyakaigan at Tei in Konan City, Kochi (33.5203443, 133.7536717) that benthos previously colonized and dominated by macroalgae have now been replaced by coral populations (the figure modified from Mezaki and Kubota 2012, also see Table 4.2)

Table 4.2 Percentage occurrence and change of dominant coral species between 1993 and 2011 in Kochi, Japan (also see Fig. 4.4)

No.	1993	Percentage of occurrence	2011	Percentage of occurrence
1	*Acropora solitaryensis*	30.5%	*Acropora hyacinthus*	46.1%
2	*Cyphastrea serailia*	15.7%	*Acropora japonica*	14.8%
3	*Leptastrea pruinosa*	8.0%	*Acropora solitaryensis*	9.9%
4	*Acropora japonica*	5.6%	*Cyphastrea serailia*	6.9%
5	*Coscinaraea columna*	4.8%	*Psammocora superficialis*	2.9%
6	*Cyphastrea chalcidicum*	4.4%	*Pavona decussata*	2.2%
7	*Acropora pruinosa*	4.3%	*Montastrea valenciennesi*	1.8%
8	*Acropora hyacinthus*	4.1%	*Acropora muricata*	1.7%
9	*Psammocora profundacella*	3.5%	*Acropora pruinosa*	1.5%
10	*Hydnophora exesa*	2.7%	*Leptastrea pruinosa*	1.3%

Fig. 4.5 Coral *Acropora pruinosa*, which has been increasingly found in temperate *Sargassum* and kelp beds in Ooshima, Yawatahama City (33.3794814, 132.3435057)), in southwestern Ehime, Shikoku. As seawater gets warmer, this coral is seen encroaching and emerging in locations (**a**–**d**) where benthos before were solely populated by macroalgae (kelp, seaweed in Kuroshio Town, Kochi) (**e**, **f**). In addition, increases in coral density and cover have given stiff competition to macroalgae for benthic space (**g**)

corals in high latitudes may be more likely to come from species and populations already present high latitude corals, and not from tropical or subtropical coral species. Overall, discerning between increases in abundance and range shifts is difficult in the absence of molecular data. Given all the unique features of high-latitude marine environments when compared to subtropical and tropical coral reefs (see previous section), the presence of unique and adapted zooxanthellate coral populations or even species should not be surprising.

4.1.4 Climate Change and Future High-Latitude Coral Communities in East Asia: Perspectives

There is global evidence of ongoing expansion of coral populations because of sea surface temperature increases (Precht and Aronson 2004; Greenstein and Pandolfi 2008). Gradual warming due to climate change seems to enable population expansions of tropical to subtropical fauna into temperate communities (Tuckett et al. 2017). The southern region of mainland Japan (Honshu west of Tokyo, northern Shikoku, and Kyushu) harbours many high-latitude coral communities, and it has been projected that these regions will host both tropical/subtropical and high-latitude coral communities in the future (Yara et al. 2011). However, climate change in the Anthropocene may be occurring faster than zooxanthellate coral species can adapt, and it remains unclear as to how many species will continue to survive as the oceans warm (Hoegh-Guldberg et al. 2017). Failure to acclimate or adapt may lead to extinction for some species (see previous section; Dietzel et al. 2021; Muir et al. 2022).

One option for corals is to range-shift or migrate towards locations with relatively cooler environments compared to the hot conditions in future subtropical and tropical locations. Such locations at higher latitudes have been termed as refugia (see Soares 2020). However, migration or range-shifting in benthic organisms such as comparatively slow-growing and slow-to-adapt corals (Hoegh-Guldberg et al. 2017) may be difficult (Bates et al. 2013, 2014; Hargreaves et al. 2015), and influenced by interactions of factors including seawater temperature variations (Yara et al. 2011) and rising ocean acidification (Yara et al. 2012).

For example, observations have shown that it is difficult for potential range-shifting tropical species to survive the wide annual seawater temperature ranges in Kochi, Japan (see (Keshavmurthy et al. 2014), especially during the winter. Another example is a tropical *Acropora* sp. that occurred in Kochi between 2009 and 2011 but died in the winter of 2011 from a sudden drop in seawater temperatures (minimum 13.4 °C) (Mezaki 2014; Takao et al. 2015). The effects of cold seawater events on fish and coral have also been discussed in Leriorato and Nakamura (2019), highlighting how sudden shifts in seawater temperatures can limit the refuge potential of high-latitude locations.

We hypothesize that for tropical and subtropical zooxanthellate corals to naturally relocate poleward requires rapid acclimation and/or adaptation to local environmental conditions, specifically with regards to seawater temperature variations. Results from recent studies have shown that a given coral species is locally and seasonally adapted to the environment it thrives in (Lee et al. 2018; Keshavmurthy et al. 2021, 2022). Molecular studies have also demonstrated that high-latitude populations are distinct from lower latitude populations for many coral species, and not recent arrivals to these regions (Suzuki et al. 2016; Nakabayashi et al. 2019; Nakamura et al. 2021).

Asides from rising temperatures, there are several other environmental factors in the high-latitude regions of East Asia that zooxanthellate coral species must contend with now and into the future. Unlike tropical and subtropical locations, the environment and weather patterns are more dynamic and often less predictable in high latitudes. Notably, while global warming will bring overall warmer temperatures to the northwest Pacific (Yara et al. 2014), this may also bring increased instability into temperature and weather patterns. For example, the Kuroshio Current is known to occasionally meander in its course along mainland Japan, and when this occurs in winter, the results can be devastating for local zooxanthellate corals (Suzuki et al. 2013). In particular, for many high-latitude corals, their distributions are limited by cold water tolerance (Higuchi et al. 2015), and a single cold-shock event during winter can wipe out species with warmer-water affinities (Kemp et al. 2011, 2016), or even native species not adapted to such extreme events. Meanders may also influence spawning timing (Nozawa 2012), making potential future restoration efforts harder to plan. Global warming may increase the strength of typhoons in this region (Emanuel, 2005; Trenberth, 2005; Webster et al., 2005; Emanuel et al. 2008; Tu et al. 2009), and increasingly northwards courses for typhoons (Tu et al. 2009). As coral communities are not protected by large coral reef formations, this could mean more mechanical damage from larger waves, which is already a major component of loss of coral in the region (Cant et al. 2022). Furthermore, as most coral communities in this region are attached to rocky shorelines, they are in the direct vicinity of land, meaning that they experience potentially more low salinity (True 2012; Coles and Jokiel 2018) and high runoff/smothering events (Fabricius 2005) from typhoons and heavy rains, and if typhoons become more extreme, so will these negative influences. As well, increasing acidification has also been posited as a potential limiting factor in this region (Yara et al. 2012). Finally, lower amounts of available light for photosynthesis of endosymbiotic Symbiodiniaceae at higher latitudes have also been posited as another potential limiting factor in poleward expansion (Muir et al. 2015; see also Madin et al. 2016).

It is predicted that most of the world's coral reefs will become marginalized due to the changing climate (Hoegh-Guldberg et al. 2007; Soares 2020), and therefore it is necessary to understand the ecophysiology of as many zooxanthellate coral species as possible in such environments to provide useful analogues for understanding the ecology of coral reefs subjected to changing conditions (Perry and Larcombe 2003). Although the possibility of poleward migration and refugia needs to be modelled on a species-by-species basis, the overall currently available data do not suggest this to be likely for most zooxanthellate coral species.

However, very few focused molecular and physiological studies have been undertaken to determine if tropical and subtropical fauna (especially corals) will be able to adjust, adapt, and/or acclimatize to high-latitude environmental conditions. It is necessary to define these capacities of corals to understand how each coral species present across latitudinal and seasonal gradients will respond under current to future climate changes. With projected seawater temperatures rising by 1.8–4.0 °C by 2050 under IPCC scenarios (IPCC 2021), and pH values decreasing due to ocean acidification (Caldeira and Wickett 2003), the responses of corals and their associated organisms, ranging from microbes to large reef-associated organisms, and their interactions with local stressors must all be considered in order to effectively plan conservation efforts (Orr et al. 2005; Doney et al. 2009; Yara et al. 2011, 2012). Thus, to better elucidate the responses of zooxanthellate corals to predicted future environmental changes, focused multidisciplinary research approaches must be undertaken across tropical, subtropical, and high-latitude coral communities (Castillo and Helmuth 2005; Wicks 2009; Freeman 2015; Higuchi et al. 2015; Samiei et al. 2015; Camp et al. 2018; Kavousi et al. 2020). Such research will help us understand the mechanisms present in corals to face and survive changes in the face of increases in the frequency and severity of extreme temperature events (Rooke et al. 2017).

Moving forward, as the potential futures of zooxanthellate corals are explored, a clear definition or explanation of many of the terms used by scientists may need clarification or refinement. For example, one definition of 'tropicalization' has been thought of as warm-water species increasingly colonizing high-latitude environments and replacing native species, resulting in high-latitude locations becoming 'tropicalized'. 'Tropicalization' was defined early on by Bianchi and Morri (2003) as an 'increased occurrence of warm-water biota' in the Mediterranean, suggesting the tropicalization of the Mediterranean. The definition of 'tropicalized' is when a species become acclimatized and/or adapted to the tropical environment and can exist and thrive. Thus, a location cannot become 'tropicalized', rather it can be 'tropical-like' and host such organisms, which can acclimatize and/or adapt to the environment of that location due to conditions become suitable or similar to the invading specie's previous location. In East Asia, there is no doubt that high-latitude marine environments are becoming warmer and gaining characteristics of tropical or subtropical environments as the climate continues to change (IPCC 2021). Due to increases in average seawater temperature (winter and summer), certain marginal (high-latitude, non-reefal) locations have and will experience the arrival of tropical fauna and flora. Continued existence of these tropical species at high-latitude sites can happen if arriving species are plastic enough to tolerate wide temperature ranges between seasons, as well as the effects of factors such as typhoons and salinity changes. Thus, range-shifted organisms will become 'marginalized' when they reach high-latitude locations. 'Marginalized' in this sense means when flora and fauna (from tropical and subtropical locations) range-shift or migrate to marginal (high-latitude, non-reefal) locations and become acclimatized and/or adapted to thrive in such locations.

As scientists, clearly defining the aspects of 'tropicalization' being investigated in each study should be explained, e.g., explaining the range-shifting of tropical/subtropical species as opposed to functional or more general changes in marine communities. At the same time, discerning range shifts from increases in abundance of already present local populations needs much more research effort to better understand the overall prognosis for zooxanthellate corals. It is possible that native corals in high latitudes have better chances of surviving the warming of their native environment than range-expanding species, which is another risk facing range-shifting and poleward migration of species (Sorte et al. 2010; Cobben et al. 2012; Valladares et al. 2014). Critically, these native high-latitude species are already adapted to wide temperature ranges (Keshavmurthy et al. 2021).

Rather than acting as refugia for corals from tropical and subtropical locations, we propose that high-latitude areas in East Asia will act instead as 'oases' for the already present native coral populations and a limited number of range-shifted coral species that can adapt to the variable and unique environment of the region (e.g., Fifer et al. 2021). There are many such locations in high-latitude East Asia that may be considered 'oases', not only for *Acropora* species (see Fig. 4.2), but also other species such as *Pavona decussata* (e.g., Fig. 4.6; also, Nomura 2004; Mezaki et al. 2014). Thus, instead of the 'tropicalization' proposed by Bianchi and Morri (2003), high-latitude benthic communities may become more 'tropicalized' from a functional point of view (e.g., loss of functional species—Vergés et al. 2019; Agostini et al. 2021), with higher abundances of 'tropical'-like zooxanthellate coral species that were already present in the region, and decreased abundances of temperate macroalgae (Haraguchi and Sekida 2008; Haraguchi et al. 2009; Tanaka et al. 2012). Although it is possible for certain macroalgal and fish species to migrate and range-shift (Nakamura et al. 2013; Vergés et al. 2014), it may be rather infeasible (for example, availability of light; Muir et al. 2015) for most zooxanthellate coral species to do so (Mizerek et al. 2021).

Supporting this idea, coral communities on the high-latitude coastal reefs of eastern Australia typically comprise widely distributed generalists, as well as stress-tolerant coral species with horizontal morphologies (Sommer et al. 2014). Other studies have noted increased abundances of generalist anthozoan species under conditions considered marginal for most zooxanthellate coral species (Inoue et al. 2013; Rodriguez et al. 2020; Reimer et al. 2021).

For zooxanthellate coral diversity, the worst-case future scenario is that existing high-latitude coral species lose their ability to produce healthy juveniles (Keshavmurthy et al. 2014) or to recruit due to increasing summer seawater

Fig. 4.6 Occurrences of large coral colonies at various locations in Shikoku, Japan. Large colony of *Porites lutea* (**a**, **b**) in Oshima, Mugi Town, Tokushima (33.6367559, 134.4847409); *Porites* sp. (**c**) in Konan City, Kochi (33.5203443, 133.7536717); *Lithophyllon undulatum* (**d**) and *Stylocoeniella guentheri* (2 m in area) (**e**) in Takegashima, Kaiyo Town, Tokushima (33.5420455, 134.3163478); *Acropora muricata* and *Pavona decussata* (**f**, **g**) respectively in Tatsukushi, Tosashimizu City, Kochi (32.7748762, 132.8689544), and *Plesiastrea verispora* (**h**) in Sunokawa, Ainan Town, Ehime can be found to be increasing in past decade

temperatures, increased temperature fluctuations, and, to some extent, acidification (Nakabayashi et al. 2019) and more extreme weather events (see Beyer et al. 2018). In addition, recruitment via external population sources (e.g., subtropical, or tropical regions) may also not be possible (Nakabayashi et al. 2019; Nakamura et al. 2021), resulting in the presently existing high-latitude coral communities representing the coda of such ecosystems. Similar results have been reported in other disturbed environments (Hoegh-Guldberg et al. 2019), and it is not improbable that this could happen in high-latitude locations.

4.2 Conclusions

At high latitudes in eastern Asia, in some locations, previously dominant macroalgal and kelp communities are being succeeded by corals such as *Al. japonica* in Jeju Island, Korea

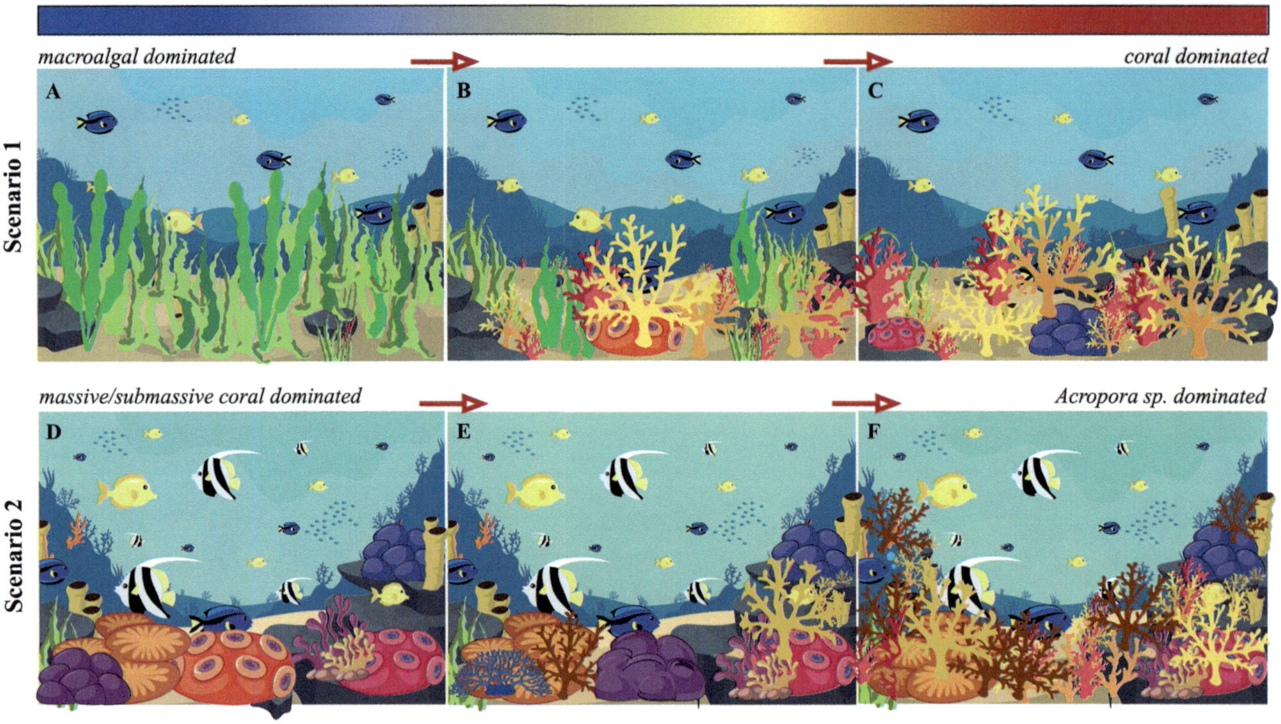

Fig. 4.7 Schematic diagram showing possible emergence and succession in high-latitude locations. Scenario 1 depicts locations previously dominated by macroalgae (for example locations with *Alveopora japonica* dominance in Jeju, Korea; locations previously dominated by macroalgae in Kochi, Japan) and slowly being overtaken by coral populations through succession. Scenario 2 depicts locations previously devoid of coral populations or composed mainly of massive or sub-massive coral species (for example Kochi, Japan) being populated by endemic and previously cryptic *Acropora* sp. populations. Illustrations were drawn using free to download and use vector images from https://www.vecteezy.com/free-vector/nature, Nature Vectors by Vecteezy

and *Acropora* spp. in Shikoku, Japan (scenario 1, Fig. 4.7a–c). However, rather than poleward range shifts, it appears increasing abundances of previously less abundant local coral species may be responsible for these changes. Thus, high-latitude locations generally dominated by massive and/or sub-massive corals are now being overtaken by *Acropora* species (Fig. 4.7d–f), representing perhaps functional shift. Even if and when some coral species do range-shift (proven by molecular ecology and taxonomic studies), it is more plausible that high-latitude locations in East Asia will become 'oases' for existing coral species and populations already present in these locations rather than to poleward-migrating tropical and subtropical corals. The proliferation of previously rare or cryptic species (e.g., acroporid corals) with morphologies and ecophysiologies adapted to the particular environmental features of high latitudes could form the majority of the future benthos in these areas (Scenario 2, Fig. 4.7 d–f). Due to slow coral skeletal growth, wide temperature ranges, and typhoons, the coral communities in high-latitude locations in higher latitude eastern Asia may favour such adapted species (Kleypas et al. 1999; Yamano et al. 2001; Takao et al. 2015).

Previous research has often assumed that coral populations in high-latitude locations represent the range edges of species from tropical and subtropical locations. However, recent molecular data do not show this, instead often indicating the presence of regional temperate endemic lineages (see Nakabayashi et al. 2019, Fifer et al. 2021). Considering that high-latitude locations are 'oases' for these temperate coral species, such regions should be designated as areas for conservation priority in order to protect endemic species and lineages. To better facilitate future conservation, it is necessary to conduct more studies on high-latitude coral communities, particularly on those endemic species and lineages by including eco-physiological, molecular ecological, and taxonomic (molecular and morphological) datasets to understand whether future coral communities will be dominated by the succession of these local species or the poleward range-shifting of lower latitude species, or a mixture of both scenarios.

Definition

Tropical-like: when, due to increases in average seawater temperatures (winter and summer), certain marginal (high-latitude, non-reefal) locations experience arrival and presence of tropical fauna and flora. This

(continued)

can happen when arriving species are plastic enough to tolerate wide temperature ranges between seasons, as well as effects of factors such as typhoons and salinity changes.

Tropicalized: when species become acclimatized and/or adapted to tropical environments. A location cannot become 'tropicalized', rather it can become 'tropical-like' and host organisms that can acclimatize and/or adapt to the environment of that location, due to conditions become suitable or similar to the invading organism's original location.

Marginalized: when flora and fauna from tropical and subtropical locations range-shift or migrate to marginal (high-latitude, non-reefal) locations and become acclimatized and/or adapted to thrive in such locations.

Acknowledgements The authors thank the editors of this edition of the Corals of the World series, and Profs. Takeuchi and Yamashiro for giving us an opportunity to contribute. We also thank members of the Coral Lab at the Biodiversity Research Center, Academia Sinica, Taiwan, and the Shellfish Research & Aquaculture Lab at the Department of Marine Life Science, Jeju National University, for their support. Comments from Dr. Danwei Huang on an earlier version helped improved this work.

References

Abdo DA, Bellchambers LM, Evans SN (2012) Turning up the heat: increasing temperature and coral bleaching at the high latitude coral reefs of the Houtman Abrolhos Islands. PLoS One 7:e43878. https://doi.org/10.1371/journal.pone.004387

Adjeroud M, Kayal M, Penin L (2017) Importance of recruitment processes in the dynamics and resilience of coral reef assemblages. In: Rosssi S, Bramanti L, Gori A, Orejas C (eds) Marine animal forests: the ecology of benthic biodiversity hotspots. Springer, Cham, pp 549–569

Agostini S, Harvey BP, Milazzo M, Wada S, Kon K, Floc'h N, Komatsu K, Kuroyama M, Hall-Spencer JM (2021) Simplification, not "tropicalization", of temperate marine ecosystems under ocean warming and acidification. Glob Chang Biol 27:4771–4784

Baird AH, Sommer B, Madin JS (2012) Poleward range expansion of *Acropora* spp. along the east coast of Australia. Coral Reefs 31:1063

Bates AE, McKelvie CM, Sorte CJB, Morley SA, Jones NAR, Mondon JA, Bird TJ, Quinn G (2013) Geographical range, heat tolerance and invasion success in aquatic species. Proc Royal Soc B Biol Sci 280: 20131958. https://doi.org/10.1098/rspb.2013.1958

Bates AE, Pecl GT, Frusher S, Hobday AJ, Wernberg T, Smale DA, Sunday JM, Hill NA, Dulvy NK, Colwell RK, Holbrook NJ, Fulton EA, Slawinski D, Feng M, Edgar GJ, Radford BT, Thompson PA, Watson RA (2014) Defining and observing stages of climate-mediated range shifts in marine systems. Glob Environ Chang 26: 27–38

Beger M, Sommer B, Harrison PL, Smith SDA, Pandolfi JM (2014) Conserving potential coral reef refuges at high latitudes. Divers Distrib 20:245–257

Bennett S, Wernberg T, Connell SD, Hobday AJ, Johnson CR, Poloczanska ES (2016) The 'Great Southern Reef': social, ecological and economic value of Australia's neglected kelp forests. Mar Freshw Res 67:47–56

Beyer HL, Kennedy EV, Beger M et al (2018) Risk-sensitive planning for conserving coral reefs under rapid climate change. Conserv Lett 11. https://doi.org/10.1111/conl.12587

Bianchi CN, Morri C (2003) Global Sea warming and "tropicalization" of the Mediterranean Sea: biogeographic and ecological aspects. Biogeographia 24:319–327

Brown BE, Cossins AE (2011) The potential for temperature acclimatisation of reef corals in the face of climate change. In: Dubinsky Z, Stambler N (eds) Coral reefs: an ecosystem in transition. Springer, New York, pp 421–434

Caldeira K, Wickett M (2003) Anthropogenic carbon and ocean pH. Nature 425:365

Camp EF, Schoepf V, Mumby PJ, Hardtke LA, Rodolfo-Metalpa R, Smith DJ, Suggett DJ (2018) The future of coral reefs subject to rapid climate change: lessons from natural extreme environments. Front Mar Sci 5:4. https://doi.org/10.3389/fmars.2018.00004

Cant J, Cook K, Reimer JD, Mezaki T, Nakamura M, O'Flaherty C, Salguero-Gómez R, Beger M (2022) Transient amplification enhances the persistence of tropicalising coral populations in marginal high latitude environments. Ecography:e06156. https://doi.org/10.1111/ecog.06156

Carpenter KE, Abrar M, Aeby G, Aronson RB, Banks S, Bruckner A, Chiriboga A, Cortés J, Delbeek JC, DeVantier L, Edgar GJ, Edwards AJ, Fenner D, Guzmán HM, Hoeksema BW, Hodgson G, Johan O, Licuanan WY, Livingstone SR, Lovell ER, Moore JA, Obura DO, Ochavillo D, Polidoro BA, Precht WF, Quibilan MC, Reboton C, Richards ZT, Rogers AD, Sanciangco J, Sheppard A, Sheppard C, Smith J, Stuart S, Turak E, Veron JEN, Wallace C, Weil E, Wood E (2008) One-third of reef-building corals face elevated extinction risk from climate change and local impacts. Science 321:560–563

Castillo KD, Helmuth BST (2005) Influence of thermal history on the response of *Montastraea annularis* to short-term temperature exposure. Mar Biol 148:261–270

Chen CA (1999) Analysis of scleractinian distribution in Taiwan indicating a pattern congruent with sea surface temperatures and currents: examples from *Acropora* and Faviidae corals. Zool Stud 38:119–129

Chen CA, Keshavmurthy S (2009) Taiwan as a connective stepping-stone in the Kuroshio Traiangle and the conservation of coral ecosystems under the impacts of climate change. Kuroshio Sci 3: 15–22

Cobben MMP, Verboom J, Opdam PFM, Hoekstra RF, Jochem R, Smulders MJM (2012) Wrong place, wrong time: climate change-induced range shift across fragmented habitat causes maladaptation and declined population size in a modelled bird species. Glob Chang Biol 18:2419–2428

Coles SL, Jokiel PL (2018) Effects of salinity on coral reefs. In: Connell D, Hawker DW (eds) Pollution in tropical aquatic systems. CRC Press, pp 147–166

Cook CB, Logan A, Ward J, Luckhurst B, Berg CJ Jr (1990) Elevated temperatures and bleaching on a high latitude coral reef: the 1988 Bermuda event. Coral Reefs 9:45–49

Cook KM, Yamagiwa H, Beger M, Masucci GD, Ross S, Lee HYT, Stuart-Smith RD, Reimer JD (2022) A community and functional comparison of coral and reef fish assemblages between four decades of coastal urbanisation and thermal stress. Ecol Evol 12:e8736. https://doi.org/10.1002/ece3.8736

Coyer JA, Ambrose RF, Engle JM, Carroll JC (1993) Interactions between corals and algae on a temperate zone rocky reef: mediation by sea urchins. J Exp Mar Biol Ecol 167:21–37

Cruz ICS, Waters LG, Kikuchi RKP, Leão ZMAN, Turra A (2018) Marginal coral reefs show high susceptibility to phase shift. Mar Pollut Bull 135:551–561

Dalton SJ, Roff G (2013) Spatial and temporal patterns of eastern Australia subtropical coral communities. PLoS One 8:e75873. https://doi.org/10.1371/journal.pone.0075873

de Palmas S, Denis V, Ribas-Deulofeu L, Loubeyres M, Woo S, Hwang SJ, Song JI, Chen CA (2015) *Symbiodinium* spp. associated with high-latitude scleractinian corals from Jeju Island, South Korea. Coral Reefs 34:919–925

Denis V, Mezaki T, Tanaka K, Kuo C-Y, Palmas SD, Keshavmurthy S, Chen CA (2013) Coverage, diversity, and functionality of a high-latitude coral community (Tatsukushi, Shikoku Island, Japan). PLoS One 8:e54330. https://doi.org/10.1371/journal.pone.0054330

Denis V, Ribas-Deulofeu L, Loubeyres M, Palmas SD, Hwang S-J, Woo S, Song J-I, Chen CA (2014) Recruitment of the subtropical coral *Alveopora japonica* in the temperate waters of Jeju Island, South Korea. Bull Mar Sci 91:85–96

Denis V, Ribas-Deulofeu L, Loubeyres M, Palmas SD, Hwang S-J, Woo S, Song J-I, Chen CA (2015) Recruitment of the subtropical coral *Alveopora japonica* in the temperate waters of Jeju Island, South Korea. Bull Mar Sci 91:85–96. https://doi.org/10.5343/bms.2014.1032

Dietzel A, Bode M, Connolly SR, Hughes TP (2021) The population sizes and global extinction risk of reef-building coral species at biogeographic scales. Nat Ecol Evol 5:663–669

Doney SC, Fabry VJ, Feely RA, Kleypas JA (2009) Ocean acidification: the other CO_2 problem. Mar Sci 1:169–192

Duarte GAS, Villela HDM, Deocleciano M, Silva D, Barno A, Cardoso PM, Vilela CLS, Rosado P, Messias CSMA, Chacon MA, Santoro EP, Olmedo DB, Szpilman M, Rocha LA, Sweet M, Peixoto RS (2020) Heat waves are a major threat to turbid coral reefs in Brazil. Front Mar Sci 7:179. https://doi.org/10.3389/fmars.2020.00179

Emanuel K (2005) Increasing destructiveness of tropical cyclones over the past 30 years. Nature 436:686–688

Emanuel K, Sundararajan R, Williams JGK (2008) Hurricanes and global warming. Results from downscaling IPCC AR4 simulations. Bull Am Meteorol Soc 89:347–369

Fabricius KE (2005) Effects of terrestrial runoff on the ecology of corals and coral reefs: review and synthesis. Mar Pollut Bull 50: 125–146

Fifer JE, Yasuda N, Yamakita T, Bove CB, Davies SW (2021) Genetic divergence and range expansion in a western North Pacific coral. Sci Total Environ 813:152423

Fordyce AJ, Ainsworth TD, Heron SF, Leggat W (2019) Marine heatwave hotspots in coral reef environments: physical drivers, ecophysiological outcomes, and impact upon structural complexity. Front Mar Sci 6:498. https://doi.org/10.3389/fmars.2019.00498

Foster K, Foster G, Al-Cibahy AS, Al-Harthi S, Purkis SJ, Riegl BM (2012) Environmental setting and temporal trends in southeastern gulf coral communities. In: Riegl BM, Purkis SJ (eds) Coral reefs of the Gulf. Springer, Dordrecht, pp 51–70. https://doi.org/10.1007/978-94-007-3008-3

Freeman LA (2015) Robust performance of marginal Pacific coral reef habitats in future climate scenarios. PLoS One 10:e0128875–e0128816. https://doi.org/10.1371/journal.pone.0128875

Fujita D (2010) Current status and problems of Isoyake in Japan. Bull Fish Res Agen 32:33–42

Greenstein BJ, Pandolfi JM (2008) Escaping the heat: range shifts of reef coral taxa in coastal Western Australia. Glob Chang Biol 14: 513–528

Haraguchi H, Sekida S (2008) Recent changes in the distribution of *Sargassum* species in Kochi, Japan. Kuroshio Sci 2:41–46

Haraguchi H, Tanaka K, Imoto Z, Hiraoka M (2009) The decline of *Ecklonia cava* in Kochi, Japan and the challenge in marine afforestation. Kuroshio Sci 3:49–54

Hargreaves AL, Bailey SF, Laird RA (2015) Fitness declines towards range limits and local adaptation to climate affect dispersal evolution during climate-induced range shifts. J Evol Biol 28:1489–1501

Harii S, Omori M, Yamakawa H, Koike Y (2001) Sexual reproduction and larval settlement of the zooxanthellate coral *Alveopora japonica* Eguchi at high latitudes. Coral Reefs 20:19–23

Harriott VJ, Banks SA (1995) Recruitment of scleractinian corals in the Solitary Islands marine reserve, a high latitude coral-dominated community in eastern Australia. Mar Ecol Prog Ser 123:155–161

Harriott VJ, Harrison PL, Banks SA (1995) The coral communities of Lord Howe Island. Mar Freshw Res 46:457–446

Heery EC, Hoeksema BW, Browne NK, Reimer JD, Ang PO, Huang D, Friess DA, Chou LM, Loke LHL, Saksena-Taylor P, Alsagoff N, Yeemin T, Sutthacheep M, Vo ST, Bos AR, Gumanao GS, Hussein MAS, Waheed Z, Lane DJW, Johan O, Kunzmann A, Jompa J, Suharsono Taira D, Bauman AG, Todd PA (2018) Urban coral reefs: degradation and resilience of hard coral assemblages in coastal cities of East and Southeast Asia. Mar Pollut Bull 135:654–681. https://doi.org/10.1016/j.marpolbul.2018.07.041

Higuchi T, Agostini S, Casareto BE, Suzuki Y, Yuyama I (2015) The northern limit of corals of the genus *Acropora* in temperate zones is determined by their resilience to cold bleaching. Sci Rep 5:18467. https://doi.org/10.1038/srep18467

Ho MJ, Dai CF (2014) Coral recruitment of a subtropical coral community at Yenliao Bay, northern Taiwan. Zool Stud 53:5. https://doi.org/10.1186/1810-522X-53-5

Hoegh-Guldberg O, Mumby PJ, Hooten AJ, Steneck RS, Greenfield P, Gomez E, Harvell CD, Sale PF, Edwards AJ, Caldeira K, Knowlton N, Eakin CM, Iglesias-Prieto R, Muthiga N, Bradbury RH, Dubi A, Hatziolos ME (2007) Coral reefs under rapid climate change and ocean acidification. Science 318:1737–1742

Hoegh-Guldberg O, Poloczanska ES, Skirving W, Dove S (2017) Coral reef ecosystems under climate change and ocean acidification. Front Mar Sci 4:158. https://doi.org/10.3389/fmars.2017.00158

Hoegh-Guldberg O, Pendleton L, Kaup A (2019) People and the changing nature of coral reefs. Reg Stud Mar Sci 30:100699

Hoey A, Pratchett M, Cvitanovic C (2011) High macroalgal cover and low coral recruitment undermines the potential resilience of the world's southernmost coral reef assemblages. PLoS One 6(10): e25824. https://doi.org/10.1371/journal.pone.0025824

Hoey A, Howells E, Johansen J, Hobbs J-P, Messmer V, McCowan D, Wilson S, Pratchett M (2016) Recent advances in understanding the effects of climate change on coral reefs. Diversity 8:12–22

Hong H-K, Keshavmurthy S, Kang C-K, Hwang K, Park SR, Cho S-H, Choi K-S (2015) *Alveopora japonica* repopulation of a bare substrate off Jeju Island, Korea. Bull Mar Sci 91:477–478

Hughes TP, Day JC, Brodie J (2015) Securing the future of the Great Barrier Reef. Nat Clim Chang 5:508–511

Hughes TP, Barnes ML, Bellwood DR, Cinner JE, Cumming GS, Jackson JBC, Kleypas J, van de Leemput IA, Lough JM, Morrison TH, Palumbi SR, van Nes EH, Scheffer M (2017a) Coral reefs in the Anthropocene. Nature 546:82–90

Hughes TP, Kerry JT, Álvarez-Noriega M, Álvarez-Romero JG, Anderson KD, Baird AH, Babcock RC, Beger M, Bellwood DR, Berkelmans R, Bridge TC, Butler IR, Byrne M, Cantin NE, Comeau S, Connolly SR, Cumming GS, Dalton SJ, Diaz-Pulido G, Eakin CM, Figueira WF, Gilmour JP, Harrison HB, Heron SF, Hoey AS, Hobbs J-PA, Hoogenboom MO, Kennedy EV, Kuo C, Lough JM, Lowe RJ, Liu G, McCulloch MT, Malcolm HA, McWilliam MJ, Pandolfi JM, Pears RJ, Pratchett MS, Schoepf V, Simpson T, Skirving WJ, Sommer B, Torda G, Wachenfeld DR, Willis BL, Wilson SK (2017b) Global warming and recurrent mass bleaching of corals. Nature 543:373–377

Hughes TP, Anderson KD, Connolly SR, Heron SF, Kerry JT, Lough JM, Baird AH, Baum JK, Berumen ML, Bridge TC, Claar DC, Eakin CM, Gilmour JP, Graham NAJ, Harrison H, Hobbs J-PA, Hoey AS,

Hoogenboom M, Lowe RJ, McCulloch MT, Pandolfi JM, Pratchett M, Schoepf V, Torda G, Wilson SK (2018a) Spatial and temporal patterns of mass bleaching of corals in the Anthropocene. Science 359:80–83

Hughes TP, Kerry JT, Baird AH, Connolly SR, Dietzel A, Eakin CM, Heron SF, Hoey AS, Hoogenboom MO, Liu G, McWilliam MJ, Pears RJ, Pratchett MS, Skirving WJ, Stella JS, Torda G (2018b) Global warming transforms coral reef assemblages. Nature 556: 492–496

Inoue S, Kayanne H, Yamamoto S, Kurihara H (2013) Spatial community shift from hard to soft corals in acidified water. Nat Clim Chang 3:683. https://doi.org/10.1038/nclimate1855

IPCC (2021) Summary for policymakers. In: Masson-Delmotte V, Zhai P, Pirani A, Connors SL, Péan C, Berger S, Caud N, Chen Y, Goldfarb L, Gomis MI, Huang M, Leitzell K, Lonnoy E, Matthews JBR, Maycock TK, Waterfield T, Yelekçi O, Yu R, Zhou B (eds) Climate change 2021: the physical science basis. Contribution of working group I to the sixth assessment report of the intergovernmental panel on climate change. Cambridge University Press, Cambridge, pp 3–32. https://doi.org/10.1017/9781009157896.001

Iwase F (2004) Shikoku Island. In: Ministry of the Environment and Japanese Coral Reef Society (ed) Coral reefs of Japan. Ministry of the Environment, Tokyo, pp 259–269

Jeon BH, Yang KM, Kim JH (2015) Changes in macroalgal assemblage with sea urchin density on the east coast of South Korea. Algae 30: 139–146

Kang RS (2010) A review of destruction of seaweed habitats along the coast of the Korean Peninsula and its consequences. Bull Fish Res Agen 32:25–31

Kang JH, Jang JE, Kim JH, Kim S, Keshavmurthy S, Agostini S, Reimer JD, Chen CA, Choi K-S, Park SR, Lee HJ (2020) The origin of the subtropical coral *Alveopora japonica* (Scleractinia: Acroporidae) in high-latitude environments. Front Ecol Evol 8:12. https://doi.org/10.3389/fevo.2020.00012

Kavousi J, Denis V, Sharp V, Reimer JD, Nakamura T, Parkinson JE (2020) Unique combinations of coral host and algal symbiont genotypes reflect intraspecific variation in heat stress responses among colonies of the reef-building coral, *Montipora digitata*. Mar Biol 167:23

Kemp DW, Oakley CA, Thornhill DJ, Newcomb LA, Schmidt GW, Fitt WK (2011) Catastrophic mortality on inshore coral reefs of the Florida Keys due to severe low-temperature stress. Glob Chang Biol 17:3468–3477. https://doi.org/10.1111/j.1365-2486.2011.02487.x

Kemp DW, Colella MA, Bartlett LA, Ruzicka RR, Porter JW, Fitt WK (2016) Life after cold death: reef coral and coral reef responses to the 2010 cold water anomaly in the Florida Keys. Ecosphere 7(6): e01373. https://doi.org/10.1002/ecs2.1373

Keshavmurthy S, Fontana S, Mezaki T, González L d C, Chen CA (2014) Doors are closing on early development in corals facing climate change. Sci Rep 4:5633. https://doi.org/10.1038/srep05633

Keshavmurthy S, Beals M, Hsieh HJ, Choi K-S, Chen CA (2021) Physiological plasticity of corals to temperature stress in marginal coral communities. Sci Total Environ 758:143628

Keshavmurthy S, Chen T-R, Liu P-J, Wang J-T, Chen CA (2022) Learning from the past is not enough to survive present and future bleaching threshold temperatures. Sci Total Environ 852:158379

Kim D (2006) A study on the restoration of marine forests using artificial reef in the barren grounds along the coast of Jeju. PhD thesis, Jeju National University, Korea

Kim SW, Chung M, Park H-S (2015) Tropical fish species thriving in temperate Korean waters. Mar Biodivers 2:147–148

Kim H, Moon B, Kim M, Kwon M (2020) Dynamic mechanisms of summer Korean heat waves simulated in a longterm unforced Community Climate System Model version 3. Atmos Sci Lett 21. https://doi.org/10.1002/asl.973

Kleypas JA, McManus JW, Meñez LAB (1999) Environmental limits to coral reef development: where do we draw the line? Am Zool 39: 146–159

Kumagai NH, Molinos JG, Yamano H et al (2018) Ocean currents and herbivory drive macroalgae-to-coral community shift under climate warming. Proc Natl Acad Sci 115:201716826. https://doi.org/10.1073/pnas.1716826115

Kuo C-Y, Keshavmurthy S, Huang Y-Y, Ho M-J, Hsieh HJ, Xiao A-T, Lo W-C, Hsin Y-C, Chen CA (2023) Transitional coral ecosystem of Taiwan in the era of changing climate. In: Takeuchi I, Yamashiro H (eds) Coral reefs of eastern Asia under anthropogenic impacts. Coral reefs of the world, vol 17, Springer Nature Switzerland AG, Cham, pp 7–34

Lee STM, Keshavmurthy S, Fontana S, Takuma M, Chou W-H, Chen CA (2018) Transcriptomic response in *Acropora muricata* under acute temperature stress follows preconditioned seasonal temperature fluctuations. BMC Res Notes 11:119. https://doi.org/10.1186/s13104-018-3230-z

Lee K-T, Lee H-M, Subramaniam T, Yang H-S, Park SR, Kang C-K et al (2022) Dominance of the scleractinian coral *Alveopora japonica* in the barren subtidal hard bottom of high-latitude Jeju Island off the south coast of Korea assessed by highresolution underwater images. PLoS One 17(11):e0275244. https://doi.org/10.1371/journal.pone.0275244

Leong RC, Marzinelli EM, Low J, Bauman AG, Lim EWX, Lim CY, Steinberg PD, Guest JR (2018) Effect of coral-algal interactions on early life history processes in *Pocillopora acuta* in a highly disturbed coral reef system. Front Mar Sci 5:385. https://doi.org/10.3389/fmars.2018.00385

Leriorato JC, Nakamura Y (2019) Unpredictable extreme cold events: a threat to range-shifting tropical reef fishes in temperate waters. Mar Biol 166:110

Mezaki T (2012) First record of distribution of *Acropora* sp. with cochleariform radial coralites from Shirigai, Otsuki, Kochi prefecture, Japan. Kuroshio Biosphere 8:23–26 + 1pl. in Japanese

Mezaki T (2014) Coral migration to the high latitude area in east and North Asia. In: Kimura T, Tun K, Chou LM (eds) Status of coral reefs in east Asian seas region. JWRC, MOE Japan, pp 31–33

Mezaki T, Kubota S (2012) Changes of hermatypic coral community in costal sea area of Kochi, high-latitude, Japan. Aquabiol 201(34): 332–337. in Japanese

Mezaki T, Keshavmurthy S, Chen CA (2014) An old and massive colony of *Pavona decussata* is sexually active at high latitude (32° N) in Japan. Coral Reefs 33:97

Miller MW, Hay ME (1996) Coral-seaweed-grazer-nutrient interactions on temperate reefs. Ecol Monogr 66:323–344

Misaki H (2017) Spawning patterns of high latitude scleractinian corals from 1989 to 2012 at Kushimoto, Wakayama, Japan. Nanki-seibutsu (in Japanese) 59:55–60

Mizerek TL, Madin JS, Benzoni F, Huang D, Luiz OJ, Mera H, Schmidt-Roach S, Smith SDA, Sommer B, Baird AH (2021) No evidence for tropicalization of coral assemblages in a subtropical climate change hot spot. Coral Reefs 40:1451–1461

Muir PR, Wallace CC, Done T, Aguirre JD (2015) Limited scope for latitudinal extension of reef corals. Science 348:1135–1138

Muir PR, Obura DO, Hoeksema BW, Sheppard C, Pichon M, Richards ZT (2022) Conclusions of low extinction risk for most species of reef-building corals are premature. Nat Ecol Evol 14:1–2

Mydlarz LD, McGinty ES, Harvell CD (2010) What are the physiological and immunological responses of coral to climate warming and disease? J Exp Biol 213:934–945

Nakabayashi A, Yamakita T, Nakamura T, Aizawa H, Kitano YF, Iguchi A, Yamano H, Nagai S, Agostini S, Teshima KM, Yasuda N (2019) The potential role of temperate Japanese regions as refugia for the coral *Acropora hyacinthus* in the face of climate change. Sci Rep 9:1892. https://doi.org/10.1038/s41598-018-38333-5

Nakamura Y, Feary DA, Kanda M, Yamaoka K (2013) Tropical fishes dominate temperate reef fish communities within western Japan. PLoS One 8(12):e81107. https://doi.org/10.1371/journal.pone.0081107

Nakamura M, Nomura K, Hirabayashi I, Nakajima Y, Nakajima T, Mitarai S, Yokochi H (2021) Management of scleractinian coral assemblages in temperate non-reefal areas: insights from a long-term monitoring study in Kushimoto, Japan (33°N). Mar Biol 168:140

Nojima S (2004) Kyushu. In: Ministry of the Environment and Japanese Coral Reef Society (ed) Coral reefs of Japan. Ministry of the Environment, Tokyo, pp 272–280

Nomura K (2004) Kii Peninsula. In: Ministry of the Environment and Japanese Coral Reef Society (ed) Coral reefs of Japan. Ministry of the Environment, Tokyo, pp 252–256

Nomura K, Mezaki T (2005) Reef building corals from Otsuki, Kochi prefecture, Japan. Kuroshio Biosphere 2:29–41 + 2pls. in Japanese

Nozawa Y (2012) Annual variation in the timing of coral spawning in a high-latitude environment: influence of temperature. Biol Bull 222: 192–202

Okano T (2013) Biodiversity in the islands of Kagoshima. In: Kawai K, Terada R, Kuwahara S (eds) The islands of Kagoshima. Kagoshima University Research Center for the Pacific Islands, pp 136–145

Orr JC, Fabry VJ, Aumont O, Bopp L, Doney SC, Feely RA, Gnanadesikan A, Gruber N, Ishida A, Joos F (2005) Anthropogenic ocean acidification over the twenty-first century and its impact on calcifying organisms. Nature 437:681–686

Park J-S, Keshavmurthy S, Subramaniam T, Park S-R, Kang C-K, Choi K-S (2019) Annual gametogenesis patterns in two high-latitude corals, *Alveopora japonica* and *Oulastrea crispata*, from Jeju Island, South Korea. Estuar Coasts 43:1–10

Perry CT, Larcombe P (2003) Marginal and non-reef-building coral environments. Coral Reefs 22:427–432

Peteiro C (2017) Alginates and their biomedical applications. Springer Ser Biomater Sci Eng:27–66. https://doi.org/10.1007/978-981-10-6910-9_2

Pipithkul S, Ishizu S, Shimura A, Yokochi H, Nagai S, Fukami H, Yasuda N (2021) High clonality and geographically separated cryptic lineages in the threatened temperate coral, *Acropora pruinosa*. Front Mar Sci 8:668043. https://doi.org/10.3389/fmars.2021.668043

Pontasch S (2014) Living on the edge: protective mechanisms underlying thermal tolerance in high latitude. PhD thesis, Victoria University of Wellington, New Zealand

Precht WF, Aronson RB (2004) Climate flickers and range shifts of reef corals. Front Ecol Environ 2:307–314

Reimer JD, Ono S, Sinniger F, Tsukahara J (2008) Distribution of zooxanthellate zoanthid species (Zoantharia: Anthozoa: Hexacorallia) in southern Japan limited by cold temperatures. Galaxea J Coral Reef Stud 10:57–67

Reimer JD, Kim S, Arai S, Keshavmurthy S, Choi KS (2018) First records of zooxanthellate *Zoanthus* (Anthozoa: Hexacorallia: Zoantharia) from Korea and Japan (east) sea. Mar Biodivers 48: 1269–1273

Reimer JD, Fujii T, Kise H, Yanagi K, Cook K, Cant J, Koeda K, Koido T, Kitamura T, Mezaki T (2020) A *Goniopora stokesi* community at Tatsugasako, Otsuki, Kochi, Japan: a new northernmost specimen-based record. Plankton Benthos Res 15:185–187

Reimer JD, Fourreau CJ, Yamagiwa H, Poliseno A (2021) First record of *Nanipora* (Lithotelestidae: Helioporacea: Octocorallia: Anthozoa) from the Yaeyama Islands. Fauna Ryukyuana 63:1–5

Richards Z, Kirkendale L, Moore G, Hosie A, Huisman J, Bryce M, Marsh L, Bryce C, Hara A, Wilson N, Morrison S, Gomez O, Ritchie J, Whisson C, Allen M, Betterridge L, Wood C, Morrison H, Salotti M, Hansen G, Slack-Smith S, Fromont J (2016) Marine biodiversity in temperate Western Australia: multi-taxon surveys of Minden and Roe reefs. Diversity 8:7. https://doi.org/10.3390/d8020007

Riegl B (2003) Climate change and coral reefs: different effects in two high-latitude areas (Arabian Gulf, South Africa). Coral Reefs 22: 433–446

Riegl B, Piller WE (2003) Possible refugia for reefs in times of environmental stress. Int J Earth Sci 92:520–531

Rivera HE, Aichelman HE, Fifer JE, Kriefall NG, Wuitchik DM, Wuitchik SJS, Davies SW (2021) A framework for understanding gene expression plasticity and its influence on stress tolerance. Mol Ecol 30:1381–1397

Roberts CM, McClean CJ, Veron JEN et al (2002) Marine biodiversity hotspots and conservation priorities for tropical reefs. Science 295: 1280–1284. https://doi.org/10.1126/science.1067728

Rodriguez MV, Segumalian CS, Lalas JA, Maningas JM (2020) Octocorals outcompete scleractinian corals in a degraded reef. IOP Conf Ser Earth Environ Sci 420:012027

Rooke AC, Burness G, Fox MG (2017) Thermal physiology of native cool-climate, and non-native warm-climate pumpkinseed sunfish raised in a common environment. J Thermal Biol 64:48–57

Samiei JV, Saleh A, Mehdinia A, Shirvani A, Kayal M (2015) Photosynthetic response of Persian Gulf acroporid corals to summer versus winter temperature deviations. PeerJ 3:e1062. https://doi.org/10.7717/peerj.1062

Sato A, Kuwahara H, Hashimoto O (2010) Efforts by fishers and support activities to conserve and rehabilitate seaweed beds: adaptive management of the fishery resources and habitats in Japan, vol 8, Fish for the People: Southeast Asian Fisheries Development Center, pp 25–31

Schleyer MH, Floros C, Laing SCS, Macdonald AHH, Montoya-Maya PH, Morris T, Porter SN, Seré MG (2018) What can South African reefs tell us about the future of high-latitude coral systems? Mar Pollut Bull 136:491–507

Schoepf V, Stat M, Falter JL, McCulloch MT (2015) Limits to the thermal tolerance of corals adapted to a highly fluctuating, naturally extreme temperature environment. Sci Rep 5:17639. https://doi.org/10.1038/srep17639

Schoepf V, Carrion SA, Pfeifer SM, Naugle M, Dugal L, Bruyn J, McCulloch MT (2019) Stress-resistant corals may not acclimatize to ocean warming but maintain heat tolerance under cooler temperatures. Nat Commun 10:4031. https://doi.org/10.1038/s41467-019-12065-0

Serisawa Y, Imoto Z, Ishikawa T, Ohno M (2004) Decline of the *Ecklonia cava* population associated with increased seawater temperatures in Tosa Bay, southern Japan. Fish Sci 70:189–191

Shiu J-H, Keshavmurthy S, Chiang P-W, Chen H-J, Lou S-P, Tseng C-H, Hsieh HJ, Chen CA, Tang S-L (2017) Dynamics of coral-associated bacterial communities acclimated to temperature stress based on recent thermal history. Sci Rep 7:14933. https://doi.org/10.1038/s41598-017-14927-3

Soares M d O (2020) Marginal reef paradox: a possible refuge from environmental changes? Ocean Coast Manage 185:105063. https://doi.org/10.1016/j.ocecoaman.2019.105063

Sommer B, Harrison PL, Beger M, Pandolfi JM (2014) Trait-mediated environmental filtering drives assembly at biogeographic transition zones. Ecology 95:1000–1009

Sommer B, Sampayo EM, Beger M, Harrison PL, Babcock RC, Pandolfi JM (2017) Local and regional controls of phylogenetic structure at the high-latitude range limits of corals. Proc R Soc B 284:20170915–20170910. https://doi.org/10.1098/rspb.2017.0915

Sorte CJB, Williams SL, Carlton JT (2010) Marine range shifts and species introductions: comparative spread rates and community impacts. Glob Ecol Biogeogr 19:303–316

Steneck RS, Johnson CR (2014) Kelp forests: dynamic patterns, processes, and feedbacks. In: Bertness MD, Bruno JF, Silliman BR, Stachowicz JJ (eds) Marine community ecology and conservation. Sinauer Associates, Inc., Boston, MA, pp 315–336. ISBN 9781605352282 [Research Book Chapter]

Sugihara K, Yamano H, Choi K-S, Hyeong K (2014) Zooxanthellate scleractinian corals of Jeju Island, Republic of Korea. In: Nakano S, Yahara T, Nakashizuka T (eds) Integrative observations and assessments. Ecological research monographs. Springer, Tokyo, pp 111–130. https://doi.org/10.1007/978-4-431-54783-9_6

Suzuki G, Yatsuya K, Muko S (2013) Bleaching of tabular *Acropora* corals during the winter season in a high-latitude community (Nagasaki, Japan). Galaxea J Coral Reef Stud 15:43–44

Suzuki G, Keshavmurthy S, Hayashibara T, Wallace CC, Shirayama Y, Chen CA, Fukami H (2016) Genetic evidence of peripheral isolation and low diversity in marginal populations of the *Acropora hyacinthus* complex. Coral Reefs 35:1419–1432

Takao S, Yamano H, Sugihara K, Kumagai NH, Fujii M, Yamanaka Y (2015) An improved estimation of the poleward expansion of coral habitats based on the inter-annual variation of sea surface temperatures. Coral Reefs 34:1125–1137

Tanaka K, Taino S, Haraguchi H, Prendergast G, Hiraoka M (2012) Warming off southwestern Japan linked to distributional shifts of subtidal canopy-forming seaweeds. Ecol Evol 2:2854–2865

Terazono Y, Nakamura Y, Imoto Z, Hiraoka M (2012) Fish response to expanding tropical *Sargassum* beds on the temperate coasts of Japan. Mar Ecol Prog Ser 464:209–220

Thompson PL, MacLennan MM, Vinebrooke RD (2018) Species interactions cause non-additive effects o,f multiple environmental stressors on communities. Ecosphere 9:e02518. https://doi.org/10.1002/ecs2.2518

Trenberth K (2005) Uncertainty in hurricanes and global warming. Science 308:1753–1754

True JD (2012) Salinity as a structuring force for near shore coral communities. In, Proceedings of the 12th International Coral Reef Symposium, vol 9, Cairns, Australia, p e13

Tu J-Y, Chou C, Chu P-S (2009) The abrupt shift of typhoon activity in the vicinity of Taiwan and its association with western North Pacific-East Asian climate change. J Climate 22:3617–3628

Tuckett CA, Wernberg T (2018) High latitude corals tolerate severe cold spell. Front Mar Sci 5:14. https://doi.org/10.3389/fmars.2018.00014

Tuckett CA, de Bettignies T, Fromont J, Wernberg T (2017) Expansion of corals on temperate reefs: direct and indirect effects of marine heatwaves. Coral Reefs 36:947–956

Underwood AJ, Fairweather PG (1989) Supply-side ecology and benthic marine assemblages. Trends Ecol Evol 4:16–20

Valladares F, Matesanz S, Guilhaumon F, Araújo MB, Balaguer L, Benito-Garzón M, Cornwell W, Gianoli E, Kleunen M, Naya DE, Nicotra AB, Poorter H, Zavala MA (2014) The effects of phenotypic plasticity and local adaptation on forecasts of species range shifts under climate change. Ecol Lett 17:1351–1364

van der Zande RM, Achlatis M, Bender-Champ D, Kubicek A, Dove S, Hoegh-Guldberg O (2020) Paradise lost: end-of-century warming and acidification under business-as-usual emissions have severe consequences for symbiotic corals. Glob Chang Biol 26:2203–2219

Vergés A, Steinberg PD, Hay ME, Poore AG, Campbell AH, Ballesteros E, Heck KL, Booth DJ, Coleman MA, Feary DA, Figueira W, Langlois R, Marzinelli EM, Mezerek T, Mumby PJ, Nakamura Y, Roughan M, van Sebille E, Gupta AS, Smale DA, Tomas F, Wernberg T, Wilson SK (2014) The tropicalization of temperate marine ecosystems: climate-mediated changes in herbivory and community phase shifts. Proc R Soc B 281:20140846. https://doi.org/10.1098/rspb.2014.0846

Vergés A, McCosker E, Mayer-Pinto M, Coleman MA, Wernberg T, Ainsworth T, Steinberg PD (2019) Tropicalisation of temperate reefs: implications for ecosystem functions and management actions. Funct Ecol 33:1000–1013. https://doi.org/10.1111/1365-2435.13310

Veron JEN (2000) Corals of the world, vol 3. Australian Institute of Marine Science, Townsville

Veron JEN, Done TJ (1979) Corals and coral communities of Lord Howe Island. Aust J Mar Freshwat Res 30:203–236

Veron JEN, Minchin PR (1992) Correlations between sea surface temperature, circulation patterns and the distribution of hermatypic corals of Japan. Cont Shelf Res 12:835–857

Vieira C, Keshavmurthy S, Ju S-J, Hyeong K, Seo I, Kang C-K, Hong H-K, Chen CA, Choi K-S (2016) Population dynamics of a high-latitude coral *Alveopora japonica* Eguchi from Jeju Island, off the southern coast of Korea. Mar Freshw Res 67:594–604

Webster PJ, Holland GJ, Curry JA, Chang HR (2005) Changes in tropical cyclone number, duration, and intensity in a warming environment. Science 309:1844–1846

Wernberg T, Thomsen MS, Tuya F, Kendrick GA (2011) Biogenic habitat structure of seaweeds change along a latitudinal gradient in ocean temperature. J Exp Mar Biol Ecol 400:264–271

Wernberg T, Bennett S, Babcock RC, de Bettignies T, Cure K, Depczynski M, Dufois F, Fromont J, Fulton CJ, Hovey RK, Harvey ES, Holmes TH, Kendrick GA, Radford B, Santana-Garcon J, Saunders BJ, Smale DA, Thomsen MS, Tuckett CA, Tuya F, Vanderklift MA, Wilson S (2016) Climate-driven regime shift of a temperate marine ecosystem. Science 353:169–172

Wicks LC (2009) Persistence of corals in marginal habitats: the role of the environment and symbiont diversity and ecophysiology. PhD thesis. Victoria University of Wellington, New Zealand

Wicks LC, Hill R, Davy SK (2010) The influence of irradiance on tolerance to high and low temperature stress exhibited by *Symbiodinium* in the coral, *Pocillopora damicornis*, from the high-latitude reef of Lord Howe Island. Limnol Oceanogr 55: 2476–2486

Yabe H, Sugiyama T (1931) Reef-building coral fauna of Japan. Proc Imperial Acad 7:357–360

Yamano H, Hori K, Yamauchi M, Yamagawa O, Ohmura A (2001) Highest-latitude coral reef at Iki Island, Japan. Coral Reefs 20:9–12

Yamano H, Sugihara K, Nakai T, Yamagata O (2004) Iki Islands. In: Ministry of the Environment and Japanese Coral Reef Society (ed) Coral reefs of Japan. Ministry of the Environment, Tokyo, pp 242–244

Yamano H, Sugihara K, Nomura K (2011) Rapid poleward range expansion of tropical reef corals in response to rising sea surface temperatures. Geophys Res Lett 38:4. https://doi.org/10.1029/2010GL046474

Yamano H, Sugihara K, Watanabe T, Shimamura M, Hyeong K (2012) Coral reefs at 34°N, Japan: exploring the end of environmental gradients. Geology 40:835–838

Yara Y, Fujii M, Yamanaka Y, Okada N, Yamano H, Oshima K (2009) Projected effects of global warming on coral reefs in seas close to Japan. Galaxea J Coral Reef Stud 11:131–140

Yara Y, Oshima K, Fujii M, Yamano H, Yamanaka Y, Okada N (2011) Projection and uncertainty of the poleward range expansion of coral habitats in response to sea surface temperature warming: a multiple climate model study. Galaxea J Coral Reef Stud 13:11–20

Yara Y, Vogt M, Fujii M, Yamano H, Hauri C, Steinacher M, Gruber N, Yamanaka Y (2012) Ocean acidification limits temperature-induced poleward expansion of coral habitats around Japan. Biogeosciences 9:4955–4968

Yara Y, Fujii M, Yamano H, Yamanaka Y (2014) Projected coral bleaching in response to future sea surface temperature rises and the uncertainties among climate models. Hydrobiologia 733:19–29. https://doi.org/10.1007/s10750-014-1838-0

Succession and Spread of Coral Diseases and Coral-Killing Sponges with Special Reference to Microbes in Southeast Asia and Adjacent Waters

5

Hideyuki Yamashiro, Naohisa Wada, and Sen-Lin Tang

Abstract

Corals and coral-associated organisms are threatened by repeated bleaching events and anthropogenic stresses, resulting in mass mortality and property loss (including foods, natural breakwater reefs, tourism resources, and landscapes). In addition to physical and chemical disturbances, the biological destruction caused by infectious diseases and encrusting/boring organisms exceeds our expectations. This chapter describes the present situation of coral diseases and coral-killing sponges in the region of Southeast Asia to East Asia. These areas maintain a wide array of coral reefs with highly diverse organisms, but there is limited information available compared to the Caribbean and Great Barrier Reef. Recent research focused on the pathological and microbiological perspectives of reefs in the Southeast Asian area are summarized in this chapter.

Keywords

Threat · Prevalence · Pathogenesis · Global warming

5.1 Introduction

Coral reefs are valuable ecosystems that host a variety of marine organisms and produce food for people living in oligotrophic coastal areas. Small and low-lying islands depend on coral reefs for protection, providing a natural breakwater against rough waves and storms formed in the open ocean. The majority of the coral reef islands in Southeast Asia are surrounded by coral reefs. Contrary to their increasing protective value, most coral reefs are facing threats caused by human activities. These threats include global warming and ocean acidification due to high carbon dioxide emissions and other various anthropogenic impacts, such as increased sediments and chemical pollutants, including herbicides and fertilizers. In particular, both coral bleaching and diseases are highly visible phenomena facilitated by ocean warming (Hughes et al. 2003; Bruno and Selig 2007) which is globally increasing annually (IPCC 2021). On the local scale, human-induced activities such as sedimentation, overfishing, excess nutrient loads (eutrophication), and the addition of various chemicals have degraded coral reefs.

With decreasing coral populations, coral reefs are predicted to be replaced by algae or sponges (Hoegh-Guldberg et al. 2007; Carpenter et al. 2008; Mumby and Steneck 2008; Bell et al. 2013; Wulff 2016; Hughes et al. 2017; Bruno et al. 2019). Elevated stress levels enhance the susceptibility of corals to diseases and infections. The reef framework depends on a delicate balance between bioaccretion and bioerosion processes (Alvarado et al. 2017). A reduction in corals decreases the populations of other organisms and degrades the limestone reef structure (Glynn and Manzello 2015). The degraded structures attract destructive grazers of corals and limestone (fish and sea urchins), borers (sponges, mollusks, sipunculans, and polychaetes), and endolithic organisms (algae, fungi, and bacteria) which further degrade the calcium carbonate limestones. The number of boring sponges increases in disturbed areas (Carballo et al. 2013) and some boring sponges (such as *Cliona*) overgrow the living coral and dissolve the coral calcium carbonate skeleton (e.g., Glynn and Manzello 2015; Huang et al. 2021).

In the Atlantic and Pacific regions, the number of coral diseases has been increasing since the 1990s (e.g., Sutherland et al. 2004), including in remote island areas of Hawaii and the US Pacific (Aeby 2006; Vargas-Angel 2009). Coral reefs are threatened by infectious diseases which instigate mass losses of corals and coral-associated organisms (e.g., Harvell

H. Yamashiro (✉)
Sesoko Station, Tropical Biosphere Research Center, University of the Ryukyus, Motobu, Okinawa, Japan
e-mail: hyama@lab.u-ryukyu.ac.jp

N. Wada · S.-L. Tang
Biodiversity Research Center, Academia Sinica, Taipei, Taiwan

I. Takeuchi, H. Yamashiro (eds.), *Coral Reefs of Eastern Asia under Anthropogenic Impacts*, Coral Reefs of the World 17,
https://doi.org/10.1007/978-3-031-27560-9_5

et al. 1999, 2002). Initial reports of coral disease-like phenomenon were skeletal anomalies ("tumors or neoplasms" in the early literature), these appeared as disordered and rapidly degraded coral surface skeletal structures (Squires 1965). However, skeletal anomalies are not considered to be coral diseases because the abnormal shapes are not lethal to the coral itself. The first recorded infectious coral disease was black band disease (BBD), identified in Caribbean coral reefs (Antonius 1973). Thereafter, diverse coral diseases have emerged and threatened coral reef ecosystems worldwide, similar to the crown-of-thorns starfish. Environmental factors such as high water temperatures, nutrients, sediments, chemicals, and changes in the biotic composition have enhanced coral diseases.

Symbiosis with the dinoflagellate "zooxanthellae" (*Symbiodinium* and other genera closely related to *Symbiodinium*) is crucial for coral nutrition, promoting coral growth and reproduction. Coral-associated microbiomes can be positive or negative relationships. Coral holobiont microbiomes support coral health, producing improved nutrition and immunological defense (e.g., Shiu et al. 2017). Conversely, some microorganisms are pathogenic to corals, resulting in coral injury or death (e.g., Patterson et al. 2002). Coral diseases are caused by a wide range of microorganisms, mostly bacteria.

Southeast Asia and East Asia comprise tropical, subtropical, and temperate regions. Coral reefs have developed around numerous islands. Zooxanthellate corals occur in the temperate region of Japan. In addition, the Kuroshio Current brings warm waters to the north. These combined factors enable us to observe the effects of global heat stress and local stress on corals which cause bleaching and diseases and compare the continuous variations in their effects on reefs of different islands and temperature gradients.

5.2 Coral Diseases

Coral diseases have variable causal agents (including bacteria, protozoa, flatworms, but most remain unknown), severity (from small effects, minor injuries to lethal effects), progression rates (from slow to rapid), and affected areas of coral reefs (from limited to widespread). Coral reefs are affected by biological features (including microbes and macroalgae), chemicals (including nutrients), physical conditions (such as temperature, light, water flow, and turbidity), and human activities (for example, ships, leisure activities, eutrophication, and chemicals), potentially increasing the prevalence of coral diseases.

It is difficult to identify and assess diseases by the appearance of the affected corals. In the simplest terms, diseases should be assessed using Koch's postulates, such as isolating the pathogen, culture of the pathogen, infection of healthy corals, and the reinfection should match the original disease. However, when assuming a bacterium is a pathogen, most marine bacteria are uncultivable (99.9%), making it extremely difficult to elucidate the species. Furthermore, there is a risk of misdiagnosis when applying cultivable bacteria at high concentrations in isolation to host coral, even when the bacteria support Koch's postulate.

The Japanese Environmental Agency has instigated a long-term monitoring program of various ecosystems, including coral reefs. Coral reef surveys have occurred regularly from 2003 to 2020 (and are planned to continue for 100 years) in Yaeyama (Sekisei Lagoon). The results show coral diseases have been increasing in prevalence and geographic distribution (Biodiversity Center of Japan 2003, 2014, 2015, 2016, 2017, 2018, 2019, 2020).

This chapter aims to describe each coral disease/syndrome. Due to common misconceptions, other factors which are often confused with coral diseases/syndromes are not addressed in this chapter. These factors include: predation (starfish, snails, fish), grazing, bleaching (heat-induced, focal/non-focal), skeletal deformations (associated with other organisms), competition (with diatoms, algae, and cyanobacteria), and crustose coralline white syndrome. In addition, coral infectious diseases observed in aquariums, such as brown jelly, rapid tissue necrosis (Borneman 2001; Wada et al. 2012, 2015), tissue loss, and recovery of branching *Porites* (Asoh 2008) in Japan are not included in this chapter.

5.2.1 Lethal Coral Diseases: Rapid Tissue Loss Diseases

The diverse effects of infectious diseases on corals range in lethality, infection rate, and transmission rate, and vary with time and place. Presently, some coral diseases are lethal and widely distributed, impacting corals and coral-associated organisms. Lethal coral diseases cause rapid tissue loss. Examples include black band disease, white syndrome, and brown band. Other tissue loss diseases have more limited effects. For example, ulcerative white spots, skeletal eroding band, pink line syndrome, and atramentous coral tissue necrosis. The less destructive tissue loss diseases are discussed in the compromised disease/syndrome section of this chapter.

5.2.2 Black Band Disease (BBD)

Black band disease (BBD) was the first infectious coral disease identified in the Caribbean Sea (Antonius 1973). The black color and mat formation ensure BBD is easy to distinguish from other infectious diseases (Fig. 5.1a), with

Fig. 5.1 (**a**) Black band disease (BBD) on a massive coral, *Leptastrea purpurea*. (**b**) An enlarged view of BBD on an encrusting coral, *Montipora informis*, circles highlight the coral polyps. (**c**) White syndrome (WS) on *Acropora cytherea*. (**d**) An enlarged view of WS showing the degrading coral tissues. (**e**) Brown band (BrB) on *Acropora* sp., insets provide micrographs of the ciliates engulfing the zooxanthellae (left) and an image using fluorescent microscopy showing the autofluorescence of zooxanthellae (right). (**f**) Skeletal eroding band (SEB) on *M. informis*, the inset is the sessile ciliates that cause SEB. All photographs: Sesoko Island, Japan

the only exception of skeletal eroding band (SEB) (Fig. 5.1f). BBD has been spreading worldwide and is now reported from the Indo-Pacific to the Great Barrier Reef. BBD is a multi-microbial disease with a thin black band containing filamentous cyanobacteria (Fig. 5.1b), sulfate-reducing bacteria, and sulfur-oxidizing bacteria (Sato et al. 2016). Therefore, the mat has both aerobic and anaerobic environments. BBD produces toxic H_2S and associated bacteria-producing enzymes (proteases) to digest coral proteins. The initial infection is considered another filamentous bacterium called the cyanobacterial patch (Sato et al. 2010), which is gradually replaced by a black band forming cyanobacteria. Water-borne infections are potentially transmitted by BBD (Zvuloni et al. 2009).

BBD was initially observed in Japan (Yaeyama Islands) in 2003 (Biodiversity Center of Japan 2013). BBD is a commonly observed coral disease in Japanese coral reefs. In Okinawa Island (Japan), infections vary between species, species affected include *Acropora*, *Astreopora*, *Montipora*, *Gardineroseris*, *Pachyseris*, and *Leptastrea* (Yamashiro pers. obs.). In Okinawa Island, BBD has also been reported on *Pachyseris* in the upper mesophotic site (Kubomura et al. 2018). BBD is in the process of distribution expansion in Japan, dispersing from Yonaguni Island (westernmost) to

Kakeroma Island in the Amami Islands (Yamashiro unpublished data). These islands are close to the northern limit of the coral reefs (especially, Yakushima and Tanegashima Islands) (Yamano et al. 2001). The progression rates of BBD disease are estimated to be 2.9 mm d^{-1} in Akajima, Japan (Wada et al. 2017).

5.2.3 White Syndrome (WS)

White syndrome (WS) has been spreading widely in many coral reefs, similar to BBD. WS is considered a common and severe infectious disease (Fig. 5.1c, d). The causative agent was previously suggested as *Vibrio* (Luna et al. 2007; Sussman et al. 2008; Arboleda and Reichardt 2010; Ushijima et al. 2012, 2014). However, recent periodical measurements have suggested *Rhodobacteraceae* as the candidate bacteria causing WS after 18 months of repeated sampling of the diseased fronts (Pollock et al. 2017). WS is highly chronic when compared to BBD, with a progress rate of several cm/day. Outbreaks of WS often result in the sudden extinction of local coral communities (Yamashiro et al. 2016).

High water temperature enhances the susceptibility of corals to disease infections and the progress rate of the diseases. For example, BBD quickly spreads in the warm waters (seawater temperatures 25 °C or higher) of the Caribbean Sea (Kuta and Richardson 1997, 2002), and WS also disperses rapidly in the tropical waters of the GBR in Australia (Bruno et al. 2007). In contrast, an outbreak of WS occurred only on *Acropora* corals in Miyazaki, a temperate region of Japan where no coral reef develops (Yamashiro et al. 2016). Interestingly, WS-infected corals were still observed in winter, when water temperatures dropped to 15 °C in Miyazaki Japan.

WS has been reported in a wide range of coral reefs, including reefs in Africa, the Red Sea, the Indian Ocean, the Pacific Ocean (up to Hawaii), and the GBR in Australia (summarized by Bourne et al. 2015). Similarly, WS is distributed in most coral reefs in Southeast Asia, including Indonesia (Lamb et al. 2017; Subhan et al. 2020), Malaysia (Akmal and Shahbudin 2020), Taiwan (Huang et al. 2021), and Guam in Micronesia (Myers and Raymundo 2009).

5.2.4 Brown Band (BrB)

Brown band (BrB) (Willis et al. 2004) is characterized by a brown band along the progressing disease front (Fig. 5.1e). BrB is composed of an aggregation of numerous ciliates that engulf coral zooxanthellae (Bourne et al. 2008) (Fig. 5.1e). The elongated ciliates forming BrB have been identified as members of the class Oligohymenophorea, subclass Scuticociliatia (Bourne et al. 2008), and *Porpostoma guamense* n. sp. (Lobban et al. 2011). The progress of BrB is quite rapid, at a rate of 1.2 cm a day (Haapkylä et al. 2009). The prevalence of BrB is not high or frequent when compared to BBD and WS. However, the brown color is generally indistinguishable from coral tissue. The rapid progression and consumption rate suggest further research on BrB is required.

5.2.5 Compromised Disease/Syndrome

Other diseases/syndromes that cause relatively moderate tissue loss or bleaching of coral tissues are discussed in this section.

Growth anomalies (GAs) (previously called "tumors") are easy to identify in the field due to the affected coral's distinct abnormal development (hypercalcification) (Fig. 5.2a, b). GAs prevail in a wide range of coral reefs, including the East Pacific region (Table 5.1). GAs were the first reported coral disease, but the etiology of GAs remains unknown. The effect of GAs on coral health is not serious, except it decreases the reproductive output of the affected coral (Yamashiro et al. 2000).

Skeleton eroding band (SEB) is similar to BBD with its black color and shape. SEB is distinguished by the dotted appearance of the white zone behind the infection front (Antonius 1999) (Fig. 5.1f). SEB is caused by the colonial heterotrich ciliate *Halofolliculina corallasia* Antonius & Lipscomb 2000 (Antonius 1999). In addition to the sessile *H. corallasia* in SEB, there are motile ciliates that cause tissue loss in the Indo-Pacific and Caribbean coral reefs (e.g., Randall et al. 2015).

Ulcerative white spots (UWS) were first described as *Porites* Ulcerative White Spots (PUWS) and predominantly occurred on *Porites* (Raymundo et al. 2003) in the Philippines. Recently it has been identified in other coral genera and other regions (Myers and Raymundo 2009; Weil et al. 2012; Subhan et al. 2020; Akmal and Shahbudin 2020, Table 5.1). UWS is identified by the round to oval white spot with an exposed coral skeleton inside (Fig. 5.2c). Similar diseases to UWS include white spot syndrome on *Turbinaria* in the temperate region of Japan (Yamashiro and Fukuda 2009) and white spots on massive *Porites* caused by the coral-inhabiting copepod *Xenomorgus varius* (Kim and Yamashiro 2007) (Fig. 5.2d, f). *Vibrio* spp. are the suggested causative agents of UWS (Arboleda and Reichardt 2010).

Atramentous necrosis (AtN) Anthony et al. (2008) is characterized by a white to grayish film-like mass. AtN affects *Montipora* corals and has been observed in the prefectures of Miyazaki and Okinawa, Japan (Fig. 5.2e), the GBR, Australia (Jones et al. 2004), and Indonesia (Subhan et al. 2020). Typically, the diseased bacterial films are loosely attached and easily swept off (Anthony et al. 2008). AtN is

Fig. 5.2 (**a**) Growth anomalies (GAs) on tabular *Acropora cytherea*. (**b**) GAs on encrusting *Montipora* sp., showing the elongated GAs progressing toward the edge of the colony. (**c**) Ulcerative white spot (UWS) on *Porites* sp. The spot sizes vary, and the insert reveals the bare coral skeleton inside the UWS. (**d**) White spots on *P. lobata* caused by a gall-forming copepod, the spot sizes are uniform and covered by bleached coral tissue. (**e**) Atramentous necrosis (AtN) on *Montipora* sp. (**f**) White spot syndrome on *Turbinaria peltata* (Photograph by M. Fukuda). (**a**–**e**) Sesoko Island, Okinawa, Japan. (**f**) Miyazaki, Japan

considered the second most common cause of coral tissue mortality. AtN is believed to include an opportunistic bacterial mass. Further research is required into the progression and etiology of AtN, the environmental factors involved, and detailed microbial analyses.

Pink coloration (pigmented response, PR) is identified as lines and spots in massive *Porites* colonies. PR is abundant in the *Porites* colonies surrounding Okinawa Island, Japan (Kubomura et al. 2021). The tiny pink spots on *Porites* (called trematodiasis) are caused by parasitic flatworms (Aeby 2003). Further pathological investigation of the internal cysts is required for concise identification. Trematodiasis, pink line syndrome (PLS), and other pink-colored tissues are considered to be caused by the tissue's own immune response which produces red fluorescent proteins when stressed by epibionts (Palmer et al. 2009). Larger pink pigmentations and purple pigmentations (Kubomura et al. 2021) are also frequently observed, regardless of the level of stresses or competition, such as injury, algal touch, or other organisms.

There are a variety of other diseases or related phenomena in Southeast and East Asia. These include white band disease in the Philippines (Antonius 1985), white plaque disease in Thailand (Roder et al. 2014), *Waminoa* flatworm infestations, and a BBD-like syndrome in Taiwan (Huang et al. 2021).

Coral diseases have been observed in most coral reefs around the world, including the Caribbean Sea, Great Barrier

Table 5.1 Coral diseases in Eastern Asia and related regions

Disease	Country	Reference
Black band disease (BBD)	Japan	Biodiversity Center of Japan (2004); Weil et al. (2012); Wada et al. (2017); Kubomura et al. (2018); Hutabarat et al. (2018)
	Taiwan	Huang et al. (2021)
	Malaysia	Akmal and Shahbudin (2020)
	Indonesia	Haapkylä et al. (2009); Muller et al. (2012); Johan et al. (2015); Lamb et al. (2017); Subhan et al. (2020)
	Philippines	Antonius (1985)
	Guam, Micronesia	Myers and Raymundo (2009)
	Palau, Micronesia	Sussman et al. (2006); Page et al. (2009)
White syndrome (WS)	Japan	Biodiversity Center of Japan (2004); Weil et al. (2012); Wada et al. (2017)
	Japan, temperate	Yamashiro et al. (2016)
	Taiwan	Huang et al. (2021)
	Malaysia	Akmal and Shahbudin (2020)
	Indonesia	Haapkylä et al. (2009); Muller et al. (2012); Lamb et al. (2017); Subhan et al. (2020)
	Guam, Micronesia	Myers and Raymundo (2009)
Brown band disease (BrB)	Japan	Weil et al. (2012)
	Indonesia	Haapkylä et al. (2009); Lamb et al. (2017); Subhan et al. (2020)
	Guam, Micronesia	Myers and Raymundo (2009); Lobban et al. (2011)
	Philippines	Raymundo and Weil (2015)
Skeletal eroding band (SEB)	Japan	Wada et al. (2017)
	Guam, Micronesia	Myers and Raymundo (2009)
	Indonesia	Haapkylä et al. (2009); Page et al. (2009); Lamb et al. (2017)
	Malaysia	Akmal and Shahbudin (2020)
Growth anomalies (GAs)	Japan	Yamashiro et al. (2000); Yamashiro (2004); Irikawa et al. (2011); Yasuda et al. (2012); Weil et al. (2012); Wada et al. (2017)
	Taiwan	Huang et al. (2021)
	Indonesia	Lamb et al. (2017); Subhan et al. (2020)
	Guam, Micronesia	Myers and Raymundo (2009)
	Palau, Micronesia	Page et al. (2009)
	Malaysia	Akmal and Shahbudin (2020)
	Philippines	Kaczmarsky (2006)
Ulcerative White Spots (UWS)	Japan	Weil et al. (2012)
	Taiwan	Huang et al. (2021)
	Malaysia	Akmal and Shahbudin (2020)
	Guam, Micronesia	Myers and Raymundo (2009)
	Indonesia	Haapkylä et al. (2009); Subhan et al. (2020)
	Philippines	Kaczmarsky (2006)
White spot syndrome	Japan, temperate	Yamashiro and Fukuda (2009)

Reef, Red Sea, and Indo-Pacific region, even in isolated or remote islands. To study the prevalence and abundance of coral diseases/syndromes, quantitative field surveys are required. Previous abundance studies have been undertaken in Malaysia (Akmal and Shahbudin 2020), Japan (Weil et al. 2012; Wada et al. 2017), Indonesia (Akmal and Shahbudin 2020), the Philippines (Raymundo and Weil 2015), and Guam (Myers and Raymundo 2009). High water temperatures drive the prevalence of some coral diseases, such as BBD (Bruno and Selig 2007; Zvuloni et al. 2009; Sato et al. 2009). Climate change may cause an increase in coral diseases in the near future. Higher water temperatures might increase the activity of microbes related to coral diseases, decreasing coral health, respectively.

In addition to increasing sea surface temperatures, several other environmental factors seem to initiate or drive coral diseases. High concentrations of nutrients such as sewage effluent or upwelling aid BBD (Kaczmarsky et al. 2005; Rodríguez and Cróquer 2008), and coral injury (tourist activity) aids BrB (Lamb and Willis 2011). Dredging-associated sedimentation and turbidity facilitate the prevalence of white syndrome, sediment-associated tissue necrosis, and pigmentation response; however, these activities do not correlate with BBD, GAs, and SEB infections (Pollock et al. 2016).

Biological factors also drive or affect some coral diseases. These include predators (crown-of-thorn starfish, fish, and snails), vectors (fish, snails, and fireworms) (Sussman et al. 2003; Nicolet et al. 2018 with a review list), and algal contact (Nugues et al. 2004). Some microbes, mostly bacteria, are associated with coral diseases. These bacteria originate from human activity and enter by inflows from land or fish farms (e.g., Garren et al. 2009). Thus, any survey of coral diseases requires comprehensive analyses of the coral diseases, environment, anthropogenic effects, and coral-associated organisms.

5.3 Sponges

Some demosponges have become a threat to coral and coral ecosystems. Several coral-encrusting (or coral-killing) sponges are spreading in coral reefs. This section summarizes the most coral-threatening sponges that are predicted to increase in the distribution or proliferate in the near future. *Clathria* (*Microciona*) *aceratoobtusa* (Carter 1887) (Poecilosclerida, Microcionidae) and *Mycale* (*Mycale*) *grandis* Gray, 1867 (Poecilosclerida, Mycalidae) are considered location competitors with corals and are referred to as coral-killing sponges in Indo-Pacific coral reefs (Table 5.2).

5.3.1 *Terpios* Sponge

Terpios hoshinota are grayish to blackish thin (<1 mm) encrusting sponges that overgrow live corals (Fig. 5.3a, b). Since the initial report in Guam and the Mariana Islands in 1973 (Bryan 1973), *T. hoshinota* has been spreading widely in the Indo-Pacific coral reefs of Japan, Taiwan, Southeast China Sea, India, Maldives, Indonesia, Australia, and Papua New Guinea (Table 5.2). *T. hoshinota* is associated with numerous symbiotic cyanobacteria (Aini and Yamashiro 2022). *T. hoshinota* is distributed in shallow water with high light; it does not grow in reduced light conditions (Thinesh et al. 2017). A small snail closely associated with this sponge has also been reported (Yamashiro et al. 2021). Linear progression rates are 11.5–23.0 mm month^{-1} (Bryan 1973; Plucer-Rosario 1987; Elliott et al. 2016; Thinesh et al. 2017), and 28 mm month^{-1} (Aini et al. 2021). Dispersal is faster in warm months (summer) and lower in winter and when disturbed by typhoons (Aini et al. 2021). *T. hoshinota* releases larvae at midnight (Hirose et al. 2022) (Fig. 5.3d–g). Self and non-self recognition are determined by both direct and indirect contact assays (Hirose et al. 2021). Syue et al. (2021) demonstrated the *T. hoshinota* sponge does not kill corals for food or nutrients. *T. hoshinota* uses corals as a substrate and demonstrated using the stable isotope technique, on Green Island, Taiwan (Syue et al. 2021).

5.3.2 *Cliona* Sponge

Cliona sponges are diverse and comprise many species that are distributed widely from deep sea to shallow areas, including coral reefs of the Caribbean and Indo-Pacific (Fig. 5.4a, b). *Cliona* excavates calcium carbonate substratum and lives inside the limestone. *Cliona* produces the enzyme carbonic anhydrase (CA) to assist in the biochemical dissolution of calcareous substances (Hatch 1980). *Cliona* can effectively remove calcareous matter as small chips by etching the surface with filopodia that secretes CA and mechanically removes the chips through oscula (Rützler and Rieger 1973).

Herein, we define *Cliona* sp. in coral reefs as coral-covering type sponges. The sponge hosts the symbiotic algae zooxanthella. Therefore, the sponge is a brown-like zooxanthellate coral. Unlike *Terpios* (<1 mm), the tissue of *Cliona* sp. is thick (cm) with a robust structure. In Taiwan, coral-encrusting *Cliona* sp. in combination with *T. hoshinota* have become a threat to coral reefs on Green Island and Orchid Island (Huang et al. 2021). *Cliona viridis* threatens coral reefs in India (Ashok et al. 2018).

5.3.3 *Chalinula* Sponge

Chalinula nematifera (de Laubenfels 1954) (originally described as *Nara nematifera* when collected from the Marshall Islands) is a thin and bright purple sponge (it is commonly called the magenta ribbon sponge) (Kelly et al. 2016). *C. nematifera* overgrows hard substrates, including live corals (Fig. 5.4c–f). The soft tissues are very delicate and slimy and contain no photosynthetic symbionts. It is distributed from the GBR, Micronesia, Marshall Islands, The Philippines, Maldives (Kelly et al. 2016), and Japan (Reimer et al. 2022). Although this species has been recorded on the Pacific Mexican coast, it seems to have been translocated from the Pacific-Indian Ocean (Ávila and Carballo 2009). Recently, *C. nematifera* is considered a coral-killing sponge in the Mexican Pacific coast and the

Table 5.2 Coral-killing sponges in the Indo-Pacific and related region

Genus	Country	Reference
Terpios	Japan	Yamaguchi (1986); Reimer et al. (2010, 2011); Yomogida et al. (2017); Aini et al. (2021); Hirose et al. (2021, 2022)
	Taiwan	Liao et al. (2007); Wang et al. (2012); Hsu et al. (2013); Huang et al. (2021)
	Guam, Micronesia	Bryan (1973); Plucer-Rosario (1987);
	Malaysia	Hoeksema et al. (2014); Akmal and Shahbudin (2020)
	Indonesia	de Voogd et al. (2013); van der Ent et al. (2015)
	Southeast China Sea	Shi et al. (2012); Yang et al. (2018)
	Papua New Guinea	Ekins et al. (2017)
	Australia	Fujii et al. (2011); Fromont et al. (2019)
	Maldives	Montano et al. (2015)
Cliona	Taiwan	Huang et al. 2021
	India	Ashok et al. (2018)
Chalinula	Indonesia	Rossi et al. (2015); Turicchia et al. (2018)
	Vietnam	Quang (2020)
	Japan	Reimer et al. (2022)
	GBR, Micronesia, Marshall Islands, Philippines, Maldives	Kelly et al. (2016)
	Pacific Mexican coast	Ávila and Carballo (2009)
Clathria	Yemen	Benzoni et al. (2008)
	India	Ashok et al. (2020)
	Taiwan	Huang et al. (2015)
Mycale	Taiwan	Huang et al. (2015)
	Hawaii	Coles and Bolick (2007)

Western–central Pacific, in areas such as Indonesia (Rossi et al. 2015; Turicchia et al. 2018) and Vietnam (Quang 2020). The progression rate was 0.17–0.22 cm d^{-1} (50 mm $month^{-1}$) in Sulawesi Island, Indonesia (Rossi et al. 2015). This sponge is easy to find and identify with its bright purple color and specific white threads radiating or aligned in the same direction. Presently, *C. nematifera* is not a threat in some places such as Japan. However, it has the potential to become a threat in the near future, similar to *Terpios*. Monitoring should be conducted in the coral reefs of Japan, due to the increasing dispersal of this sponge. To better understand *C. nematifera*, we require further studies on the reproduction, progression, symbionts, and morphological/molecular traits.

5.3.4 Other Coral-Killing Sponges

The red sponge *Clathria* (*Microciona*) sp. overgrows *Porites lutea* in Yemen at an average rate of 1 cm $month^{-1}$ (Benzoni et al. 2008). Other coral-killing sponges include *C.* (*Microciona*) *aceratoobtusa* (Carter 1887) in southeast India (Ashok et al. 2020), and *Clathria* spp. in the Penghu Archipelago, Taiwan (Huang et al. 2015). The orange keyhole sponge (*Mycale grandis*) in Hawaii aggressively overgrows *Porites* and *Montipora* corals (Coles and Bolick 2007). Conversely, *Mycale grandis* occurs naturally in Indonesia and Australia. Therefore, this sponge may be invasive when introduced. It has become dominant over corals in south Kaneohe Bay, Hawaii (Coles and Bolick 2007). *Mycale* (*Zygomyclae*) *parishii* (Bowerbank 1875) is reported in the Penghu Archipelago, Taiwan (Huang et al. 2015).

It is predicted in future (coral) reefs will be replaced by algae and sponges and there will be a decrease in corals due to global warming, ocean acidification, and anthropogenic stresses such as sedimentation, excess nutrients, and a variety of introduced chemicals that negatively affect corals. This regime shift will not occur suddenly, it will require the repeated degradation of coral reefs and reef organisms compounding over time. The recent increases in reports of coral-killing sponges are providing us with a warning, yet information on these sponges is scarce when compared to coral diseases. Continuous monitoring of coral-killing sponges is required to describe their prevalence and progression for the future conservation of coral reefs. Furthermore, fundamental scientific research on sponge reproduction and population dynamics is needed with a focus on global and local environmental changes.

5.4 Coral Holobiont

Corals harbor an array of microorganisms, including endosymbiotic dinoflagellate algae (of the family *Symbiodiniaceae*), bacteria, archaea, fungi, ciliates, and viruses to sustain their health and homeostasis (which are referred to as the coral holobiont) (reviewed in Rohwer et al.

Fig. 5.3 Images of *Terpios hoshinota*. (**a**) Massive coral of *Porites* sp. covered by *T. hoshinota*, Ogimi, Okinawa, Japan. (**b**) Foliose coral, *Montipora aequituberculata* covered by *T. hoshinota*, Kikai Island, Kagoshima, Japan. (**c**) The sponge overgrowing the branching coral, *M. digitata*, with many extending threads. (**d**) Numerous larvae on the surface tissue of the sponge. (**e**) Sponge larvae caught in a nylon mesh cup. (**f**) Sponge larva. (**g**) Settled sponge larvae developing a canal system. (**c–g**) Nakijin, Okinawa, Japan

2002; Bourne et al. 2016). It is commonly recognized that *Symbiodiniaceae* is a major contributor to the host coral by providing crucial nutrients, whereas the algae symbiont resides intracellularly in the epidermis of coral tissues (Muscatine et al. 1981). Other members of the holobiont, especially bacteria, also provide mutualistic benefits, such as the assimilation of limited nutrients in an ambient environment and the production of antibiotics (Rohwer et al. 2002; Ritchie 2006; Olson et al. 2009). If any of these holobiont members are disrupted or altered by environmental stress, the health of the entire coral is compromised (reviewed in Bourne et al. 2016; Mera and Bourne 2018). For instance, *Symbiodiniaceae* loss from coral tissues can lead to bleaching and death in corals under thermal stress (Brown et al. 1995). Dramatic shifts in the bacterial community are also regularly observed when the coral is stressed (e.g., Bourne et al. 2007; Shiu et al. 2017). These dramatic shifts eventually affect the health of the coral, reducing the growth and survival of the host, and increasing disease occurrences (Guerra et al. 2014; Teplitski et al. 2016; Zaneveld et al. 2016). In addition, thermal stress increases the vulnerability of corals to opportunistic pathogenic bacteria within the coral microbiome and decreases the relative abundance of healthy core bacterial symbionts (Zaneveld et al. 2016). Coral diseases are often a complex interaction of various holobiont members, affecting the fitness of the holobiont array, such as bacteria. Core

Fig. 5.4 (**a**, **b**) *Cliona* sp. (brown), Green Island, Taiwan. (**a**) Bar: 1 m. (**b**) An enlarged view showing the oscula. (**c**–**f**) *Chalinula nematifera*, Okinawa, Japan. (**c**) *C. nematifera* (bright purple) encrusting branching *Acropora* and foliose *Montipora* coral. (**d**) *Pocillopora* sp. is fully covered by the sponge. The image was taken without lighting at the 25 m depth, Minna Island, Okinawa, Japan (photograph by Ayaka H). (**e**, **f**) An enlarged view highlighting the oscula and white threads

bacterial symbionts in healthy corals show significant decreases in abundance in diseased corals (Vega Thurber et al. 2009; Meyer et al. 2014; Gignoux-Wolfsohn et al. 2017). One study suggested the microbiome of a diseased coral depends on the existing healthy microbiome, the disease history of the infected coral, the source of the disease, and the rate of the disease progression (Gignoux-Wolfsohn et al. 2017). If the origin of the coral microbiome is a factor in the future health and survival of corals then there is an urgent need for research into the microbiome of healthy corals, including diversity, core symbionts, and their functions. This data will enable us to understand the physiology of healthy coral and the mechanisms of coral disease.

5.5 Causes of Coral Diseases by Biotic Agents

The causes of coral diseases are complex due to biotic and/or abiotic stressors and remain poorly understood. However, a diverse array of studies have investigated the coral reefs of the world. Biotic stresses are instigated by microorganisms, including bacteria, fungi, viruses, and protists (reviewed in Sutherland et al. 2004). To date, the etiologies of a few coral diseases have been determined by their fulfilling Koch's postulates (Table 5.3). However, the traditional assessment approach, based on Koch's postulates, is difficult to apply to coral diseases (Richardson 1998).

Table 5.3 Pathogens of coral diseases that satisfy Koch's postulates

Disease	Location	Tested coral species	Potential pathogen (s)	References
Vibrio shiloi induced bleaching	Israel	*Oculina patagonica*	*Vibrio mediterranei*[a]	Kushmaro et al. (1996, 1997)
Aspergillosis	USA and Bahamas	*Gorgonia ventalina*	*Aspergillus sydowii*	Geiser et al. (1998)
Vibrio coralliilyticus induced bleaching and coral tissue lysis	Tanzania	*Pocillopora damicornis*	*V. coralliilyticus*	Ben-Haim and Rosenberg (2002); Ben-Haim and Zicherman-Keren (2003)
Yellow band disease	USA	*Motastraea faveolata*	*Vibrio* consortium	Cervino et al. (2008)
Porites uncreative white spot	Philippines	*Porites cylindrica*	*Vibrio* sp.	Arboleda and Reichardt (2010)
Pocillopora White syndrome (WS)	Italy	*Poci. damicornis*	*V. harveyi*	Luna et al. (2010)
Montipora WS	Hawaii	*Montipora capitata*	*V. owensii, and V. coralliilyticus*	Ushijima et al. (2012, 2014)
Porites WS[b]	China	*Por. andrewsi*	*V. alginolyticus*	Zhenyu et al. (2013)
Porites white patch syndrome	South Africa	*Por. lutea*	*V. tubiashii*	Séré et al. (2015)
Acropora WS	Hawaii	*Acropora cytherea*	*V. coralliilyticus*	Ushijima et al. (2016)

[a]Originally classified as *V. shilonii* (synonym, *V. shiloi*)
[b]Original paper describes as *Porites andrewsi* White Syndrome (PAWS)

The *Vibrio* genus is a prominent coral pathogen (Table 5.3), and a possible causative agent for coral diseases throughout the world (reviewed in Bourne et al. 2009; Arotsker and Kushmaro 2015; Mera and Bourne 2018). However, *Vibrio* bacteria can be present in low densities in the microbiome of healthy corals (e.g., Arboleda and Reichardt 2009; Hong et al. 2009; Kvennefors et al. 2010). *Vibrio coralliilyticus* (a pathogen that infects *Pocillopora damicornis*; see Table 5.3) has been relatively well studied. There have been 88-related publications since 2003 (based on data from ScienceGate database, https://www.sciencegate.app/). *Vibrio coralliilyticus* is present in *V. coralliilyticus*-induced bleaching and coral tissue lysis and also in white syndromes (WS) (Sussman et al. 2008; Ushijima et al. 2014, 2016) and BBD (Arotsker et al. 2009). Both the chemotaxis and chemokinesis by *V. coralliilyticus* target coral mucus through chemical cues (Garren et al. 2014, 2016). *V. coralliilyticus* affects coral by assisting the production of extracellular protease (VcpB zinc metalloprotease) to photoinactivate *Symbiodiniaceae* and coral tissue lysis which is activated by high temperatures (temperature-dependent virulence response) (Sussman et al. 2009; Kimes et al. 2012; van de Water et al. 2018). Other species of the *Vibrio* genus have also been involved in the pathogenesis of corals in the Caribbean Sea and the Indo-Pacific Ocean, including the South China Sea (Table 5.3). *V. mediterranei* (synonym, *V. shionii* and *V. shiloi*) successfully fulfilled Koch's postulates in *Oculina patagonica* to become the first identified infectious disease pathogen of corals (Kushmaro et al. 1996, 1997). Since 2004, *V. mediterranei* has no longer been detected in either healthy or bleached corals when assessed using culture-based methods or by fluorescence in situ hybridization techniques (Koren and Rosenberg 2006; Ainsworth and Hoegh-Guldberg 2008). However, *V. mediterranei* is responsible for coral bleaching when in collaboration with *V. coralliilyticus* (Mills et al. 2013; Meyer et al. 2014; Rubio-Portillo et al. 2020). Four other *Vibrio* species (*V. harveyi*, *V. owensii*, *V. alginolyticus*) are also considered coral pathogens with WS (with *Pocillopra*, *Montipora*, and *Porites* corals, respectively) (Luna et al. 2010; Zhenyu et al. 2013; Ushijima et al. 2014).

In addition to the genus *Vibrio*, other bacteria are also implicated as biotic causative agents of coral diseases. These include *Serratia marcescens* (which causes white pox disease in the Caribbean Sea) (Sutherland et al. 2016), *Pseudoalteromonas pirati* (which causes *Montipora* white syndrome in Hawaii) (Beurmann et al. 2017), and *Thalassomonas loyana* (which causes white plaque-like disease in the Red Sea) (Thompson et al. 2006). BBD lesions have been characterized as complex microbial consortiums, called polymicrobial diseases because they can include cyanobacteria (the dominant group), sulfur-cycling bacteria (including sulfate-reducing bacteria and sulfide-oxidizing bacteria), other heterotrophic bacteria, archaea, fungi, and eukaryotes (reviewed in Sato et al. 2016). A phylogenetic analysis using the 16S rRNA gene revealed the dominant cyanobacteria are tightly clustered in monophyletic groups (Sato et al. 2016). The dominant cyanobacteria include the *Oscillatoria* sp.-related cyanobacterium RMS in the Indo-Pacific, *Pseudoscillatoria coralii* in the Red Sea, and *Roseofilum reptotaenium* in the Caribbean (Sussman et al. 2006; Rasoulouniriana et al. 2009; Casamatta et al. 2012).

In non-axenic culture, *R. reptotaenium* can cause progressive BBD lesions in healthy corals within days of infection (Casamatta et al. 2012). Although the group has been suggested to be involved in BBD pathogenesis as a major key player, an infection occurred only in the presence of sulfate-reducing bacteria as a necessary secondary pathogen (Brownell and Richardson 2014). Polymicrobial diseases can be caused by complex interactions within a microbial consortium. Although unable to fulfill Koch's postulates, the virulence of BBD may be due to the lesion becoming anoxic and promoting sulfide accumulation by sulfate-reducing bacteria which is a very harmful and toxic biogeochemical cocktail for the underlying coral tissues (Carlton and Richardson 1995; Kuta and Richardson 1997; Glas et al. 2012; Sato et al. 2016).

5.6 Pathological Studies in Eastern Asia

Pathological studies in eastern Asia have investigated coral diseases, including WS, *V. coralliilyticus* infectious disease, BBD, GAs, and bleaching corals. In Japan, GAs (Yamashiro et al. 2000, 2001; Irikawa et al. 2011; Yasuda et al. 2012; Yasuda and Hidaka 2012) and cyanobacterial blooms (Yamashiro et al. 2014) have been described in detail (Wada et al. 2018). Therefore, this section aims to update earlier information from previous pathological studies, especially those undertaken in eastern Asia (including China, Hong Kong, Japan, South Korea, and Taiwan).

5.6.1 Coral Pathogens in Eastern Asia

Zhenyu et al. (2013) identified *V. alginolyticus* as the cause of a new coral disease, *Porites andrewsi* White Syndrome (PAWS) in the north Xisha Archipelago, South China Sea (Table 5.3). PAWS was identified by bright white lesions on *P. andrewsi*, which dominated corals in the lagoons at depths of 1–3 m. Newly dead colonies became darker (dull white) within 4–7 days. After 10 days, algae began growing on the surface of the dead colonies. The initial transmission experiments in the field (Fig. 5.5) confirmed some PAWS lesions transmitted to healthy corals, resulting in an average tissue loss of 5.40 cm^2 from the site where the lesion was placed in direct contact with a healthy fragment (Zhenyu et al. 2013). The study concluded an infection occurred between days 5 and 14, most likely occurring after approximately 7–10 days of the PAWS lesion being placed on the healthy coral. The second experiment involved mixed bacterial challenges (with a five-strain mixture) which confirmed one of the five strain mixtures (the T_1 group containing XSBZ01, XSBZ03, XSBZ07, XSBZ14, and XSBZ22) caused coral tissue loss 6 days after inoculation. Conversely, the other bacterial mixtures did not affect the coral with the coral maintaining visibly healthy signs (the experiment involved immersing coral nubbins in the bacterial mixture before transferring the corals into seawater). The third experiment aimed to determine whether two single strains (XSBZ03 and XSBZ14) from the five-strain mixture (T_1 group) were the PAWS pathogen. This experiment involved single-strain bacterial challenges to fulfill Koch's postulates (the experiment was carried out by the coral nubbins being immersed in the bacterial inoculation solution during the entire experimental period). This experiment identified *V. alginolyticus* as the PAWS pathogen using the 16S rRNA gene. In addition, two sensitive detection methods for identifying *V. alginolyticus* were developed (Li et al. 2018; Yu et al. 2019). Interestingly, one method can detect viable *V. alginolyticus* when bound to single-stranded DNA aptamers using flow cytometry (Yu et al. 2019).

Higuchi et al. (2013) also identified a bacteria isolated from bleached *Montipora digitata* and the surrounding seawater in Okinawa, Japan. The research used bacterial challenges to investigate the corals in heat stress. First, five bacterial strains were collected. *V. coralliilyticus* [AB490821] and *V. harveyi* [AB490822] were obtained from seawater, *Paracoccus carotinifaciens* [AB490820] from bleached *M. digitata*, and *Pseudoalteromonas* sp. [AB691769] and *Sulfitobacter* sp. [AB691770] from the water surrounding the bleached *M. digitata*. The five bacterial strains were mixed and immersed in the nubbins of healthy *M. digitata* at two different temperatures (27 °C and 32 °C). The coral nubbins at 27 °C showed no signs of bleaching, even when the mixed bacteria were challenged. The high temperature (32 °C) treated coral nubbins with a mixed bacterial challenge became bleached, while the corals without the bacterial treatment showed moderate signs of bleaching and loss of coloration (Higuchi et al. 2013). Thus, the most affected nubbins were those with the added bacterial mixture at 32 °C; these nubbins displayed potential necrosis signs, especially on the coenosarc. Second, Higuchi et al. (2013) confirmed that only the addition of *Sulfitobacter* sp. (AB691770) resulted in bleached corals at 32 °C. In this instance, the maximum quantum efficiency of photosystem II [Fv/Fm ratio] also decreased by approximately 24% when compared with control coral nubbins. Conversely, the other four strains did not provoke any signs of bleaching after 4 days of incubation at a high temperature (Higuchi et al. 2013).

In a BBD pathology study, Hutabarat et al. (2018) also attempted to isolate the dominant cyanobacteria from BBD-affected colonies (*Montipora* sp. and *Goniopora* sp.) from two Okinawa islands, Japan (Sesoko Island and Miyako Island, respectively). All of the cyanobacterial cultures from BBD were dark brown color. The morphological structures of all of the BBD-related cyanobacteria had similar size ranges

Fig. 5.5 Images of the transition stages (day 0, day 14, and day 28 after bacterial inoculation) in the field transmission experiment of *Porites andrewsi* White Syndrome (PAWS). The upper (**a**) and lower (**b**) panels showing the healthy colony of *P. andrewsi* in contact with the PAWS-diseased coral fragment and with a healthy coral fragment, respectively, both previously collected from other *P. andrewsi* colonies. Red arrows highlight the contact fragments (source: Zhenyu et al. 2013)

(cell lengths ranging from 2.58 to 4.80 μm; cell widths ranging from 3.10 to 4.80 μm; Fig. 5.6a). The size ranges are more like the Australian strain (*R. reptotaenium* AO1, Buerger et al. 2016) and Caribbean strain (*R. reptotaenium* 100-1 and 101-1, Casamatta et al. 2012) than the Red Sea strain (*P. coralii* BGP10-4S, Rasoulouniriana et al. 2009). Cyanobacterial-specific primers targeting the 16S rRNA gene were used to further examine the cyanobacteria from both locations in Japan. Three cyanobacterial isolates originating from Sesoko Island ($n = 1$, CDS) and Miyako Island ($n = 2$, CDM1 and CDM2) were identical for 860 bp of the gene. The three isolates were affiliated with the BBD group retrieved from the Caribbean and Indo-Pacific (Fig. 5.6b).

5.6.2 Bacterial Community in Coral Diseases

Several bacterial community analyses of coral diseases have characterized the communities by their abundance, including BBD and GAs, in eastern Asia (Chiu et al. 2012; Yang et al. 2014; Ng et al. 2015).

A BBD mat on *Echinopora lamellosa* was sampled from Yongxing Island in the South China Sea, and a full-length 16S rRNA gene was sequenced using the cloning library method (Yang et al. 2014). A total of 81 sequences were obtained, a total of 64 of the sequences were identified as cyanobacteria (79.01% of the total number). The remaining sequences belonged to *Bacteroidetes* (3.95%), *Actinobacterium* (1.22%), *Verrucomicrobia* (1.22%), and *Proteobacteria*, including *α-proteobacteria* (12.3 5%) and *γ-proteobacteria* (2.25%). The cyanobacterial sequences in the BBD mat clustered into 31 operational taxonomic units (OTUs) at the phylum level. The most abundant OTU SD-39 (which included 15 sequences) and the second most abundant OTU SD-95 (which contained seven sequences) were similar to uncultured cyanobacteria (DQ269106 and AM159754), which were derived from red algae in Australia (95% similarity) and from a marine sponge (97% similarity), respectively. However, some OTUs in the phylogenetic analysis were also similar to the BBD-associated cyanobacteria from the Caribbean and Red Sea (Yang et al. 2014).

GA-affected coral tissues did not affect healthy coral tissue in *Acropora cytherea* or *Porites australiensis* (Irikawa et al. 2011; Yasuda et al. 2012). Conversely, Kaczmarsky and Richardson (2007) reported contrasting results, with successful transmission in *Porites* corals and suggested the cause was pathogenic microorganisms. In 2010 and 2011, GA-associated bacterial communities were characterized in *Platygyra carnosus* samples in Hong Kong (Chiu et al. 2012; Ng et al. 2015). Surprisingly, in Hong Kong, more than 60% of the *P. carnosus* colonies developed GAs (Fig. 5.7, Chiu et al. 2012). Although the culture-dependent method showed no difference in the bacterial communities between the GAs and the healthy regions of *P. carnosus* (Chiu et al. 2012), 16S

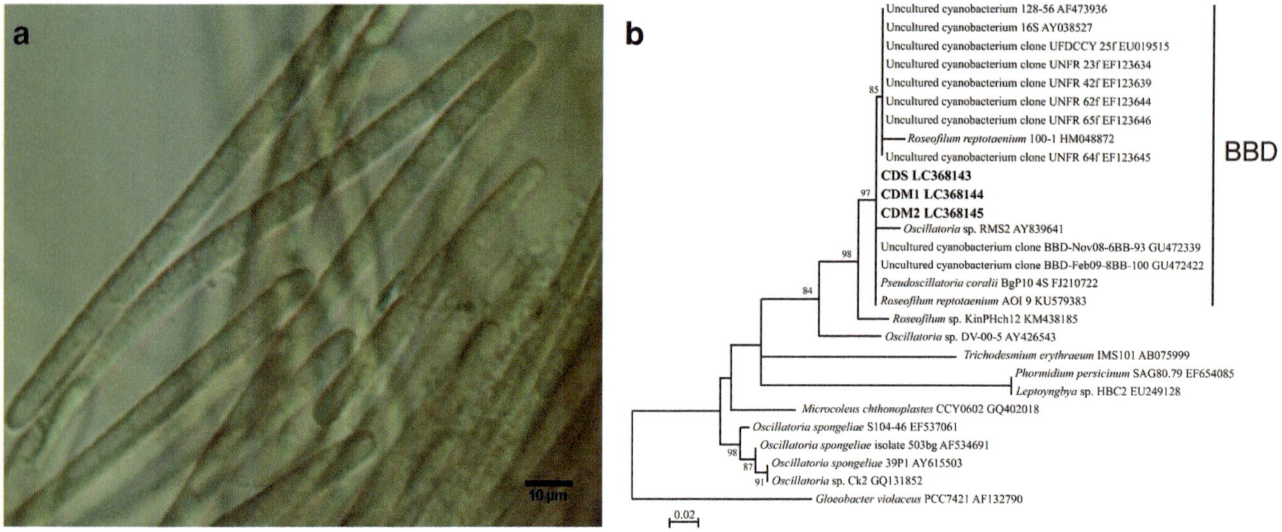

Fig. 5.6 Image of BBD-related cyanobacteria using blight field (**a**) and the phylogenetic tree (**b**) of the cyanobacterial partial 16S rRNA gene sequences (860 bp) based on a maximum likelihood analysis (with 1000 bootstrap replications). Adapted by permission from Springer Nature: Hutabarat et al. (2018) J Appl Phycol 30:3197–3203. Copyright (2018) Springer Science+Business Media B.V.

rRNA amplicon sequencing revealed at the phylum level, the GAs-affected tissues contained high numbers of *Acidobacteria* and *Gemmatimonadetes*, and low numbers of *Spirochaetes* (Ng et al. 2015). In addition, Ng et al. (2015) identified certain specific OTUs were exclusively associated with GA-affected tissues, including *Rhodospirillaceae*, *Rhizobiales*, *γ-proteobacteria*, and the *Cytophaga-Flavobacterium-Bacteroidetes* (CFB) group. This result suggests they are potential pathogens or opportunistic colonizers that proliferate during the development of GAs. Subsequently, Zhang et al. (2017) performed a transcriptome analysis of GAs-affected tissues sampled from Hong Kong and focused on the gene expression of *P. carnosus* and *Symbiodiniaceae*. When compared to healthy tissues, the GAs-affected tissues contained 109 differentially expressed genes (61 upregulated and 48 downregulated genes) in *P. carnosus* and 31 genes (16 upregulated and 15 downregulated genes) in *Symbiodiniaceae* (Zhang et al. 2017). Of the 109 differentially expressed genes in *P. carnosus*, 32 gene ontology (GO) functional terms were enriched. Their functions include cell differentiation, osteoblast differentiation, ossification, branching morphogenesis of epithelial tubes, regulating tissue remodeling, and sterol metabolic processes, indicating that GAs are malignant neoplasms (Zhang et al. 2017). Of the 31 differentially expressed genes in *Symbiodiniaceae*, only five GO cellular components were enriched (including reproduction, organic nitrogen metabolism, and the metabolism of plant pigments), indicating that GAs affect several aspects of symbiont physiology (Zhang et al. 2017). Moreover, some immunological genes (such as malignant brain tumors 1 (DMBT1), bacterial permeability-increasing protein (BPI), ADP-ribosylation factor 1 (Arf1), and autophagy) were identified in the formation of GAs, suggesting the formation is a response to both bacterial and viral attacks (Zhang et al. 2017). However, conclusive evidence has yet to be provided to prove the theory of pathogenic microorganisms.

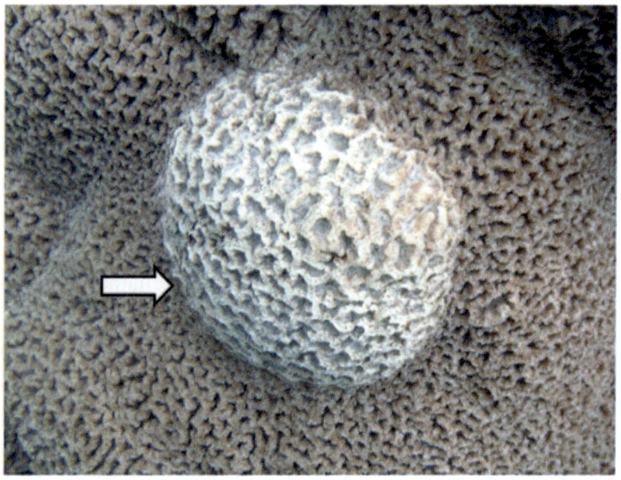

Fig. 5.7 Image highlighting growth anomalies (GAs) in *Platygyra carnosus*. Arrow indicates GAs-affected tissues (source: Ng et al. 2015)

5.6.3 *Vibrio corallilyticus* Infection with Corals

There are two representative studies that have treated corals with *V. coralliilyticus* under thermal stress. These include *Acropora diditifera* in Japan and *Galaxea fascicularis* in China (Tang et al. 2020; Takagi et al. 2020).

Takagi et al. (2020) delineated the *A. diditifera* (*Symbiodiniaceae*-aposymbiotic juvenile polyps) immune system response to *V. coralliilyticus* invasion by transcriptome analysis. To confirm the bacterial infection on the polyp, Takagi et al. (2020) transformed the DsRed2 fluorescent protein-coding gene into a cell of *V. coralliilyticus* strain P1 (LMG23696) and visually inspected the outcome. The *V. coralliilyticus* was densely localized around the actinopharynx of the polyp (Takagi et al. 2020, Fig. 5.8). The host gene expression was analyzed at 5, 30, 60, and 180-min post-infection (mpi). The analysis focused on the early stages of the *V. coralliilyticus* infection, and compared the treated polyps with uninfected polyps, in thermal stress conditions (30 °C). The notable results of the 30 and 60 mpi revealed the GO functional genes for both the "extracellular exosome (which is related to the mediator of cell-to-cell communication and also the transport of antimicrobial peptides)" and the "extracellular matrix" were upregulated. Conversely, the GO functional genes for the "innate immune response" were downregulated and included the Toll-like receptor (TLR1), the nucleotide-binding oligomerization domain- (NOD-) containing proteins (NOD1 and NOD2), and the NOD-like (NLR) family CARD domain-containing proteins (NLRC4 and NLRC5). In addition, in the GO category, the "oxidation-reduction process" related to redox homeostasis was also upregulated, including in the mitochondrion, the mitochondrial respiratory chain complex I, and the mitochondrial inner membrane. This result suggests the mitochondrial reactive oxygen species (mROS) may be involved in bacterial defense and facilitate antibacterial innate immune signaling. In summary, Takagi et al. (2020) revealed the invasion of *V. coralliilyticus* suppresses host gene expressions related to the innate immune response. Simultaneously, the coral responds to the altered mitochondrial and protein metabolism by exosome release and extracellular matrix remodeling to eliminate the infected pathogen, in the early stages of interaction between the coral pathogen and the coral polyp (Takagi et al. 2020).

Tang et al. (2020) infected *V. coralliilyticus* into *G. fascicularis* corals in the South China Sea. First, healthy *G. fascicularis* were heat treated (temperature, N.D.) and a total of 101 bacterial isolates were obtained from the gastric cavity and the taxonomy was determined using the full-length 16S rRNA gene. In the gastric cavity, the genus *Vibrio* was the most abundant bacteria (38% of the total isolates), followed by *Alteromonas* (11%), *Pseudoalteromonas* (8%), *Nautella* (8%), *Neptuniibacter* (7%), *Halomonas* (4%), and *Erythrobacter* (4%). One strain, *V. coralliilyticus* (SCSIO 43001) from *Vibrio* was further investigated. The testing involved *G. fascicularis* being inoculated with the strain (final concentration, 5×10^6–1×10^7 CFU ml) every 12 h for 2 days at 26 °C, and kept with no-inoculation at 32 °C for 2 days. The results showed the bacterial treatment resulted in the area of the coenosarc becoming noticeably bleached in the inoculated corals while the control corals (in sterile water without any bacteria) still appeared relatively healthy (Fig. 5.9). Various *Vibrios* species were isolated from the bleached area, including *V. coralliilyticus*, *V. tubiashii*, *V. alginolyticus*, *V. owensii,* and *V. mediterranei* (as *V. shilonii*). In addition, Tang et al. (2020) confirmed the antagonistic effect by screening the specimens using other isolates from the gastric cavity with the *V. coralliilyticus* strain SCSIO 43001 (more details are provided in Sect. 5.7).

5.6.4 Further Pathological Research for Coral Diseases

Since the first report of a coral disease in 1973, the identification of coral diseases has been based on the visual inspection of macroscopic lesions on the coral surface and coral diseases are named using the appearance of the lesion (Moriarty et al. 2020). This has resulted in many confusing and often contradictory disease diagnoses which is difficult to systematically overcome in the field. In addition, the accuracy of coral health assessments is yet to be determined. These aspects limit our diagnosis of coral diseases and make the elucidation of etiology difficult. Currently, coral cell lines have been established (Kawamura et al. 2021). This technique is expected to be a breakthrough that aids our understanding of the etiology of coral diseases, especially the interaction between potential pathogens and coral host cells, and the mechanisms of coral diseases.

5.7 Potential Beneficial Bacteria to Reduce Coral Disease

Bacteria are among the most diverse and abundant organisms in the coral holobiont. Recent studies in Taiwan have suggested that some bacterial isolates sampled from corals might drive the holobiont roles that are involved in dimethylsulfoniopropionate (DMSP) reduction, nitrogen fixation, prevention of host mitochondrial dysfunction, and promotion of host gluconeogenesis in corals (Ding et al. 2016; Yang et al. 2019; Tandon et al. 2020). Several studies in China and Japan have also demonstrated that bacteria isolated from *G. fascicularis* corals can inhibit the colonization and growth of the coral pathogen *Vibrio corallilyticus* (Miura et al. 2019; Tang et al. 2020).

Tang et al. (2020) successfully induced coral bleaching by an isolated strain of *V. coralliilyticus* (SCSIO 43001) (further details are provided in Sect. 5.6.3). Subsequently, the research confirmed the three *Pseudoalteromonas* strains (*Pseudoalteromonas* sp. SCSIO 43201, and *P. elyakovii* SCSIO 43202 and SCSIO 43203) form the gastric cavity of *G. fascicularis* are capable of high levels of antibacterial activity against *V. coralliilyticus* SCSIO 43001 from the

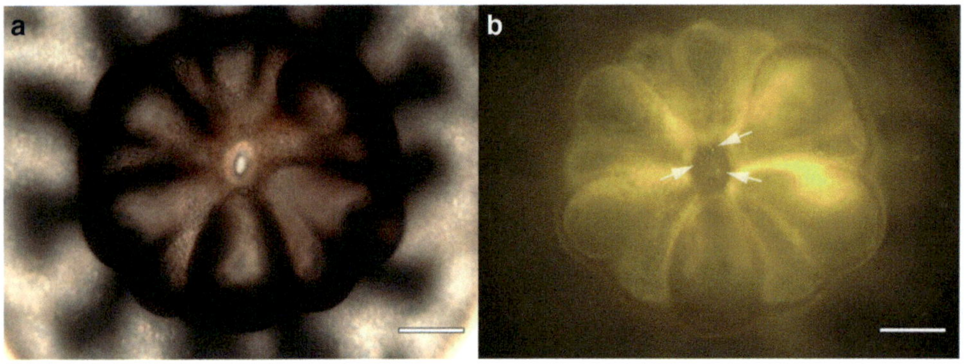

Fig. 5.8 Microscopic images (**a**–**b**) of a *Vibrio coralliilyticus* (tagging with DsRed fluorescent) invasion in *Symbiodiniaceae*-aposymbiotic juvenile polyps of *Acropora digitifera*. The images show the acquisition of the two channels, using bright field (DIC, **a**) and red fluorescent (DsRed, **b**). The white arrows highlight *V. coralliilyticus* attached to mouth of the polyp (**b**). Scale bars represent 100 μm. Reprinted by permission from Springer Nature: Takagi et al. (2020) Mar Biotechnol 22:748–759. Copyright (2020) Springer Science+Business Media, LLC

Fig. 5.9 Images of coral fragments treated without any bacterial inoculation (upper panels) and with a *V. coralliilyticus* infection (strain SCSIO 43001) (lower panels) at temperature at 26 °C for 2 days (left panels) and at 32 °C for 2 days (right panels). Reprinted by permission from Springer Nature: Tang et al. (2020) Sci China Earth Sci 63:157–166. Copyright (2020) Science China Press and Springer-Verlag GmbH Germany

same gastric cavity. Tang et al. (2020) suggested the bacterial communities in the gastric cavity could be competing, between the coral pathogens and the other bacteria. In particular, *Pseudoalteromonas* is an antagonistic bacterium which may provide a protective role in the defense against coral pathogens at elevated temperatures (Tang et al. 2020).

Miura et al. (2019) isolated bacteria from single polyps of healthy *G. fascicularis* purchased from a company in Japan. A total of 29 bacterial strains were isolated, of which 14 were identified as *Ruegeria* spp. and five were identified as *Vibrio* spp. and included one *V. coralliilyticus* strain TOK8 (99% similarity to the type strain ATCC BAA-450). All *Ruegeria* strains were tested for their antagonistic effects on the coral pathogen *V. coralliilyticus* strain P1 (LMG23696) and *V. coralliilyticus* strain TOK8. Three of the *Ruegeria* spp. (TOK22, TOK27, and TOK33 strains) were equally capable of in vitro growth at 25 °C and 30 °C, and were able to inhibit the growth of both *V. coralliilyticus* strains at 25 °C, and even at 30 °C (Fig. 5.10, Miura et al. 2019). Certain *Ruegeria* spp. were harbored in *G. fascicularis* inhabiting natural seawater in Okinawa, Japan, suggesting *Ruegeria* spp. are potential probiotic bacteria that may be used to improve coral fitness (Kitamura et al. 2021).

The coral probiotics hypothesis was established in 2006 (Reshef et al. 2006), it proposes it may maintain coral fitness and protect coral from specific harmful environmental

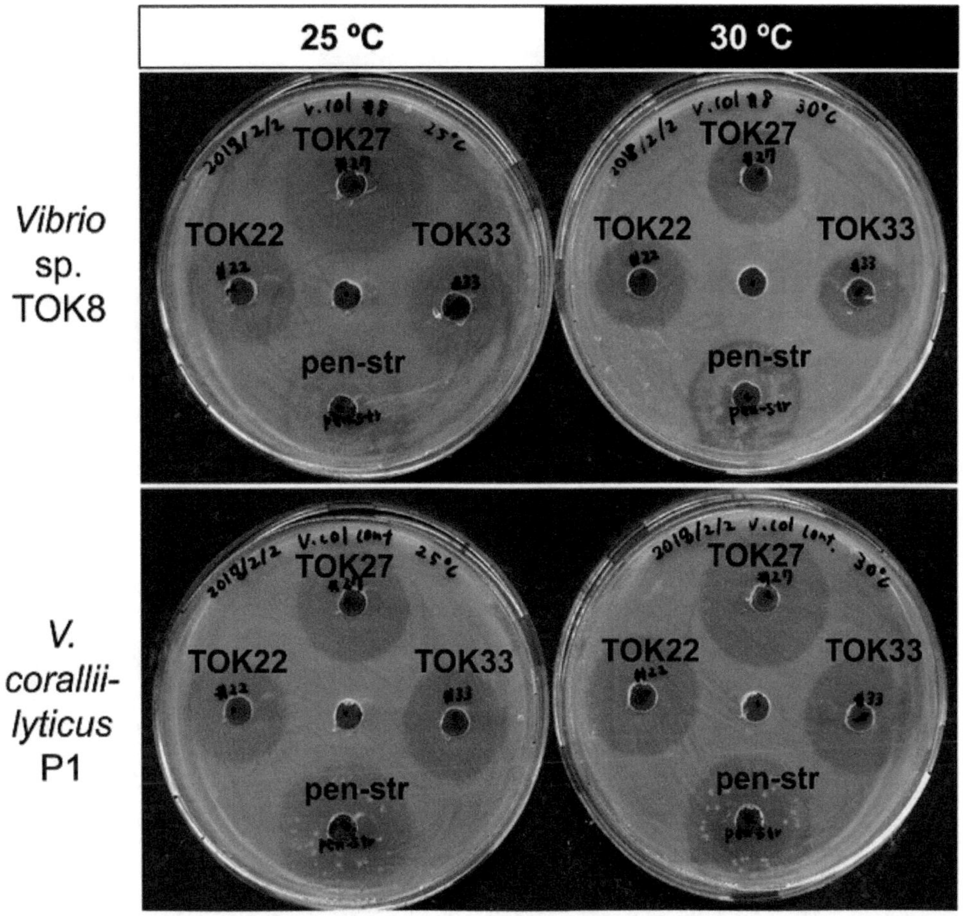

Fig. 5.10 Growth inhabitation of the *V. coralliilyticus* strains (P1 and TOK8) by three *Ruegeria* strains (TOK22, TOK27, and TOK33) at 25 °C and 30 °C. Images showing the growth inhabitation from the three *Ruegeria* strains and mixed antibiotic agents (pen-str indicates the mixture of penicillin with 10,000 U/ml and streptomycin with 10,000 μg/ml) with two *V. coralliilyticus* strains, treated at two different temperatures. Reprinted by permission from Springer Nature: Miura et al. (2019) Mar Biotechnol 21:1–8. Copyright (2018) Springer Science+Business Media, LLC

conditions and pathogens. Manipulating and engineering the microbiome is a new approach for treating diseases and enhancing the adaptability of host microorganisms (West et al. 2019). In corals, probiotics may be very difficult to apply because of the coral holobiont system and natural environment. Our priority is to accumulate knowledge to clarify the bacterial functions in coral holobionts, and it should plan to promote the adoption of the potential probiotics while taking into consideration their impact on the environment.

5.8 Microbiome-Associated Coral-Killing Sponge *Terpios hoshinota*

Sponges harbor abundant and diverse microorganisms, described as sponge holobionts (Webster and Thomas 2016). *T. hoshinota*, a coral-killing sponge, is vastly accommodated by symbiotic cyanobacteria (5 × 10^5 cells/cm^2), which fills almost all the intercellular spaces in the sponge (Fig. 5.11, Rützler and Muzik 1993; Tang et al. 2011; Hirose and Murakami 2011). A sponge-associated bacterial community analysis revealed high cyanobacterial abundance (relatively 61–98%) in the sponges covering different corals, including *Favia complanate*, *Isopora palifera*, *Millepora* sp., *Montipora efflorescens*, and *Porites lutea* (Tang et al. 2011). Current research has revealed cyanobacteria are widely and dominantly distributed in all sponge samples from diverse geographic locations in the Indian Ocean and the western Pacific Ocean. The cyanobacteria were identified as *Candidatus* Paraprochloron terpiosi in the novel genus *Paraprochloron* by phylogenomic analysis (Chen et al. 2021). Additionally, the other bacterial proportions varied between the sponge samples collected from sponges covering different corals, with 15% of the relative abundance consistently involving α-proteobacteria and γ-proteobacteria (Tang et al. 2011). Some other bacteria were randomly distributed in the sponge tissue, these were identified using a fluorescence in situ hybridization technique, while a large volume of cyanobacteria was present (Fig. 5.12).

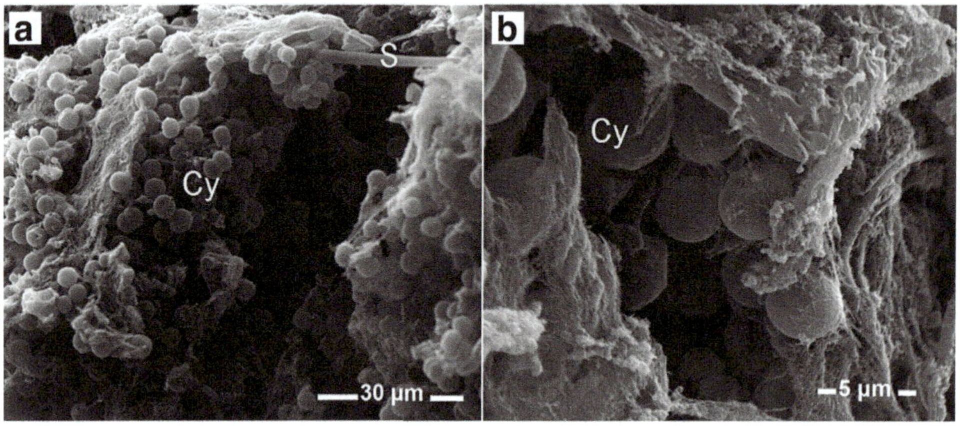

Fig. 5.11 Scanning electron micrographs showing the symbiotic cyanobacteria in the interior of the coral-killing sponge *T. hoshinota* (**a**) and an enlarged image of the cyanobacteria (**b**). Cy, cyanobacteria; S, spicule. Reprinted by permission from John Wiley and Sons: Tang et al. (2011) Environ Microbiol 13:1179–1191. Copyright (2011) Society for Applied Microbiology and Blackwell Publishing Ltd

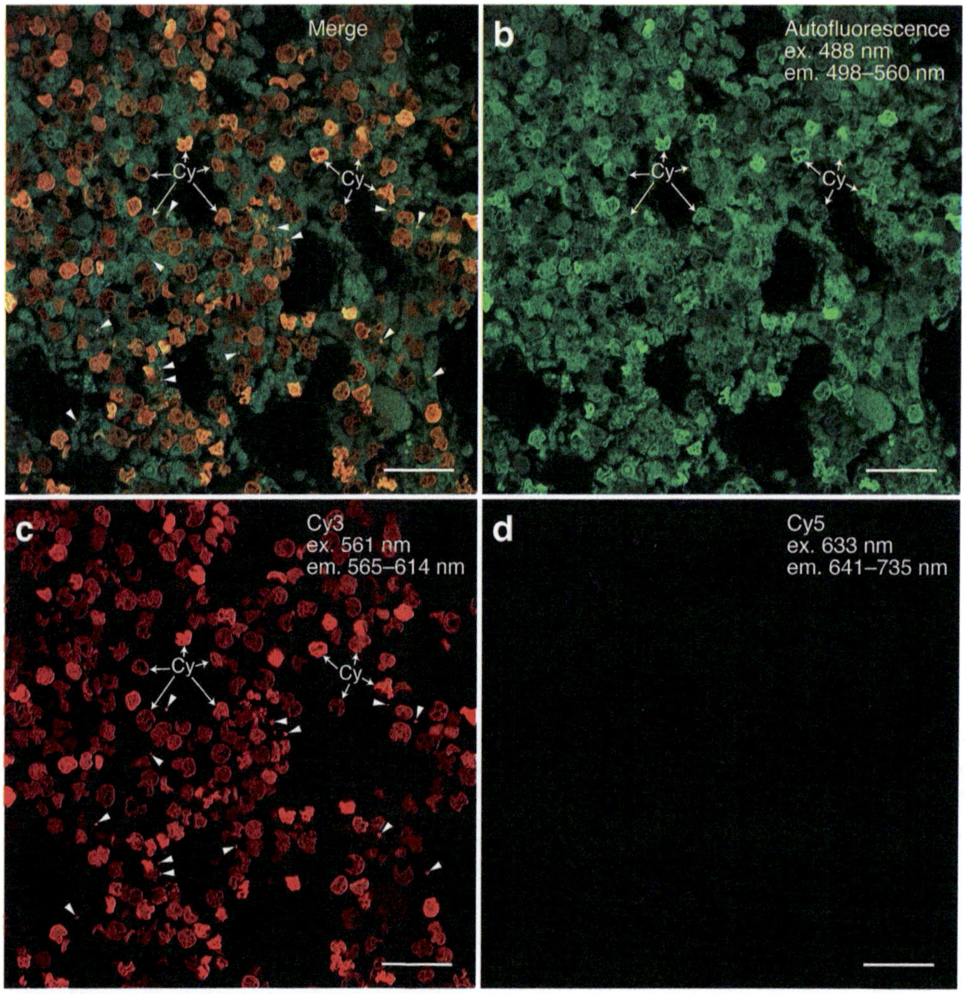

Fig. 5.12 Confocal laser scanning micrographs depicting the symbiotic cyanobacteria and other bacteria in the tissue of the coral-killing sponge *T. hoshinota* using the Fluorescence in situ hybridization (FISH) technique. FISH was performed with 10% formalin-fixed and paraffin-embedded *T. hoshinota* tissues (collected in Okinawa, Japan by Siti Nurul Aini) using a universal bacterial probe EUB338mix labeled with Cy3 and a negative control probe Non338 labeled with Cy5 according to the protocol detailed in Wada et al. (2016). The merge image (**a**) was acquired from three channels; the autofluorescence of both the sponge tissue and cyanobacteria (green, **b**), Cy3 signals (red, **c**) and Cy5 signals (magenta, **d**). The images reveal abundant symbiont cyanobacteria (arrows indicate representative cyanobacteria, Cy) and other bacteria (arrowheads) in the sponge tissue. Note the cyanobacterial autofluorescence (green) and Cy3 signals (red) from EUB338mix are manifested in orange-yellow overlap regions in the merged image (**a**). The acquired conditions are provided using a confocal microscope LSM 880 (Zeiss) for each image (**b**-**d**). Scale bars indicate 20 µm (source: this study)

Symbiotic cyanobacteria are suggested to drive some vital roles in the host *T. hoshinota*, including vertical transmission during embryogenesis (Nozawa et al. 2016). Vertical transmission during embryogenesis causes irreversible damage to *T. hoshinota* due to the low cyanobacterial biomass and inhibition of cyanobacterial photosynthesis (Thinesh et al. 2017) and forms hairy tip structures containing high densities of cyanobacteria in the forefront during coral invasion (Wang et al. 2012). Chen et al. (2021) reconstructed the genome of the cyanobacterial symbiont, Ca. P. terpiosi from sponge samples collected from Green Island (Taiwan) and revealed that Ca. P. terpiosi can process most of the genes required for photosynthesis, carbon fixation, the tricarboxylic acid cycle, and glycolysis. The Ca. P. terpiosi might uptake host-derived wastes, such as urea, and nitrogenous and carbohydrate wastes, and supply some vitamins (vitamins B_{12}, B_7, and B_2) to the host sponge cells (Chen et al. 2021).

To understand the physiology of the coral-killing sponge *T. hoshinota*, we need to investigate the role of the cyanobacterial symbiont in the host. Genome reconstruction of Ca. P. terpiosi will aid future studies to explore the detailed ecosystem inside the holobiont of *T. hoshinota* and the cyanobacterial symbiont.

Acknowledgments This work was supported by the Japan Society for the Promotion of Science (JSPS) KAKENHI (grants 19 K06091 and 21H036191A).

References

Aeby G (2003) Corals in the genus *Porites* are susceptible to infection by a larval trematode. Coral Reefs 22:216–216

Aeby GS (2006) Baseline levels of coral disease in the Northwestern Hawaiian Islands. Atoll Res Bull 543:471–488

Aini SN, Yamashiro H (2022) Densities of cyanobacterial cells, spicules, and particles in the coral-killing sponge *Terpios hoshinota* in Sesoko Island, Okinawa, Japan. Plankton Benthos Res 17:263–270

Aini SN, Tang S-L, Yamashiro H (2021) Monthly progression rates of the coral-killing sponge *Terpios hoshinota* in Sesoko Island, Okinawa, Japan. Coral Reefs 40:973–981

Ainsworth TD, Hoegh-Guldberg O (2008) Cellular processes of bleaching in the Mediterranean coral *Oculina patagonica*. Coral Reefs 27:593–597

Akmal KF, Shahbudin S (2020) Baseline assessment of coral health and disease in Tioman Island Marine Park, Malaysia. Community Ecol 21:285–301

Alvarado JJ, Grassian B, Cantera-Kintz JR, Carballo JL, Londoño-Cruz E (2017) Coral reef bioerosion in the Eastern Tropical Pacific. In: Glynn PW, Manzello DP, Enochs IC (eds) Coral reefs of the Eastern Tropical Pacific: persistence and loss in a dynamic environment. Springer, Dordrecht, pp 369–403

Anthony SL, Page CA, Bourne DG, Willis BL (2008) Newly characterized distinct phases of the coral disease 'atramentous necrosis' on the Great Barrier Reef. Dis Aquat Org 81:255–259

Antonius A (1973) New observations on coral destruction in reefs. Abs Assoc Isl Mar Lab Caribb 10:3

Antonius A (1985) Coral diseases in the Indo-Pacific: a first record. Mar Ecol 6:197–218

Antonius A (1999) *Halofolliculina corallasia*, a new coral-killing ciliate on Indo-Pacific reefs. Coral Reefs 18:300

Arboleda M, Reichardt W (2009) Epizoic communities of prokaryotes on healthy and diseased scleractinian corals in Lingayen Gulf, Philippines. Microb Ecol 57:117–128

Arboleda M, Reichardt W (2010) *Vibrio* sp. causing Porites ulcerative white spot disease. Dis Aquat Org 90:93–104

Arotsker L, Kushmaro A (2015) Vibriosis. In: Porter J, Woodley C (eds) Diseases of Coral. John Wiley & Sons, Ltd, pp 206–220

Arotsker L, Siboni N, Ben-Dov E, Kramarsky-Winter E, Loya Y, Kushmaro A (2009) *Vibrio* sp. as a potentially important member of the black band disease (BBD) consortium in *Favia* sp. corals. FEMS Microbiol Ecol 70:515–524

Ashok AM, Schönberg CHL, Raj KD (2018) A sponge of the *Cliona viridis* complex invades and excavates corals of the Gulf of Mannar, south-eastern India. Mar Freshw Res 69:874–882

Ashok AM, Calcinai B, Edward JKP (2020) The coral-killing red sponge *Clathria* (*Microciona*) *aceratoobtusa* (Porifera: Demosponigiae) invades various coral communities of Gulf of Mannar Marine National Park, southeast India. European Zool J 87:1–11

Asoh K (2008) Loss of live tissue and exposure of coral skeleton in *Porites cylindrica* at Shiraho, Ishigaki Island, Japan. Galaxea J Coral Reef Stud 10:45–45

Ávila E, Carballo JL (2009) A preliminary assessment of the invasiveness of the Indo-Pacific sponge *Chalinula nematifera* on coral communities from the tropical Eastern Pacific. Biol Invasions 11: 257–264

Bell JJ, Davy SK, Jones T, Taylor MW, Webster NS (2013) Could some coral reefs become sponge reefs as our climate changes? Glob Change Biol 19:2613–2624

Ben-Haim Y, Rosenberg E (2002) A novel *Vibrio* sp. pathogen of the coral *Pocillopora damicornis*. Mar Biol 141:47–55

Ben-Haim Y, Zicherman-Keren M (2003) Temperature-regulated bleaching and lysis of the coral *Pocillopora damicornis* by the novel pathogen *Vibrio coralliilyticus*. Appl Environ Microbiol 69: 4236–4242

Benzoni F, Calcinai B, Eisinger M, Klaus R (2008) Coral disease mimic: sponge attacks *Porites lutea* in Yemen. Coral Reefs 27:695–695

Beurmann S, Ushijima B, Videau P, Svoboda CM, Smith AM, Rivers OS, Aeby GS, Callahan SM (2017) *Pseudoalteromonas piratica* strain OCN003 is a coral pathogen that causes a switch from chronic to acute Montipora white syndrome in *Montipora capitata*. PLoS One 12:e0188319. https://doi.org/10.1371/journal.pone.0188319

Biodiversity Center of Japan (2003) The annual report of the "Monitoring Sites 1000" in Japan: coral reefs. Biodiversity Center of Japan, Ministry of the Environment of Japan, Yamanashi

Biodiversity Center of Japan (2004) The annual report of the "Monitoring Sites 1000" in Japan: coral reefs. Biodiversity Center of Japan, Ministry of the Environment of Japan, Japan. (in Japanese with English summary)

Biodiversity Center of Japan (2013) The annual report of the "Monitoring Sites 1000" in Japan: coral reefs. Biodiversity Center of Japan, Ministry of the Environment of Japan, Yamanashi

Biodiversity Center of Japan (2014) The annual report of the "Monitoring Sites 1000" in Japan: coral reefs. Biodiversity Center of Japan, Ministry of the Environment of Japan, Yamanashi

Biodiversity Center of Japan (2015) The annual report of the "Monitoring Sites 1000" in Japan: coral reefs. Biodiversity Center of Japan, Ministry of the Environment of Japan, Yamanashi

Biodiversity Center of Japan (2016) The annual report of the "Monitoring Sites 1000" in Japan: coral reefs. Biodiversity Center of Japan, Ministry of the Environment of Japan, Yamanashi

Biodiversity Center of Japan (2017) The annual report of the "Monitoring Sites 1000" in Japan: coral reefs. Biodiversity Center of Japan, Ministry of the Environment of Japan, Yamanashi

Biodiversity Center of Japan (2018) The annual report of the "Monitoring Sites 1000" in Japan: coral reefs. Biodiversity Center of Japan, Ministry of the Environment of Japan, Yamanashi

Biodiversity Center of Japan (2019) The annual report of the "Monitoring Sites 1000" in Japan: coral reefs. Biodiversity Center of Japan, Ministry of the Environment of Japan, Yamanashi

Biodiversity Center of Japan (2020) The annual report of the "Monitoring Sites 1000" in Japan: coral reefs. Biodiversity Center of Japan, Ministry of the Environment of Japan, Yamanashi

Borneman E (2001) Aquarium corals: selection, husbandry, and natural history. T.F.H Publications, Inc

Bourne D, Iida Y, Uthicke S, Smith-Keune C (2007) Changes in coral-associated microbial communities during a bleaching event. ISME J 2:350–363

Bourne DG, Boyett HV, Henderson ME, Muirhead A, Willis BL (2008) Identification of a ciliate (Oligohymenophorea: Scuticociliatia) associated with brown band disease on corals of the Great Barrier Reef. Appl Environ Microbiol 74:883–888

Bourne DG, Garren M, Work TM, Rosenberg E, Smith GW, Harvell CD (2009) Microbial disease and the coral holobiont. Trend Microbiol 17:554–562

Bourne DG, Ainsworth TD, Pollock FJ, Willis BL (2015) Towards a better understanding of white syndromes and their causes on Indo-Pacific coral reefs. Coral Reefs 34:233–242

Bourne DG, Morrow KM, Webster NS (2016) Insights into the coral microbiome: underpinning the health and resilience of reef ecosystems. Annu Rev Microbiol 70:317–340. https://doi.org/10.1146/annurev-micro-102215-095440

Brown BE, Tissier MDA, Bythell JC (1995) Mechanisms of bleaching deduced from histological studies of reef corals sampled during a natural bleaching event. Mar Biol 122:655–663. https://doi.org/10.1007/BF00350687

Brownell AC, Richardson LL (2014) Sulfate reducing bacteria as secondary and necessary pathogens in black band disease of corals. Rev Biol Trop 62:249–257

Bruno JF, Selig ER (2007) Regional decline of coral cover in the Indo-Pacific: timing, extent, and subregional comparisons. PLoS One 2: e711. https://doi.org/10.1371/journal.pone.0000711

Bruno JF, Selig ER, Casey KS, Page CA, Willis BL, Harvell CD, Sweatman H, Melendy AM (2007) Thermal stress and coral cover as drivers of coral disease outbreaks. PLoS Biol 5:e124. https://doi.org/10.1371/journal.pbio.0050124

Bruno JF, Côté IM, Toth LT (2019) Climate change, coral loss, and the curious case of the parrotfish paradigm: why don't marine protected areas improve reef resilience? Annu Rev Mar Sci 11:307–334

Bryan PG (1973) Growth rate, toxicity, and distribution of the encrusting sponge *Terpios* sp. (Hadromerida: Suberitidae) in Guam, Mariana Islands. Micronesica 9:237–242

Buerger P, Alvarez-Roa C, Weynberg KD, Baekelandt S, van Oppen MJH (2016) Genetic, morphological and growth characterisation of a new *Roseofilum* strain (Oscillatoriales, Cyanobacteria) associated with coral black band disease. PeerJ 4:e2110. https://doi.org/10.7717/peerj.2110

Carballo JL, Bautista E, Nava H, Cruz-Barraza JA, Chávez JA (2013) Boring sponges, an increasing threat for coral reefs affected by bleaching events. Ecol Evol 3:872–886

Carlton RG, Richardson LL (1995) Oxygen and sulfide dynamics in a horizontally migrating cyanobacterial mat: black band disease of corals. FEMS Microbiol Ecol 18:155–162

Carpenter KE, Abrar M, Aeby G, Aronson RB, Banks S, Bruckner A, Chiriboga A, Cortés J, Delbeek JC, DeVantier L, Edgar GJ, Edwards AJ, Fenner D, Guzmán HM, Hoeksema BW, Hodgson G, Johan O, Licuanan WY, Livingstone SR, Lovell ER, Moore JA, Obura DO, Ochavillo D, Polidoro BA, Precht WF, Quibilan MC, Reboton C, Richards ZT, Rogers AD, Sanciangco J, Sheppard A, Sheppard C, Smith J, Stuart S, Turak E, Veron JEN, Wallace C, Weil E, Wood E (2008) One-third of reef-building corals face elevated extinction risk from climate change and local impacts. Science 321:560–563

Casamatta D, Stanić D, Gantar M, Richardson LL (2012) Characterization of *Roseofilum reptotaenium* (Oscillatoriales, Cyanobacteria) gen. et sp. nov. isolated from Caribbean black band disease. Phycologia 51:489–499

Cervino JM, Thompson FL, Gomez-Gil B, Lorence EA, Goreau TJ, Hayes RL, Winiarski-Cervino KB, Smith GW, Hughen K, Bartels E (2008) The *Vibrio* core group induces yellow band disease in Caribbean and Indo-Pacific reef-building corals. J Appl Microbiol 105: 1658–1671

Chen Y-H, Chen H-J, Yang C-Y, Shiu J-H, Hoh DZ, Chiang P-W, Chow WS, Chen CA, Shih T-H, Lin S-H, Yang C-M, Reimer JD, Hirose E, Iskandar BH, Huang H, Schupp PJ, Tan CHJ, Yamashiro H, Liao M-H, Tang S-L (2021) Prevalence, complete genome, and metabolic potentials of a phylogenetically novel cyanobacterial symbiont in the coral-killing sponge, *Terpios hoshinota*. Environ Microbiol 24:1308–1325

Chiu JMY, Li S, Li A, Po B, Zhang R, Shin PKS, Qiu J-W (2012) Bacteria associated with skeletal tissue growth anomalies in the coral *Platygyra carnosus*. FEMS Microbiol Ecol 79:380–391

Coles SL, Bolick H (2007) Invasive introduced sponge *Mycale grandis* overgrows reef corals in Kāne'ohe Bay, O'ahu, Hawai'i. Coral Reefs 26:911–911

de Laubenfels M (1954) The sponges of the West-Central Pacific. Oregon State Monogr Stud Zool 7:1–306

de Voogd NJ, Cleary DFR, Dekker F (2013) The coral-killing sponge *Terpios hoshinota* invades Indonesia. Coral Reefs 32:755. https://doi.org/10.1007/s00338-013-1030-4

Ding J-Y, Shiu J-H, Chen W-M, Chiang Y-R, Tang S-L (2016) Genomic insight into the host–endosymbiont relationship of *Endozoicomonas montiporae* CL-33T with its coral host. Front Microbiol 7:251. https://doi.org/10.3389/fmicb.2016.00251

Ekins M, Willis B, Bridge T, Srinivasan M, Rowley S, Hooper JNA (2017) The coral killing sponge *Terpios hoshinota* in Kimbe Bay, Papua New Guinea. Mem Queensl Mus Nat 60:174–175

Elliott J, Patterson M, Vitry E, Summers N, Miternique C (2016) Morphological plasticity allows coral to actively overgrow the aggressive sponge *Terpios hoshinota* (Mauritius, Southwestern Indian Ocean). Mar Biodivers 46:489–493

Fromont J, Richards ZT, Wilson NG (2019) First report of the coral-killing sponge *Terpios hoshinota* Rützler and Muzik, 1993 in Western Australia: A new threat to Kimberley coral reefs? Diversity 11: 184. https://doi.org/10.3390/d11100184

Fujii T, Keshavmurthy S, Zhou W, Hirose E, Chen CA, Reimer JD (2011) Coral-killing cyanobacteriosponge (*Terpios hoshinota*) on the Great Barrier Reef. Coral Reefs 30:483. https://doi.org/10.1007/s00338-011-0734-6

Garren M, Raymundo L, Guest J, Harvell CD, Azam F (2009) Resilience of coral-associated bacterial communities exposed to fish farm effluent. PLoS One 4:e7319. https://doi.org/10.1371/journal.pone.0007319

Garren M, Son K, Raina J-B, Rusconi R, Menolascina F, Shapiro OH, Tout J, Bourne DG, Seymour JR, Stocker R (2014) A bacterial pathogen uses dimethylsulfoniopropionate as a cue to target heat-stressed corals. ISME J 8:999–1007

Garren M, Son K, Tout J, Seymour JR, Stocker R (2016) Temperature-induced behavioral switches in a bacterial coral pathogen. ISME J 10:1363–1372

Geiser DM, Taylor JW, Ritchie KB (1998) Cause of sea fan death in the West Indies. Nature 394:137–138

Gignoux-Wolfsohn SA, Aronson FM, Vollmer SV (2017) Complex interactions between potentially pathogenic, opportunistic, and resident bacteria emerge during infection on a reef-building coral. FEMS Microbiol Ecol 93:7

Glas MS, Sato Y, Ulstrup KE, Bourne DG (2012) Biogeochemical conditions determine virulence of black band disease in corals. ISME J 6:1526–1534

Glynn PW, Manzello DP (2015) Bioerosion and coral reef growth: a dynamic balance. In: Birkeland C (ed) Coral reefs in the anthropocene. Springer, pp 67–97

Guerra M, López MA, Estéves I, Zubillaga AL, Cróquer A (2014) Fourier-transformed infrared spectroscopy: a tool to identify gross chemical changes from healthy to yellow band disease tissues. Dis Aquat Org 107:249–258

Haapkylä J, Unsworth R, Seymour A, Melbourne-Thomas J, Flavell M, Willis B, Smith D (2009) Spatio-temporal coral disease dynamics in the Wakatobi Marine National Park, South-East Sulawesi, Indonesia. Dis Aquat Org 87:105–115

Harvell CD, Kim K, Burkholder JM, Colwell RR, Epstein PR, Grimes DJ, Hofmann EE, Lipp EK, Osterhaus ADME, Overstreet RM, Porter JW, Smith GW, Vasta GR (1999) Emerging marine diseases - climate links and anthropogenic factors. Science 285:1505–1510

Harvell CD, Mitchell CE, Ward JR, Altizer S, Dobson AP, Ostfeld RS, Samuel MD (2002) Climate warming and disease risks for terrestrial and marine biota. Science 296:2158–2162

Hatch WI (1980) The implication of carbonic anhydrase in the physiological mechanism of penetration of carbonate substrata by the marine burrowing sponge *Cliona celata* (demospongiae). Biol Bull 159:135–147

Higuchi T, Agostini S, Casareto BE, Yoshinaga K, Suzuki T, Nakano Y, Fujimura H, Suzuki Y (2013) Bacterial enhancement of bleaching and physiological impacts on the coral *Montipora digitata*. J Exp Mar Biol Ecol 440:54–60

Hirose E, Murakami A (2011) Microscopic anatomy and pigment characterization of coral-encrusting black sponge with cyanobacterial symbiont, *Terpios hoshinota*. Zool Sci 28:199–205

Hirose Y, Aini SN, Yamashiro H (2021) Contact reactions between individuals of the coral-killing sponge *Terpios hoshinota*. Zool Stud 60:41

Hirose Y, Aini SN, Yamashiro H (2022) Coral-killing sponge *Terpios hoshinota* releases larvae at midnight. Coral Reefs 41:149–160

Hoegh-Guldberg O, Mumby PJ, Hooten AJ, Steneck RS, Greenfield P, Gomez E, Harvell CD, Sale PF, Edwards AJ, Caldeira K, Knowlton N, Eakin CM, Iglesias-Prieto R, Muthiga N, Bradbury RH, Dubi A, Hatziolos ME (2007) Coral reefs under rapid climate change and ocean acidification. Science 318:1737–1742

Hoeksema BW, Waheed Z, de Voogd NJ (2014) Partial mortality in corals overgrown by the sponge *Terpios hoshinota* at Tioman Island, Peninsular Malaysia (South China Sea). Bull Mar Sci 90:989–990. https://doi.org/10.5343/bms.2014.1047

Hong M-J, Yu Y-T, Chen CA, Chiang P-W, Tang S-L (2009) Influence of species specificity and other factors on bacteria associated with the coral *Stylophora pistillata* in Taiwan. Appl Environ Microbiol 75: 7797–7806

Hsu C-M, Wang J-T, Chen CA (2013) Larval release and rapid settlement of the coral-killing sponge, *Terpios hoshinota*, at Green Island. Taiwan. Mar Biodiv 43:259–260. https://doi.org/10.1007/s12526-013-0176-1

Huang YM, de Voogd NJ, Cleary DFR, Li T-H, Mok H-K, Ueng J-P (2015) Biodiversity pattern of subtidal sponges (Porifera: Demospongiae) in the Penghu Archipelago (Pescadores), Taiwan. J Mar Biol Assoc U K 96:417–427

Huang C-Y, Hwang J-S, Yamashiro H, Tang S-L (2021) Spatial and cross-seasonal patterns of coral diseases in reefs of Taiwan: high prevalence and regional variation. Dis Aquat Org 146:145–156

Hughes TP, Baird AH, Bellwood DR, Card M, Connolly SR, Folke C, Grosberg R, Hoegh-Guldberg O, Jackson JBC, Kleypas J, Lough JM, Marshall P, Nyström M, Palumbi SR, Pandolfi JM, Rosen B, Roughgarden J (2003) Climate change, human impacts, and the resilience of coral reefs. Science 301:929–933

Hughes TP, Kerry JT, Álvarez-Noriega M, Álvarez-Romero JG, Anderson KD, Baird AH, Babcock RC, Beger M, Bellwood DR, Berkelmans R, Bridge TC, Butler IR, Byrne M, Cantin NE, Comeau S, Connolly SR, Cumming GS, Dalton SJ, Diaz-Pulido G, Eakin CM, Figueira WF, Gilmour JP, Harrison HB, Heron SF, Hoey AS, Hobbs J-PA, Hoogenboom MO, Kennedy EV, Kuo C, Lough JM, Lowe RJ, Liu G, McCulloch MT, Malcolm HA, McWilliam MJ, Pandolfi JM, Pears RJ, Pratchett MS, Schoepf V, Simpson T, Skirving WJ, Sommer B, Torda G, Wachenfeld DR, Willis BL, Wilson SK (2017) Global warming and recurrent mass bleaching of corals. Nature 543:373–377

Hutabarat PUB, Nguyen XH, Suda S (2018) Black band disease-related (BBD) cyanobacterium from Okinawan corals. J Appl Phycol 30: 3197–3203

IPCC (2021) Climate Change 2021: the physical science basis, The Working Group I. https://www.ipcc.ch/assessment-report/ar6/

Irikawa A, Casareto BE, Suzuki Y, Agostini S, Hidaka M, van Woesik R (2011) Growth anomalies on *Acropora cytherea* corals. Mar Pollut Bull 62:1702–1707

Johan O, Bengen DG, Zamani NP, Suharsono SMJ (2015) The distribution and abundance of black band disease and white syndrome in Kepulauan Seribu, Indonesia. HAYATI J Biosci 22:105–112. https://doi.org/10.1016/j.hjb.2015.09.001

Jones RJ, Bowyer J, Hoegh-Guldberg O, Blackall LL (2004) Dynamics of a temperature-related coral disease outbreak. Mar Ecol Prog Ser 281:63–77

Kaczmarsky LT (2006) Coral disease dynamics in the central Philippines. Dis Aqat Org 69:9–21. https://doi.org/10.3354/dao069009

Kaczmarsky L, Richardson LL (2007) Transmission of growth anomalies between Indo-Pacific *Porites* corals. J Invertebr Pathol 94:218–221

Kaczmarsky LT, Draud M, Williams EH (2005) Is there a relationship between proximity to sewage effluent and the prevalence of coral disease. Caribb J Sci 41:124–137

Kawamura K, Nishitsuji K, Shoguchi E, Fujiwara S, Satoh N (2021) Establishing sustainable cell lines of a coral, *Acropora tenuis*. Mar Biotechnol 23:373–388

Kelly M, Bell LJ, Herr B (2016) Splendid sponges of Palau. NIWA

Kim I-H, Yamashiro H (2007) Two species of poecilostomatoid copepods inhabiting galls on scleractinian corals in Okinawa, Japan. J Crustac Biol 27:319–326

Kimes NE, Grim CJ, Johnson WR et al (2012) Temperature regulation of virulence factors in the pathogen *Vibrio coralliilyticus*. ISME J 6: 835–846

Kitamura R, Miura N, Ito M, Hasan NA, Tall BD, Kothary MH, Kiss H, Munk AC, Tapia R, Green L, Detter C, Bruce DC, Brettin TS, Colwell RR, Morris PJ (2021) Specific Detection of coral-associated *Ruegeria*, a potential probiotic bacterium, in corals and subtropical seawater. Mar Biotechnol 23:576–589

Koren O, Rosenberg E (2006) Bacteria associated with mucus and tissues of the coral *Oculina patagonica* in summer and winter. Appl Environ Microbiol 72:5254–5259

Kubomura T, Yamashiro H, Reimer JD (2018) Appearance of an anomalous black band disease at upper mesophotic depths after coral bleaching. Dis Aquat Org 131:245–250

Kubomura T, Wee HB, Reimer JD (2021) Investigating incidence and possible causes of pink and purple pigmentation response in hard coral genus *Porites* around Okinawajima Island, Japan. Reg Stud Mar Sci 41:101569

Kushmaro A, Loya Y, Fine M, Rosenberg E (1996) Bacterial infection and coral bleaching. Nature 380:396

Kushmaro A, Rosenberg E, Fine M, Loya Y (1997) Bleaching of the coral *Oculina patagonica* by *Vibrio* AK-1. Mar Ecol Prog Ser 147: 159–165

Kuta K, Richardson L (1997) Black band disease and the fate of diseased coral colonies in the Florida Keys. In: Proc 8th Int Coral Reef Symp. pp. 575–578

Kuta K, Richardson L (2002) Ecological aspects of black band disease of corals: relationships between disease incidence and environmental factors. Coral Reefs 21:393–398

Kvennefors ECE, Sampayo E, Ridgway T, Barnes AC, Hoegh-Guldberg O (2010) Bacterial communities of two ubiquitous Great Barrier Reef corals reveals both site- and species-specificity of common bacterial associates. PLoS One 5:e10401. https://doi.org/10.1371/journal.pone.0010401

Lamb JB, Willis BL (2011) Using coral disease prevalence to assess the effects of concentrating tourism activities on offshore reefs in a tropical marine park. Conserv Biol 25:1044–1052

Lamb JB, Water JAJM v d, Bourne DG, Altier C, Hein MY, Fiorenza EA, Abu N, Jompa J, Harvell CD (2017) Seagrass ecosystems reduce exposure to bacterial pathogens of humans, fishes, and invertebrates. Science 355:731–733

Li H, Zhang X, Long H, Hu C, Zhou Y, Wang S, Ke S, Xie Z (2018) *Vibrio alginolyticus* 16S-23S intergenic spacer region analysis, and PCR assay for identification of coral pathogenic strain XSBZ03. Dis Aquat Org 129:71–83

Lobban CS, Raymundo LM, Montagnes DJS (2011) *Porpostoma guamensis* n. sp., a philasterine scuticociliate associated with brown-band disease of corals. J Eukaryot Microbiol 58:103–113

Luna GM, Biavasco F, Danovaro R (2007) Bacteria associated with the rapid tissue necrosis of stony corals. Environ Microbiol 9:1851–1857

Luna GM, Bongiorni L, Gili C, Biavasco F, Danovaro R (2010) *Vibrio harveyi* as a causative agent of the white syndrome in tropical stony corals. Environ Microbiol Rep 2:120–127

Mera H, Bourne DG (2018) Disentangling causation: complex roles of coral-associated microorganisms in disease. Environ Microbiol 20: 431–449

Meyer JL, Paul VJ, Teplitski M (2014) Community shifts in the surface microbiomes of the coral *Porites astreoides* with unusual lesions. PLoS One 9:e100316. https://doi.org/10.1371/journal.pone.0100316

Mills E, Shechtman K, Loya Y, Rosenberg E (2013) Bacteria appear to play important roles in both causing and preventing the bleaching of the coral *Oculina patagonica*. Mar Ecol Prog Ser 489:155–162

Miura N, Motone K, Takagi T, Aburaya S, Watanabe S, Aoki W, Ueda M (2019) *Ruegeria* sp. strains isolated from the reef-building coral *Galaxea fascicularis* inhibit growth of the temperature-dependent pathogen *Vibrio coralliilyticus*. Mar Biotechnol 21:1–8

Montano S, Chou WH, Chen CA, Galli P, Reimer JD (2015) First record of the coral-killing sponge *Terpios hoshinota* in the Maldives and Indian Ocean. Bull Mar Sci 91:97–98. https://doi.org/10.5343/bms.2014.1054

Moriarty T, Leggat W, Huggett MJ, Ainsworth TD (2020) Coral disease causes, consequences, and risk within coral restoration. Trends Microbiol 28:793–807

Muller EM, Raymundo LJ, Willis BL, Haapkylä J, Yusuf S, WilsonJR HDC (2012) Coral Health and Disease in the Spermonde Archipelago and Wakatobi, Sulawesi. J Indonesia Coral Reefs 1:147–159

Mumby PJ, Steneck RS (2008) Coral reef management and conservation in light of rapidly evolving ecological paradigms. Trends Ecol Evol 23:555–563

Muscatine LR, McCloskey LE, Marian R (1981) Estimating the daily contribution of carbon from zooxanthellae to coral animal respiration. Limnol Oceanogr 26:601–611

Myers R, Raymundo L (2009) Coral disease in micronesian reefs: a link between disease prevalence and host abundance. Dis Aquat Org 87: 97–104

Ng JCY, Chan Y, Tun HM, Leung FCC, Shin PKS, Chiu JMY (2015) Pyrosequencing of the bacteria associated with *Platygyra carnosus* corals with skeletal growth anomalies reveals differences in bacterial community composition in apparently healthy and diseased tissues. Front Microbiol 6:1142. https://doi.org/10.3389/fmicb.2015.01142

Nicolet KJ, Chong-Seng KM, Pratchett MS, Willis BL, Hoogenboom MO (2018) Predation scars may influence host susceptibility to pathogens: evaluating the role of corallivores as vectors of coral disease. Sci Rep 8:5258. https://doi.org/10.1038/s41598-018-23361-y

Nozawa Y, Huang Y-S, Hirose E (2016) Seasonality and lunar periodicity in the sexual reproduction of the coral-killing sponge, *Terpios hoshinota*. Coral Reefs 35:1071–1081

Nugues MM, Smith GW, Hooidonk RJ, Seabra MI, Bak RPM (2004) Algal contact as a trigger for coral disease. Ecol Lett 7:919–923

Olson ND, Ainsworth TD, Gates RD, Takabayashi M (2009) Diazotrophic bacteria associated with Hawaiian *Montipora* corals: diversity and abundance in correlation with symbiotic dinoflagellates. J Exp Mar Biol Ecol 371:140–146

Page CA, Baker DM, Harvell CD, Golbuu Y (2009) Influence of marine reserves on coral disease prevalence. Dis Aquat Org 87:135–150. https://doi.org/10.3354/dao02112

Palmer C, Roth M, Gates R (2009) Red fluorescent protein responsible for pigmentation in trematode-infected *Porites compressa* tissues. Biol Bull 216:68–74

Patterson KL, Porter JW, Ritche KB, Polson SW, Mueller E, Peters EC, Santavy DL, Smith GW (2002) The etiology of white pox, a lethal disease of the Caribbean elkhorn coral, *Acropora palmata*. Proc Natl Acad Sci U S A 99:8725–8730

Plucer-Rosario G (1987) The effect of substratum on the growth of *Terpios*, an encrusting sponge which kills corals. Coral Reefs 5: 197–200

Pollock FJ, Lamb JB, Field SN, Heron SF, Schaffelke B, Shedrawi G, Bourne DG, Willis BL (2016) Correction: sediment and turbidity associated with offshore dredging increase coral disease prevalence on Nearby Reefs. PLoS One 11:e0165541. https://doi.org/10.1371/journal.pone.0165541

Pollock FJ, Wada N, Torda G, Willis BL, Bourne DG (2017) White syndrome-affected corals have a distinct microbiome at disease lesion fronts. Appl Environ Microbiol 83:e02799–e02716. https://doi.org/10.1128/AEM.02799-16

Quang TM (2020) Morphology, abundance and the invasiveness of coral-killing sponge *Chalinula nematifera* (Porifera: Demosponigiae) from Con Dao National Park, Vietnam. Vietnam J Mar Sci Technol 20:95–103

Randall CJ, Jordán-Garza AG, van Woesik R (2015) Ciliates associated with signs of disease on two Caribbean corals. Coral Reefs 34:243–247

Rasoulouniriana D, Siboni N, Ben-Dov E, Kramarsky-Winter E, Loya Y, Kushmaro T (2009) *Pseudoscillatoria coralii* gen. nov., sp. nov., a cyanobacterium associated with coral black band disease (BBD). Dis Aquat Org 87:91–96

Raymundo LJ, Weil E (2015) Indo-Pacific colored-band diseases of corals. In: Porter J, Woodley C (eds) Diseases of coral. John Wiley & Sons, Ltd, pp 333–344

Raymundo LJH, Harvell CD, Reynolds TL (2003) Porites ulcerative white spot disease: description, prevalence, and host range of a new coral disease affecting Indo-Pacific reefs. Dis Aquat Org 56:95–104

Reimer JD, Nozawa Y, Hirose E (2010) Domination and disappearance of the black sponge: a quarter century after the initial *Terpios* outbreak in southern Japan. Zool Stud 50:394

Reimer JD, Mizuyama M, Nakano M, Fujii T, Hirose E (2011) Current status of the distribution of the coral-encrusting

cyanobacteriosponge *Terpios hoshinota* in southern Japan. Galaxea J Coral Reef Stud 13:35–44. https://doi.org/10.3755/galaxea.13.35
Reimer JD, Fourreau CJL, Fujii T, Ise Y, Kushida Y, Mizukami I, Nakano M (2022) Records and distribution of the coral-killing sponge *Chalinula nematifera* in the Ryukyu Islands, Japan. Plankton Benthos Res 17:249–254
Reshef L, Koren O, Loya Y, Zilber-Rosenberg I, Rosenberg E (2006) The coral probiotic hypothesis. Environ Microbiol 8:2068–2073
Richardson LL (1998) Coral diseases: what is really known? Trends Ecol Evol 13:438–443
Ritchie KB (2006) Regulation of microbial populations by coral surface mucus and mucus-associated bacteria. Mar Ecol Prog Ser 322:1–14
Roder C, Arif C, Bayer T, Aranda M, Daniels C, Shibl A, Chavanich S, Voolstra CR (2014) Bacterial profiling of white plague disease in a comparative coral species framework. ISME J 8:31–39
Rodríguez S, Cróquer A (2008) Dynamics of black band disease in a *Diploria strigosa* population subjected to annual upwelling on the northeastern coast of Venezuela. Coral Reefs 27:381–388
Rohwer F, Seguritan V, Azam F, Knowlton N (2002) Diversity and distribution of coral-associated bacteria. Mar Ecol Prog Ser 243:1–10
Rossi G, Montori S, Cerrano C, Calcinai B (2015) The coral killing sponge *Chalinula nematifera* (Porifera: Haplosclerida) along the eastern coast of Sulawesi Island (Indonesia). Ital J Zool 82:143–148
Rubio-Portillo E, Martin-Cuadrado AB, Caraballo-Rodríguez AM, Rohwer F, Dorrestein PC, Antón J (2020) Virulence as a side effect of interspecies interaction in *Vibrio* coral pathogens. mBio 11: e00201–e00220. https://doi.org/10.1128/mBio.00201-20
Rützler K, Muzik K (1993) *Terpios hoshinota*, a new cyanobacteriosponge threatening Pacific reefs. Sci Mar 57:395–403
Rützler K, Rieger G (1973) Sponge burrowing: fine structure of *Cliona lampa* penetrating calcareous substrata. Mar Biol 21:144–162
Sato Y, Bourne DG, Willis BL (2009) Dynamics of seasonal outbreaks of black band disease in an assemblage of *Montipora* species at Pelorus Island (Great Barrier Reef, Australia). Proc R Soc B 276: 2795–2803. https://doi.org/10.1098/rspb.2009.0481
Sato Y, Willis BL, Bourne DG (2010) Successional changes in bacterial communities during the development of black band disease on the reef coral, *Montipora hispida*. ISME J 4:203–214
Sato Y, Civiello M, Bell SC, Willis BL, Bourne DG (2016) Integrated approach to understanding the onset and pathogenesis of black band disease in corals. Environ Microbiol 18:752–765
Séré MG, Tortosa P, Chabanet P, Quod J-P, Sweet MJ, Schleyer MH (2015) Identification of a bacterial pathogen associated with Porites white patch syndrome in the Western Indian Ocean. Mol Ecol 24: 4570–4581
Shi Q, Liu GH, Yan HQ, Zhang HL (2012) Black disease (*Terpios hoshinota*): a probable cause for the rapid coral mortality at the northern reef of Yongxing Island in the South China Sea. Ambio 41:446–455. https://doi.org/10.1007/s13280-011-0245-2
Shiu J-H, Keshavmurthy S, Chiang P-W, Chen H-J, Lou S-P, Tseng C-H, Hsieh HJ, Chen CA, Tang S-L (2017) Dynamics of coral-associated bacterial communities acclimated to temperature stress based on recent thermal history. Sci Rep 7:14933. https://doi.org/10.1038/s41598-017-14927-3
Squires DF (1965) Neoplasia in a coral? Science 148:503–505
Subhan B, Arafat D, Rahmawati F, Dasmasela YH, Royhan QM, Madduppa H, Santoso P, Prabowo B (2020) Coral disease at Mansuar Island, Raja Ampat, Indonesia. IOP Conf Ser Earth Environ Sci 429:012027. https://doi.org/10.1088/1755-1315/429/1/012027
Sussman M, Loya Y, Fine M, Rosenberg E (2003) The marine fireworm *Hermodice carunculata* is a winter reservoir and spring-summer vector for the coral-bleaching pathogen *Vibrio shiloi*. Environ Microbiol 5:250–255
Sussman M, Bourne DG, Willis BL (2006) A single cyanobacterial ribotype is associated with both red and black bands on diseased corals from Palau. Dis Aquat Org 69:111–118
Sussman M, Willis BL, Victor S, Bourne DG (2008) Coral pathogens identified for white syndrome (WS) epizootics in the Indo-Pacific. PLoS One 3:e2393. https://doi.org/10.1371/journal.pone.0002393
Sussman M, Mieog J, Doyle J, Victor S (2009) *Vibrio* zinc-metalloprotease causes photoinactivation of coral endosymbionts and coral tissue lesions. PLoS One 4:e4511. https://doi.org/10.1371/journal.pone.0004511
Sutherland K, Porter J, Torres C (2004) Disease and immunity in Caribbean and Indo-Pacific zooxanthellate corals. Mar Ecol Prog Ser 266:273–302
Sutherland KP, Berry B, Park A, Kemp DW, Kemp KM, Lipp EK, Porter JW (2016) Shifting white pox aetiologies affecting *Acropora palmata* in the Florida Keys, 1994–2014. Philos Trans R Soc B Biol Sci 371:20150205. https://doi.org/10.1098/rstb.2015.0205
Syue S-T, Hsu C-H, Soong K (2021) Testing of how and why the *Terpios hoshinota* sponge kills stony corals. Sci Rep 11:7661. https://doi.org/10.1038/s41598-021-87350-4
Takagi T, Yoshioka Y, Zayasu Y, Satoh N, Shinzato C (2020) Transcriptome analyses of immune system behaviors in primary polyp of coral *Acropora digitifera* exposed to the bacterial pathogen *Vibrio coralliilyticus* under thermal loading. Mar Biotechnol 22: 748–759. https://doi.org/10.1007/s10126-020-09984-1
Tandon K, Lu C-Y, Chiang P-W, Wada N, Yang S-H, Chan Y-F, Chen P-Y, Chang H-Y, Chiou Y-J, Chou M-S, Chen W-M, Tang S-L (2020) Comparative genomics: dominant coral-bacterium *Endozoicomonas acroporae* metabolizes dimethylsulfoniopropionate (DMSP). ISME J 14:1290–1303
Tang S-L, Hong M, Liao M, Jane W, Chiang P, Chen C, Chen CA (2011) Bacteria associated with an encrusting sponge (*Terpios hoshinota*) and the corals partially covered by the sponge. Environ Microbiol 13:1179–1191
Tang K, Zhan W, Zhou Y, Xu T, Chen X, Wang W, Zeng Z, Wang Y, Wang X (2020) Antagonism between coral pathogen *Vibrio coralliilyticus* and other bacteria in the gastric cavity of scleractinian coral *Galaxea fascicularis*. Sci China Earth Sci 63:157–166
Teplitski M, Krediet CJ, Meyer JL, Ritchie KB (2016) Microbial interactions on coral surfaces and within the coral holobiont. In: Goffredo S, Dubinsky Z (eds) The cnidaria, past, present and future: the world of medusa and her sisters. Springer, Cham, pp 331–346
Thinesh T, Meenatchi R, Pasiyappazham JPA, Selvan M, Kiran GS, Selvin J (2017) Short-term in situ shading effectively mitigates linear progression of coral-killing sponge *Terpios hoshinota*. PLoS One 12:e0182365. https://doi.org/10.1371/journal.pone.0182365
Thompson FL, Barash Y, Sawabe T, Sharon G, Swings J, Rosenberg E (2006) *Thalassomonas loyana* sp. nov., a causative agent of the white plague-like disease of corals on the Eilat coral reef. Int J Syst Evol Microbiol 56:365–368
Turicchia E, Hoeksema BW, Ponti M (2018) The coral-killing sponge *Chalinula nematifera* as a common substrate generalist in Komodo National Park, Indonesia. Mar Biol Res 14:827–833
Ushijima B, Smith A, Aeby GS, Callahan SM (2012) *Vibrio owensii* induces the tissue loss disease *Montipora* white syndrome in the Hawaiian reef coral *Montipora capitata*. PLoS One 7:e46717. https://doi.org/10.1371/journal.pone.0046717
Ushijima B, Videau P, Burger AH, Shore-Maggio A, Runyon CM, Sudek M, Aeby GS, Callahan SM (2014) *Vibrio coralliilyticus* Strain OCN008 Is an etiological agent of acute Montipora White Syndrome. Appl Environ Microbiol 80:2102–2109
Ushijima B, Videau P, Poscablo D, Stengel JW, Beurmann S, Burger AH, Aeby GS, Callahan SM (2016) Mutation of the toxR or mshA genes from *Vibrio coralliilyticus* strain OCN014 reduces infection of the coral *Acropora cytherea*. Environ Microbiol 18:4055–4067
Van der Ent E, Hoeksema BW, de Voogd NJ (2015) Abundance and genetic variation of the coral-killing cyanobacteriosponge *Terpios hoshinota* in the Spermonde Archipelago, SW Sulawesi, Indonesia. J Mar Biol Assoc UK 96:453–463. https://doi.org/10.1017/S002531541500034X

van de Water JAJM, Chaib De Mares M, Dixon GB, Raina J-B, Willis BL, Bourne DG, van Oppen MJH (2018) Antimicrobial and stress responses to increased temperature and bacterial pathogen challenge in the holobiont of a reef-building coral. Mol Ecol 27:1065–1080

Vargas-Angel B (2009) Coral health and disease assessment in the US Pacific remote Island areas. Bull Mar Sci 84:211–227

Vega Thurber RL, Willner-Hall D, Rodriguez-Mueller B, Desnues C, Edwards RA, Angly F, Dinsdale E, Kelly L, Rohwer F (2009) Metagenomic analysis of stressed coral holobionts. Environ Microbiol 11:2148–2163

Wada M, Nishihara G, Isomura N (2012) Probing stony coral rapid tissue necrosis with optical oxygen sensors. In: 12th Intern Coral Reef Symp. Cairns, Queensland, Australia

Wada N, Ikeda S, Sekine S, Toizumi K, Nakamura R, Namba A, Uchida S, Nakamura K, Mano N (2015) *Vibrio* spp. isolated from scleractinian coral *Acropora muricata* with rapid tissue necrosis (RTN) and reared in aquarium. BioMicroWorld, Barcelona

Wada N, Pollock FJ, Willis BL, Ainsworth T, Mano N, Bourne DG (2016) *In situ* visualization of bacterial populations in coral tissues: pitfalls and solutions. PeerJ 4:e2424. https://doi.org/10.7717/peerj.2424

Wada N, Mano N, Yanagisawa Y, Mori T (2017) Occurrence of coral diseases at Akajima, Okinawa, Japan in 2010 and 2011. Galaxea J Coral Reef Stud 19:35–44. https://doi.org/10.3755/galaxea.19.1_35

Wada N, Ohdera A, Mano N (2018) Coral disease in Japan. In: Iguchi A, Hongo C (eds) Coral reef studies of Japan. Springer, Singapore, pp 41–62

Wang J-T, Chen Y-Y, Meng P-J, Sune Y-H, Hsu C-M, Wei K-Y, Chen CA (2012) Diverse interactions between corals and the coral-killing sponge, *Terpios hoshinota* (Suberitidae: Hadromerida). Zool Stud 10:150–159

Webster NS, Thomas T (2016) The sponge hologenome. mBio 7: e00135–e00116. https://doi.org/10.1128/mBio.00135-16

Weil E, Irikawa A, Casareto B, Suzuki Y (2012) Extended geographic distribution of several Indo-Pacific coral reef diseases. Dis Aquat Org 98:163–170

West AG, Waite DW, Deines P et al (2019) The microbiome in threatened species conservation. Biol Conserv 229:85–98

Willis BL, Page CA, Dinsdale EA, Bourne DG, Digby A, McKenzie VJ, Taylor MW (2004) Coral disease on the Great Barrier Reef. In: Rosenberg PE, Loya Y (eds) Coral health and disease. Springer, Berlin, Heidelberg, pp 69–104

Wulff J (2016) Sponge contributions to the geology and biology of reefs: past, present, and future. In: Hubbard DK, Rogers CS, Lipps JH, Stanley Jr George D (eds) Coral reefs at the crossroads. Springer, Dordrecht, pp 103–126

Yamaguchi M (1986) Introduction to the study of coral reefs 4 Coral reef sponges (1) Sponges as destroyers of reef-building corals. Aquabiology 8:88–92. (in Japanese)

Yamashiro H (2004) Coral diseases. In Coral Reefs of Japan. Ministry of the Environment and Japanese Coral Reef Society (ed). Ministry of the Environment, pp 56–59

Yamano H, Hori K, Yamauchi M, Yamagawa O, Ohmura A (2001) Highest-latitude coral reef at Iki Island, Japan. Coral Reefs 20:9–12

Yamashiro H, Fukuda M (2009) White spot syndrome of *Turbinaria peltata* in the temperate region of Japan. Coral Reefs 28:893

Yamashiro H, Yamamoto M, van Woesik R (2000) Tumor formation on the coral *Montipora informis*. Dis Aquat Org 41:211–217

Yamashiro H, Oku H, Onaga K, Iwasaki H, Takara K (2001) Coral tumors store reduced level of lipids. J Exp Mar Biol Ecol 265:171–179

Yamashiro H, Isomura N, Sakai K (2014) Bloom of the cyanobacterium *Moorea bouillonii* on the gorgonian coral *Annella reticulata* in Japan. Sci Rep 4:6032. https://doi.org/10.1038/srep06032

Yamashiro H, Fukuda M, Fukami H (2016) Outbreak of white syndrome in *Acropora* spp., temperate region of Japan. Mar Biodivers 46:321–322

Yamashiro H, Fukumori H, Aini SN, Hirose Y (2021) Snails associated with the coral-killing sponge *Terpios hoshinota* in Okinawa Island, Japan. Sci Rep 11:20709. https://doi.org/10.1038/s41598-021-00185-x

Yang H, Shen J, Fu F, Wang Y, Zhao N (2014) Black band disease as a possible factor of the coral decline at the northern reef-flat of Yongxing Island, South China Sea. Sci China Earth Sci 57:569–578

Yang S-Y, Chen H-J, Ho M-J, Chen Y-J, Huang Y-Y, Chow WS, Tang S-L, Jeng M-S, Chen CA (2018) Outbreak of coral-killing cyanobacteriasponge, *Terpios hoshinota*, in Taiping Island (Itu Aba), Spratlys, South China Sea. Bull Mar Sci 94:1543–1544. https://doi.org/10.5343/bms.2018.0023

Yang S-H, Tandon K, Lu C-Y, Wada N, Shih C-J, Hsiao SS-Y, Jane W-N, Lee T-C, Yang C-M, Liu C-T, Denis V, Wu Y-T, Wang L-T, Huang L, Lee D-C, Wu Y-W, Yamashiro H, Tang S-L (2019) Metagenomic, phylogenetic, and functional characterization of predominant endolithic green sulfur bacteria in the coral *Isopora palifera*. Microbiome 7:3. https://doi.org/10.1186/s40168-018-0616-z

Yasuda N, Hidaka M (2012) Cellular kinetics in growth anomalies of the scleractinian corals *Porites australiensis* and *Montipora informis*. Dis Aquat Org 102:1–11

Yasuda N, Nakano Y, Yamashiro H, Hidaka M (2012) Skeletal structure and progression of growth -anomalies in *Porites australiensis* in Okinawa, Japan. Dis Aquat Org 97:237–247

Yomogida M, Mizuyama M, Kubomura T, Reimer JD (2017) Disappearance and return of an outbreak of the coral-killing cyanobacteriosponge *Terpios hoshinota* in Southern Japan. Zool Stud 56:1–7. https://doi.org/10.6620/ZS.2017.56-07

Yu Q, Liu M, Su H, Xiao H, Wu S, Qin X, Li S, Mi H, Lu Z, Shi D, Li P (2019) Selection and characterization of ssDNA aptamers specifically recognizing pathogenic *Vibrio alginolyticus*. J Fish Dis 42: 851–858

Zaneveld JR, Burkepile DE, Shantz AA, Pritchard CE, McMinds R, Payet JP, Welsh R, Correa AMS, Lemoine NP, Rosales S, Fuchs C, Maynard JA, Thurber RV (2016) Overfishing and nutrient pollution interact with temperature to disrupt coral reefs down to microbial scales. Nat Commun 7:11833. https://doi.org/10.1038/ncomms11833

Zhang Y, Sun J, Mu H, Lun JCY, Qiu J-W (2017) Molecular pathology of skeletal growth anomalies in the brain coral *Platygyra carnosa*: a meta-transcriptomic analysis. Mar Pollut Bull 124:660–667

Zhenyu X, Shaowen K, Chaoqun H, Zhixiong Z, Shifeng W, Yongcan Z (2013) First characterization of bacterial pathogen, *Vibrio alginolyticus*, for *Porites andrewsi* white syndrome in the South China Sea. PLoS One 8:e75425. https://doi.org/10.1371/journal.pone.0075425

Zvuloni A, Artzy-Randrup Y, Stone L, Kramarsky-Winter E, Barkan R, Loya Y (2009) Spatio-temporal transmission patterns of black-band disease in a coral community. PLoS One 4:e4993. https://doi.org/10.1371/journal.pone.0004993

Succession of Ocean Acidification and its Effects on Reef-Building Corals

6

Atsushi Suzuki, Akira Iguchi, Kazuhiko Sakai, Masahiro Hayashi, and Yukihiro Nojiri

Abstract

Since 2008, we have been conducting a series of coral-rearing experiments, mainly at the Sesoko Station of the Tropical Biosphere Research Center at the University of the Ryukyus, under an overarching project, called the Acidification Impact on Calcifiers project (AICAL project). The AICAL project integrates the efforts of several individual research programs, and project members employ a custom-made, high-precision pCO_2-adjusted seawater generator (the AICAL apparatus) to study the effects of ocean acidification on marine calcifying organisms. With this system, rearing experiments can be conducted under conditions mimicking those in the preindustrial era, and the future. In this review, we summarize the results of ocean acidification experiments on corals and other organisms, with a focus on studies conducted by the AICAL project members. We examine the response of organisms to ocean acidification in a hierarchical fashion: differences among various groups of calcifying organisms, and interspecific and intraspecific variation in corals. In the case of corals, we consider not only the effects of ocean acidification, but also those caused by rising seawater temperatures and eutrophication. Our major findings are that coral calcification may have already decreased from a preindustrial level and that there are evident interspecific and intraspecific differences in tolerance against ocean acidification. These findings suggest future decrease of coral cover, accompanied by species compositional changes under climate change scenarios.

Keywords

Ocean acidification · Corals · Symbionts · Calcification

6.1 Introduction

At a time when the term "ocean acidification" had not yet been widely used, Kleypas et al. (1999) suggested that coral reef calcification might be declining due to increased anthropogenic carbon dioxide (CO_2) in the atmosphere. A certain amount of the CO_2 released into the atmosphere by human activities has already become dissolved in the ocean, causing a rise in the partial pressure of CO_2 (pCO_2) in seawater and reducing seawater pH. Scientists fear that this ocean acidification could reduce the carbonate saturation state (Ω) of seawater, inhibiting the formation of skeletons, which are composed of calcite and aragonite, by marine calcifying organisms (Orr et al. 2005).

$$\Omega = \left[Ca^{2+}\right] \left[{CO_3}^{2-}\right]/K_{sp} \tag{6.1}$$

where $[Ca^{2+}]$ and $[{CO_3}^{2-}]$ are the concentrations of calcium and carbonate ions in seawater, respectively, and K_{sp} is the solubility product of a carbonate mineral. K_{sp} is a function of temperature, salinity, and the type of carbonate mineral (aragonite or calcite). Initially, because Ω tends to decrease at cooler water temperatures, ocean acidification was expected to have seriously impacted marine ecosystems at high latitudes. However, the effects of ocean acidification on

A. Suzuki (✉) · A. Iguchi
Geological Survey of Japan, National Institute of Advanced Industrial Science and Technology (AIST), Tsukuba, Ibaraki, Japan
e-mail: a.suzuki@aist.go.jp

K. Sakai
Sesoko Station, Tropical Biosphere Research Center, University of the Ryukyus, Okinawa, Japan

M. Hayashi
Demonstration Laboratory, Marine Ecology Research Institute, Niigata, Japan

Y. Nojiri
Department of Earth and Environmental Sciences, Hirosaki University, Hirosaki, Aomori, Japan

Center for Global Environmental Research, National Institute for Environmental Studies, Tsukuba, Ibaraki, Japan

I. Takeuchi, H. Yamashiro (eds.), *Coral Reefs of Eastern Asia under Anthropogenic Impacts*, Coral Reefs of the World 17,
https://doi.org/10.1007/978-3-031-27560-9_6

calcifying organisms in tropical and subtropical coral reefs have also attracted the interest of researchers (Hoegh-Guldberg et al. 2007; Suwa et al. 2010a; Suzuki 2018; Kawahata et al. 2019). The state of ocean acidification and its biological effects have been discussed extensively in the assessment reports of the Intergovernmental Panel on Climate Change (IPCC 2021).

Since 2008, we have been conducting a series of coral-rearing experiments, mainly at the Sesoko Station of the Tropical Biosphere Research Center at the University of the Ryukyus. These experiments were performed under an overarching project called the Acidification Impact on Calcifiers project (AICAL project) which integrates the efforts of several individual research programs. AICAL project members employ a custom-made, high-precision $p\mathrm{CO_2}$-adjusted seawater generator (the AICAL apparatus) to study the effects of ocean acidification on marine calcifying organisms, especially corals. In this review, we have summarized the results of ocean acidification experiments on corals and other organisms, with a focus on studies conducted by the AICAL project members. We examine the response of organisms to ocean acidification in a hierarchical fashion: differences among various groups of calcifying organisms (corals, foraminifera, and calcareous algae), and interspecific and intraspecific variations in corals. In the case of corals, we consider not only the effects of ocean acidification, but also those caused by rising seawater temperatures and eutrophication.

6.2 Acidification of the World Ocean and Western Pacific

Ocean acidification is an observational fact that has already been confirmed using regular measurements of the carbonic acid system made by oceanographic research vessels. About 31% of CO_2 released into the atmosphere by human activities during 1994–2007 has already become dissolved in the ocean, thereby causing a rise in the partial pressure of CO_2 ($p\mathrm{CO_2}$) in seawater and reducing seawater pH (Gruber et al. 2019). Oceanic and atmospheric $p\mathrm{CO_2}$ in the western North Pacific region are increasing at 1.5–3.0 and 1.7–2.1 μatm year^{-1}, respectively (Fig. 6.1; Japan Meteorological Agency 2021). In particular, the rate of ocean acidification has been increasing in the subtropical western North Pacific (Ono et al. 2019). Studies conducted along the 137°E meridian south of Japan (Japan Meteorological Agency 2021) and the PN observation line in the East China Sea (Lui et al. 2015) have suggested that coral reef habitats around the Ryukyu Islands have already been affected by ocean acidification. Specifically, in the local coral reef environment, for example, the observation at Sesokojima (Sesoko) Island (e.g., Kurihara et al. 2019), daily and seasonal variations in seawater pH and $p\mathrm{CO_2}$ are larger than those offshore and also larger than ocean acidification trends.

The stable isotope ratio of boron (the $^{11}\mathrm{B}/^{10}\mathrm{B}$ ratio relative to that of a standard; $\delta^{11}\mathrm{B}$) in biogenic carbonates has been used as a proxy for the pH of the medium at the calcification sites of organisms. Analyses of the skeletons of large *Porites* coral colonies from Chichijima Island, one of the Ogasawara Islands, and Kikaishima (Kikai) Island, in the northern part of the Ryukyu Islands, have shown that $\delta^{11}\mathrm{B}$ showed a decreasing trend throughout the twentieth century, with the decline being particularly pronounced after 1960 (Kubota et al. 2017; Fig. 6.2). These findings are quite consistent with the results of the shipboard oceanographic observations (Japan Meteorological Agency 2021). Ocean acidification lowers the pH of the calcifying medium, suggesting that it may have already reached the coral calcification sites. Whether this pH decrease exceeds the threshold for calcification reduction is unclear, but if the decrease continues, coral calcification can be adversely affected. The fate of the corals and coral reefs around the Ryukyu Islands in the event of continued ocean acidification continues is an important question that must be addressed.

6.3 Systems for Ocean Acidification Experiments with Reef-Building Corals

In response to the rapidly growing interest in ocean acidification research and concerns about the accuracy of seawater pH and $p\mathrm{CO_2}$ measurements used in marine calcifying organism rearing experiments, Riebesell et al. (2011) published their "Guide to best practices for ocean acidification research and data reporting," which outlines experimental methods and data reporting for ocean acidification. Our series of devices used in the AICAL projects is very unique in that the $p\mathrm{CO_2}$ value of the seawater supplied is accurate, which is a major advantage, and that the equipment can be scaled to meet the water requirements of the rearing organisms. The following is a detailed description of the order of development.

In our earliest experiments, we controlled the pH of seawater by adding hydrochloric acid (HCl) and then measured changes in the calcification rate of foraminifera (Kuroyanagi et al. 2009). However, adding HCl to control the carbonic acid system of experimental seawater does not appropriately mimic the ongoing ocean acidification caused by CO_2 dissolution in seawater (e.g., Kurihara 2008). Therefore, we designed and constructed a simple pH control system composed of pH electrodes, solenoid valves, and a mixing tank (Suwa et al. 2010b; Table 6.1, Fig. 6.3). In the early stages of research, seawater pH was used to monitor experimental conditions in ocean acidification experiments, and it continues to be used for that purpose today. However, of

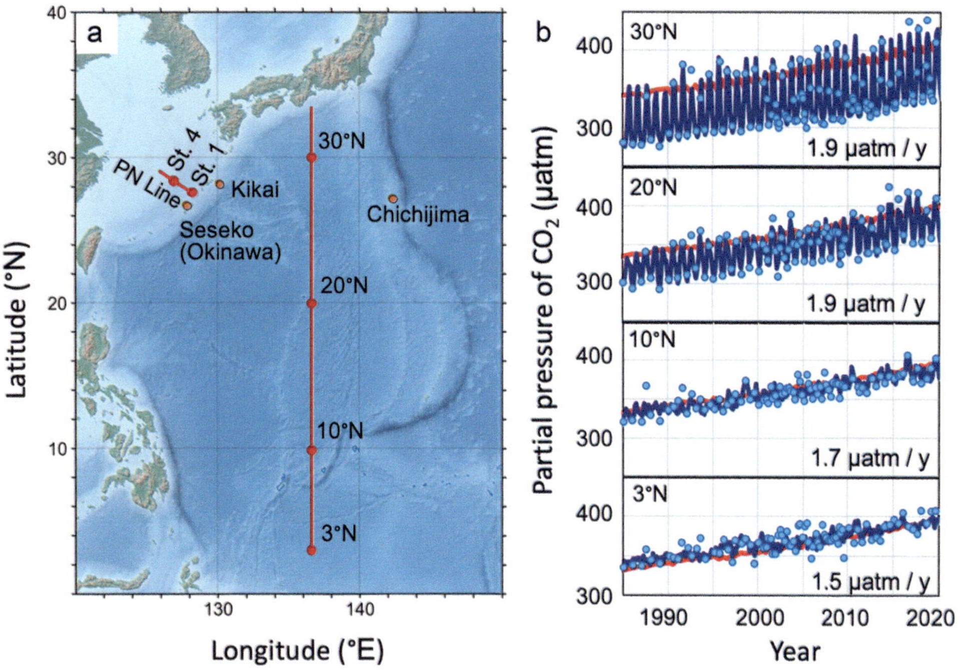

Fig. 6.1 (**a**) Locations of oceanographic observation transects used by Japan Meteorological Agency with long-term monitoring records of ocean acidification. While the transect line along 137°E is located in the Philippine Sea (Ishii et al. 2011), PN Line in the East China Sea (Lui et al. 2015) is close to Sesoko Station, University of the Ryukyus. Locations of Chichijima and Kikaishima (Kikai) islands, which are the sites for coral-based paleo-pH reconstruction, are also shown (Kubota et al. 2017). (**b**) Changes in oceanic and atmospheric pCO_2 at locations along 137°E (Japan Meteorological Agency 2021). Filled circles show pCO_2 observations of individual oceans, blue curves show estimated monthly oceanic pCO_2 values, and red curves indicate observed atmospheric pCO_2 values. The long-term oceanic pCO_2 trend at each station is shown in each panel. Panel (**a**) was made with Natural Earth (https://www.naturalearthdata.com)

the four measurable carbonate parameters (pH, total alkalinity, dissolved inorganic carbon, pCO_2), pH is the most problematic in terms of standardization and measurement stability. Some of the problems with pH measurements have been remedied by changing the standard solution used for calibration. In the past, standard solutions prescribed by the National Bureau of Standards (NBS) were used; however, these standards are only suitable for measuring the pH of freshwater and not seawater, because of its high salinity. More recently, standard solutions of artificial seawater buffered with TRIS and AMP have been used to obtain stable responses of electrodes between standards and seawater samples. pH values measured using these standard solutions are reported on the total hydrogen ion concentration scale and often written as pH_T.

Therefore, systems to control and monitor pCO_2 in seawater have been developed for use in marine calcifying organism rearing experiments to evaluate the effects of ocean acidification. Our research group has developed a series of high-precision pCO_2-adjusted-seawater generators (AICAL systems, Table 6.1, Fig. 6.4) for rearing experiments that allow the highly accurate and stable control of seawater pCO_2.

The first-generation AICAL system was designed for indoor installation (Fujita et al. 2011); thus, it is commonly referred to as an "indoor-type AICAL system" (Fig. 6.5). A mass flow controller first adjusts the CO_2 concentration of the air to the target value by mixing CO_2-free air and pure CO_2 gas, following which this air is bubbled into CO_2 dissolution and equilibration towers filled with seawater to equilibrate the seawater CO_2 content with that of air. After precisely measuring the pCO_2 of the seawater, it is supplied to the experimental tank at a rate of about 0.5 L min^{-1}, which is suitable for breeding experiments of small organisms. With this system, it is also possible to obtain seawater with a lower pCO_2 than the original seawater can be obtained; therefore, rearing experiments that mimic preindustrial or glacial conditions can also be conducted. In addition, the diurnal variation of pCO_2 in seawater can be automatically controlled.

The second-generation AICAL system produces experimental seawater by mixing highly concentrated CO_2 seawater generated by using a metered feed pump to bubble pure CO_2 gas from a liquefied CO_2 cylinder into seawater at a fixed ratio (Fig. 6.6). The system can adjust the pCO_2 in a large volume of seawater and continuously measure the pCO_2 in six tanks (Nishida et al. 2015). It measures the actual pCO_2 even when the water temperature changes; thus, it is ideal for evaluating the combined effects of water temperature and pCO_2.

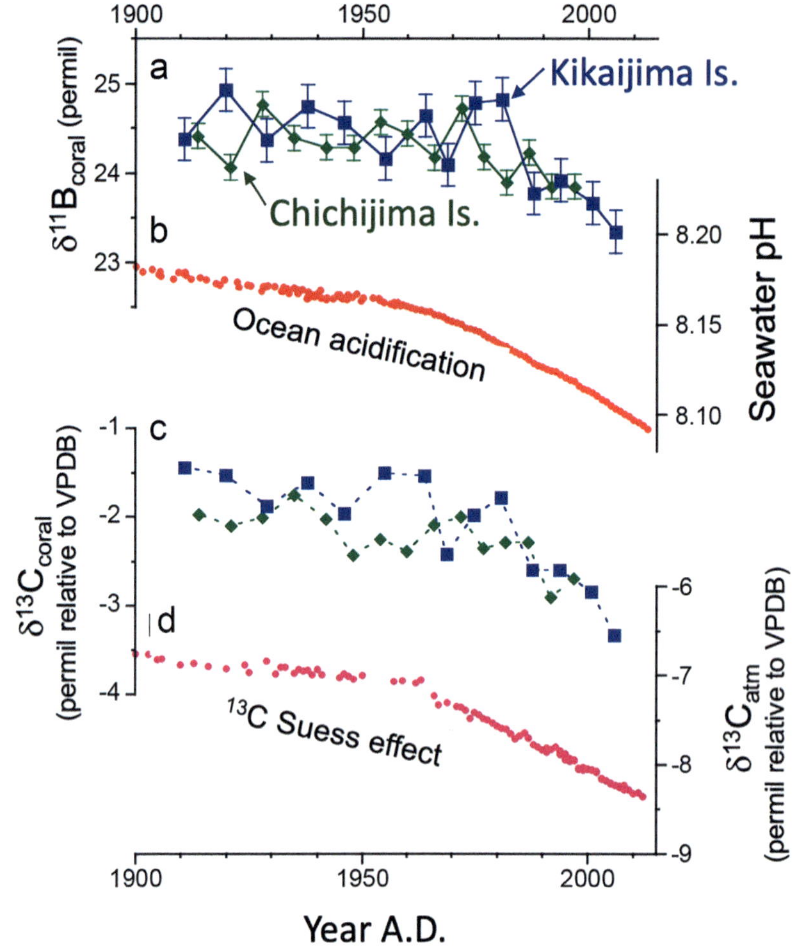

Fig. 6.2 Ocean acidification trends reconstructed from corals from the Japanese Islands and global indices (Kubota et al. 2017). (**a**, **c**) Boron and carbon stable isotope ratios ($\delta^{11}B$ and $\delta^{13}C$) of corals from Chichijima Island (green diamonds) and Kikaijima Island (blue squares). (**b**) Changes in seawater pH since the preindustrial era. (**d**) Changes in the atmospheric stable carbon isotope ratio ($\delta^{13}C_{atm}$) showing the ^{13}C Suess effect. From 1981 to 2012, $\delta\ ^{13}C_{atm}$ was measured at the Mauna Loa Observatory, Hawaii; values before 1981 were reconstructed from air trapped in the Antarctic ice sheet

The third-generation "outside AICAL system" is a simple system for outdoor installation (Fig. 6.7). It combines a five-channel gas-permeable membrane CO_2 sensor (flow-through tandem vapor–liquid equilibrator, Saito et al. 1995; Kayanne et al. 1995) with an automatic switching device and a small pCO_2 measurement device to reduce costs. The system can be scaled up to a seawater capacity of 200 L or more, and it takes advantage of the improved light conditions of outdoor aquariums (Kavousi et al. 2016). We plan to use this system for various experiments aimed at predicting ecosystem changes. When seawater with a lower pCO_2 than that of the original water is required, diluted sodium hydroxide (NaOH) solution is added.

6.4 Impacts of Ocean Acidification on Reef-Building Corals

Our research group has been conducting experiments to evaluate species-level ocean acidification responses in subtropical corals, benthic foraminifera, and calcareous algae at Sesoko Station (Table 6.2). The early life history of corals comprises spawning, fertilization, metamorphosis into larvae, settlement, acquisition of symbiotic algae by early polyps, and initiation of calcification. Experiments on *Acropora digitifera*, a common coral species around the Ryukyu Islands, have revealed that ocean acidification negatively affects sperm motility (Morita et al. 2010). Experiments conducted with a simple pH control system (Fig. 6.4) showed that a decrease in seawater pH to 7.3 had no effect on the survival of larvae in this coral species, but it was associated with a significant decrease in early polyp growth and symbiont acquisition (Suwa et al. 2010b). Subsequently, Nakamura and Morita (2012), using a first-generation AICAL system (27.1 °C, pCO_2 of 225–280, 330–395, or 790–945 µatm), confirmed that ocean acidification significantly reduced sperm motility in this coral.

Ohki et al. (2013), also using a first-generation AICAL system, conducted a 10-day exposure experiment in which they reared early *A. digitifera* polyps that were either infected or uninfected with a common green alga in seawater at 27.1 °C and five pCO_2 levels: preindustrial (279 µatm), present-day (391 µatm), and three assumed future levels (621, 842, and 1028 µatm) (Figs. 6.8 and 6.9). They found a significant decrease in polyp weight, a proxy for calcification, between preindustrial or present-day conditions and two of the

Table 6.1 Summary of experimental devices used to evaluate the effects of ocean acidification on marine calcifying organisms such as reef-building corals by the AICAL research group. A series of high-precision pCO_2-adjusted seawater generators (AICAL systems) were developed that supply CO_2 gas to the experimental seawater from cylinders of liquefied carbon dioxide

	Description (main components)	Features	References
Simple pH control system	Experimental seawater preparation system using a pH electrode and solenoid valve for automatic CO_2 injection control (pH electrode, solenoid valve, mixing tank, etc.)	Easy to use, but control accuracy is somewhat low. Suitable for small organisms and small-volume rearing experiments	Morita et al. (2010) Suwa et al. (2010b) Nakamura et al. (2011)
First-generation AICAL system (indoor AICAL system)	Generates equilibrated experimental seawater by injecting air with a set CO_2 concentration, directly measured by NDIR (CO_2 dissolution tower and equilibrium tower, mass flow controller, heatless dryer, humidification device to prevent clogging of air stones, etc.)	Can be used to set up to 5 pCO_2 levels. Control seawater supply rate to each tank is ~0.5 mL min^{-1}. pCO_2 can be set to below the atmospheric value. Suitable for small organisms and small-volume rearing experiments	Fujita et al. (2011) Hikami et al. (2011) Iguchi et al. (2012, 2014, 2015) Nakamura and Morita (2012) Ohki et al. (2013) Kato et al. (2014) Tanaka et al. (2014a) Sekizawa et al. (2017)
Second-generation AICAL system	Generates experimental seawater by adjusting the amount of concentrated CO_2 seawater added (CO_2 dissolution tower, mixing tank, metering pump, gas permeable membrane sensor, NDIR, etc.)	Up to 6 pCO_2 levels. The controlled seawater supply rate to each tank is about 5 L min^{-1}, and the pCO_2 in six tanks is continuously measured. Equipped with alkali-addition apparatus. Ideal for evaluating the combined effects of temperature and acidification. For large-volume rearing experiments	Nishida et al. (2015)
Third-generation AICAL system (outdoor AICAL system)	Generates experimental seawater by adjusting the amount of concentrated CO_2 seawater added (CO_2 dissolution tower, NDIR, mixing tank, fixed-volume pump, gas-permeable membrane sensor, automatic equilibrator switching device, small CO_2-partial-pressure measuring device)	Up to 5 levels. Controlled seawater supply rate is about 5 L min^{-1} to each tank. Alkali addition apparatus attached. Long-term large-volume experimental version of the second-generation AICAL system (shallow water recruitment experiments, etc.)	Kavousi et al. (2016)

NDIR non-dispersive infrared analyzer

assumed future conditions (842 and 1028 μatm), regardless of whether the polyps were infected with symbiotic algae. However, in four of five *A. digitifera* branch fragments, they found significant differences only between the preindustrial level and the four other levels (Fig. 6.10). These results suggest that calcification in *A. digitifera* is already significantly decreased in modern corals compared with the preindustrial period. Similarly, Iguchi et al. (2014) found a decrease in calcification since the Industrial Revolution, based on the assumed preindustrial pCO_2 level, in experiments with two other species of coral, *Porites australiensis* and *Isopora palifera*. Thus, coral calcification may already have decreased due to increasing atmospheric CO_2. Alternatively, there is interspecies variation in coral calcification responses at high pCO_2 for the future (Takahashi and Kurihara 2013), and these responses must be interpreted carefully.

Kuroyanagi et al. (2009), who acidified seawater by adding HCl, reported a decrease in foraminiferal calcification in the reef-dwelling benthic foraminifer *Marginopora kudakajimensis* with symbiotic algae. Subsequently, Fujita et al. (2011) found a decrease in foraminiferal calcification in experiments conducted with the first-generation AICAL system at a water temperature of 27.5 °C and five different pCO_2 levels, ranging from 261 μatm, mimicking preindustrial conditions, to 972 μatm, a possible future level. They observed a similar trend in *Amphisorus hemprichii*, a foraminifer species closely related to *M. kudakajimensis*. Kato et al. (2014) studied calcification in calcareous algae (*Lithophyllum kotschyanum*, *Hydrolithon samoense*), which are common in the Ryukyu Islands, by conducting a three-stage ocean acidification experiment using the first-generation AICAL system at 27.3 °C and pCO_2 levels from 263 μatm, mimicking preindustrial conditions, to 960 μatm, representing a possible future level.

6.5 Assessing the Combined Effects of Ocean Acidification and Increased Water Temperature on Reef-Building Corals

Because the first-generation AICAL system can supply four streams of pCO_2-regulated seawater, experiments that investigate the combined effects of ocean acidification and global warming can be conducted. Iguchi et al. (2015) examined the

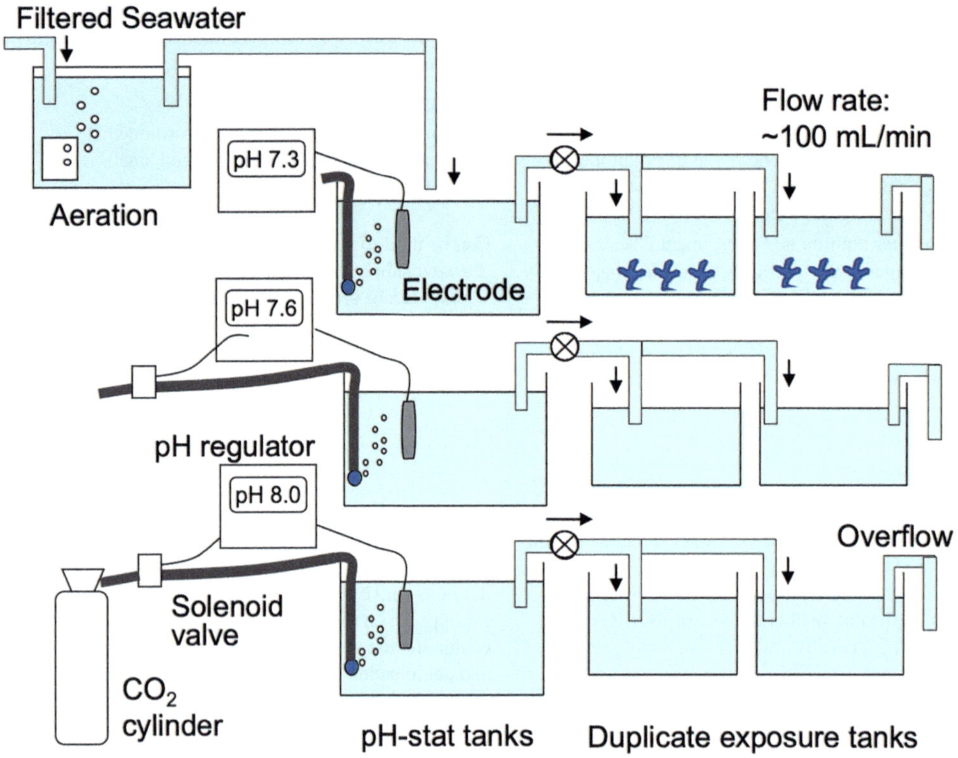

Fig. 6.3 Simple pH control system for experimental seawater preparation. The CO_2 injection volume is automatically controlled by pH electrodes and solenoid valves. It is designed for small-volume rearing experiments of small organisms

combined effects of increased water temperature and ocean acidification on the fertilization rate of *A. digitifera* corals, and they reported that an increase of approximately 4 °C in water temperature caused a significant decrease in the fertilization rate, whereas an increase in pCO_2 to about 1000 μatm had no significant effect. Earlier, Morita et al. (2010) had observed the lack of an ocean acidification effect on the fertilization rate of *A. digitifera* when the sperm concentration was somewhat high, although they found that acidified seawater caused a decrease in sperm flagellar motility. At high water temperatures, however, they observed a reduced fertilization rate even when a high sperm concentration was maintained. These results suggest that global warming effects are a greater risk to coral fertilization than ocean acidification. Considering the decrease in coral cover, it is conceivable that future increases in water temperature and ocean acidification will lead to a decrease in coral fertilization rates.

In a unique study, Tanaka et al. (2014a) examined the combined effects of ocean acidification and eutrophication on early *A. digitifera* coral polyps. Primary polyps were reared for 4 days in pretreatment tanks with high nutrient (HN treatment: $NO_3^- = 3$ μmol L^{-1}; $PO_4^{3-} = 0.2$ μmol L^{-1}) or low nutrient (LN treatment: $NO_3^- = 0.55$ μmol L^{-1}; $PO_4^{3-} = 0.02$ μmol L^{-1}) levels. Then, following exposure for another 10 days to high pCO_2 (92.5 Pa = 913 μatm) or low pCO_2 (38.8 Pa = 383 μatm) at a water temperature of 27.2 °C in the first-generation AICAL system, the calcification and chlorophyll contents of the polyps were measured. Polyps pretreated with low nutrient concentrations showed a significant decrease in calcification at high pCO_2, whereas those that had experienced high nutrient concentrations showed no decrease in calcification. The amount of oxygen produced by photosynthesis increased in all treatments. The increase in photosynthesis may have made the seawater more alkaline and led to an increase in calcification; alternatively, it may have mitigated the calcification-inhibiting effects of ocean acidification by absorbing nutrients and increasing symbiotic algal growth. The results of this experiment suggest that the nutrient environment may influence the sensitivity of early *A. digitifera* coral polyps to ocean acidification. In addition, various nutrient impact assessment experiments have been conducted in corals (Tanaka et al. 2013, 2015, 2017) and calcareous algae (Tanaka et al. 2016) at Sesoko Station (Table 6.2).

In the second-generation AICAL system, pCO_2 is continuously measured in up to six rearing tanks, and the actual pCO_2 can be obtained even when the water temperature is adjusted; therefore, it is ideal for evaluating the combined effects of water temperature and pCO_2. Nishida et al. (2015) used a second-generation AICAL system (Fig. 6.6) at the Demonstration Laboratory of the Marine Ecology Research Institute to study the temperature dependence of shell

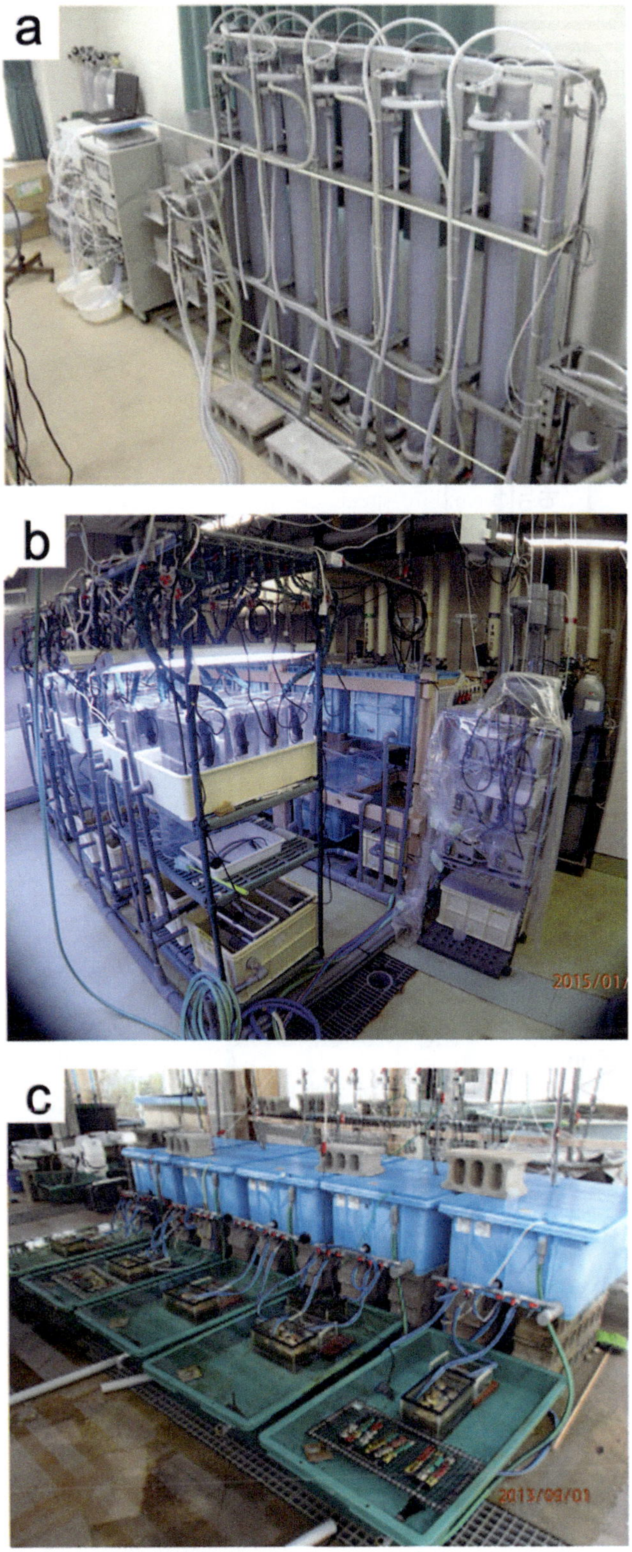

Fig. 6.4 Photographs of the (**a**) first-, (**b**) second-, and (**c**) third-generation AICAL systems. First- and third-generation AICAL systems are installed at Sesoko Station, Tropical Biosphere Research Center, University of the Ryukyus (Motobu, Okinawa Prefecture, Japan) and the second-generation AICAL system is installed at the Demonstration Laboratory of the Marine Ecology Research Institute (Kashiwazaki, Niigata Prefecture, Japan)

microstructure in the red clam (*Scapharca broughtonii*). They reported that the stable carbon isotope ratios of the shells were almost constant even though the CO_2 gas added to adjust pCO_2 values caused the carbon isotope ratio of dissolved inorganic carbon in the seawater to differ greatly among the experimental tanks. This suggests that the carbon source used for calcification is largely derived from food, rather than carbonic acid species in seawater (Nishida et al. 2015).

6.6 Inter- and Intraspecies Differences in the Responses of Reef-Building Corals to Ocean Acidification

Experiments on adult colonies of various coral species have shown that ocean acidification significantly reduces the calcification rate in *Porites* and *Isopora* corals, but not in *Acropora digitifera* (Iguchi et al. 2012, 2014; Ohki et al. 2013). These results suggest that different species show varied tolerance to ocean acidification.

Bell et al. (2022) examined the effect of five levels of pCO_2-adjusted seawater on the calcification rates of six coral species that are commonly found in the subtropical Ryukyu Islands: *Acropora nasuta*, *A. tenuis*, *Montipora digitata*, *Pocillopora damicornis*, *Porites cylindrica*, and *Galaxea fascicularis* (Fig. 6.11). They used a first-generation AICAL system to prepare pCO_2-adjusted seawater and cultured coral colonies in a common-garden manner, in which all coral species were bathed in the same tanks. The pCO_2 settings covered a relatively wide range from a preindustrial level of 300 μatm to a near-future level of 1200 μatm. In most species, calcification rates were observed to decline at the highest pCO_2 level; however, the coral species showed significantly different responses to acidified seawater, which may be attributable to interspecies differences in the physiological performance of coral–algal holobionts.

Species-wide differences in ocean acidification responses have also been observed in coral reef-dwelling benthic foraminifera. Ocean acidification experiments conducted in a first-generation AICAL system have shown that calcification was reduced in *Marginopora kudakajimensis*, but not in *Baculogypsina sphaerulata* or *Calcarina gaudichaudii* (Fujita et al. 2011; Hikami et al. 2011). These different responses may be attributable to differences in symbiotic algae or the calcification mechanism among these species.

Sekizawa et al. (2017) examined intraspecific variation in the ocean acidification responses in two branching coral species. Coral fragments prepared from 11 colonies of *Montipora digitata* and 6 colonies of *Porites cylindrica* were reared in control (pCO_2 = ~400 μatm) and high-pCO_2 (900 μatm) treatments in a first-generation AICAL

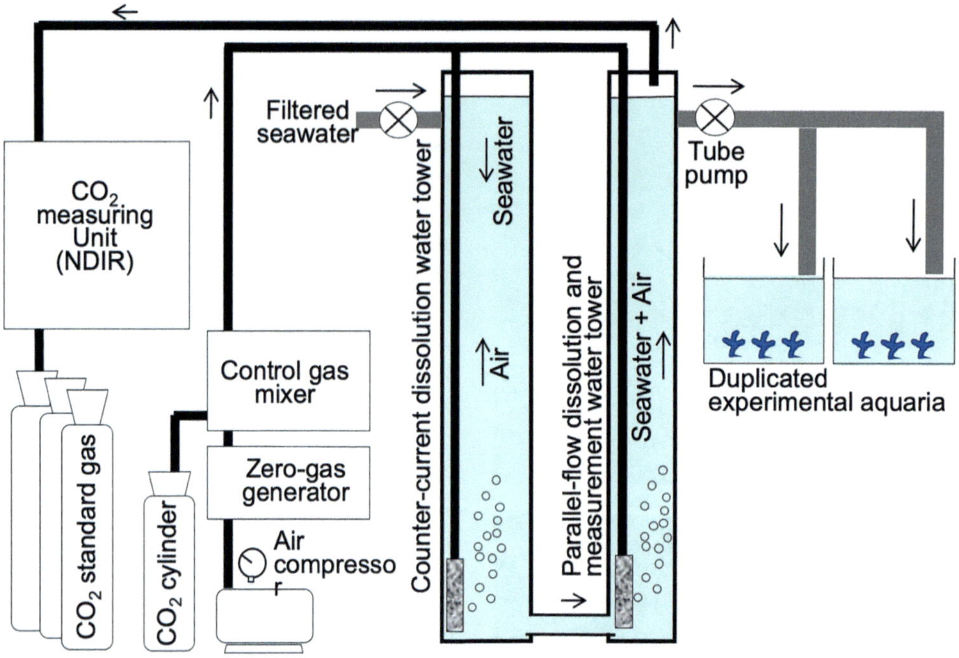

Fig. 6.5 First-generation AICAL system, also known as the "indoor AICAL system." This experimental seawater preparation system directly measures the CO_2 concentration with a non-dispersive infrared spectrometer (NDIR) and injects air controlled to the target value to achieve CO_2 equilibrium, resulting in high control accuracy. The system is characterized by sets of two dissolution towers, a counter-current dissolution tower and a parallel-flow dissolution tower. A series of five sets of towers can generate experimental seawater with five different levels of pCO_2: four levels adjusted to target pCO_2 values and one level of seawater in CO_2 equilibrium with the atmosphere. The pCO_2 level can be set to below the atmospheric value. The supply rate of the pCO_2-controlled seawater is about 0.5 L min^{-1}, which is suitable for small-volume rearing experiments of small organisms

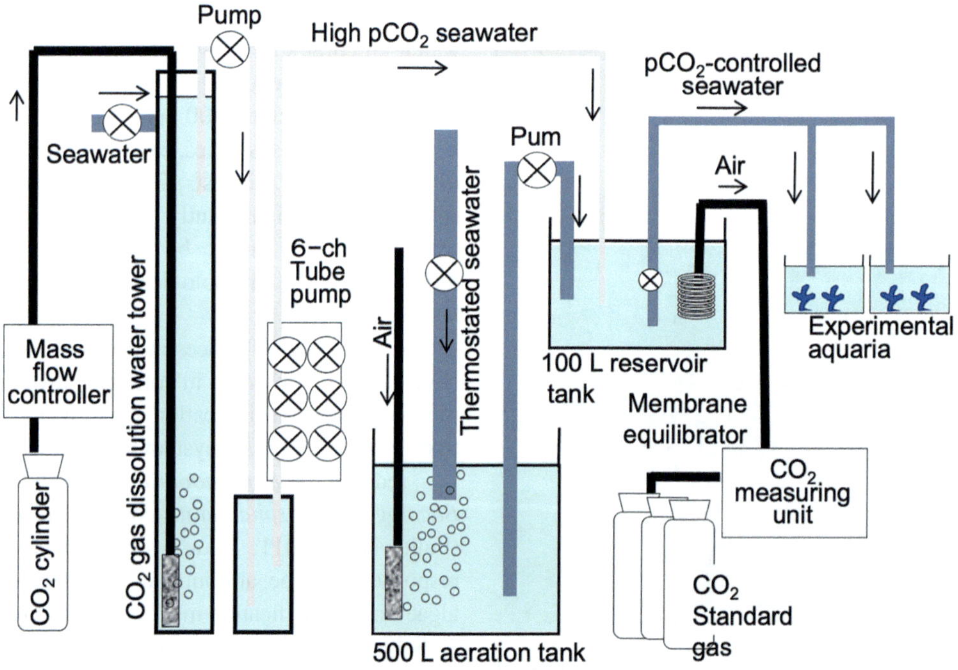

Fig. 6.6 Second-generation AICAL system. The system mixes with high CO_2 concentration with original seawater using a high-precision liquid delivery pump to produce experimental seawater with the desired pCO_2 values, and alkali can be added to set the seawater pCO_2 to below the atmospheric value. A maximum of six series of 100-L mixing tanks can be used, and each series can supply controlled seawater at a rate of about 5 L min^{-1}. The system is suitable for large-volume breeding experiments. Pre-aeration in a large water tank reduces and stabilizes the pCO_2 of the original seawater

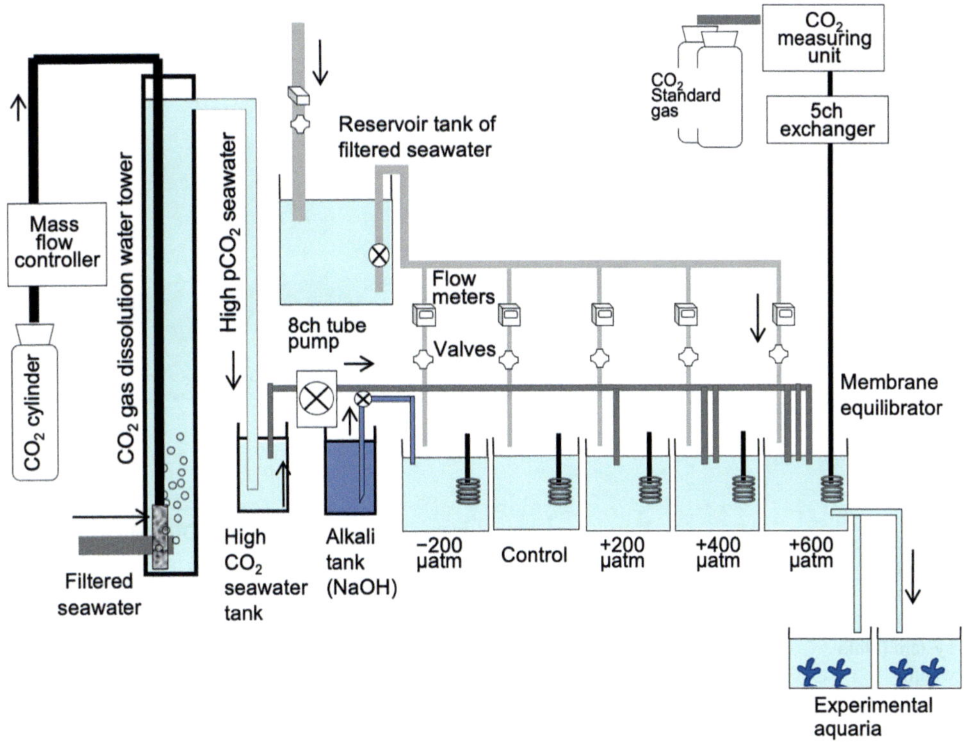

Fig. 6.7 Third-generation AICAL system for outdoor water tanks, also referred to as the "outdoor AICAL system," is a modified version of the second-generation AICAL system and is designed for long-term, large-volume experiments that can be conducted under outdoor light conditions. The water tank in which high CO_2 seawater and original seawater are mixed is enlarged compared with the second-generation system, and it also removes incoming silty sediments. The system can supply controlled seawater at a rate of approximately 5 L min^{-1} to each of a maximum of five levels. Addition of alkali allows for setting the pCO_2 level of the seawater below the atmospheric value

Table 6.2 Coral- and calcifying organism-rearing experiments conducted at Sesoko Station, Tropical Biosphere Research Center, University of the Ryukyus. The data are arranged by the environmental factors of interest. Results obtained using AICAL experimental systems (described in Table 6.1) are marked with an asterisk (*)

Environmental factor	Foraminifera (protozoan)	Corals (Phylum Cnidaria)		Calcareous algae (red algae)
		Gametes/polyps	Adults	
Temperature		Inoue et al. (2011)* Inoue et al. (2012)* Nishida et al. (2014b) Iwasaki et al. (2016) Bell et al. (2017)	Nishida et al. (2014a) Inoue et al. (2015)	
Ocean acidification	Fujita et al. (2011)* Hikami et al. (2011)*	Morita et al. (2010)* Suwa et al. (2010b)* Nakamura et al. (2011)* Nakamura and Morita (2012)* Ohki et al. (2013)* Inoue et al. (2015)* Tanaka et al. (2014a)*	Iguchi et al. (2012, 2014)* Kavousi et al. (2016)* Sekizawa et al. (2017)*	Kato et al. (2014)*
Salinity		Inoue et al. (2012)		
Nutrients		Tanaka et al. (2013)	Tanaka et al. (2015) Tanaka et al. (2017)	
Sediment-bound nutrients		Iijima et al. (2021)		
Seawater turbulence		Iwasaki et al. (2018)		
Temperature × ocean acidification		Iguchi et al. (2015)*		
Temperature × nutrients			Tanaka et al. (2014b)	Tanaka et al. (2016)
Ocean acidification × nutrients		Tanaka et al. (2014a)*		

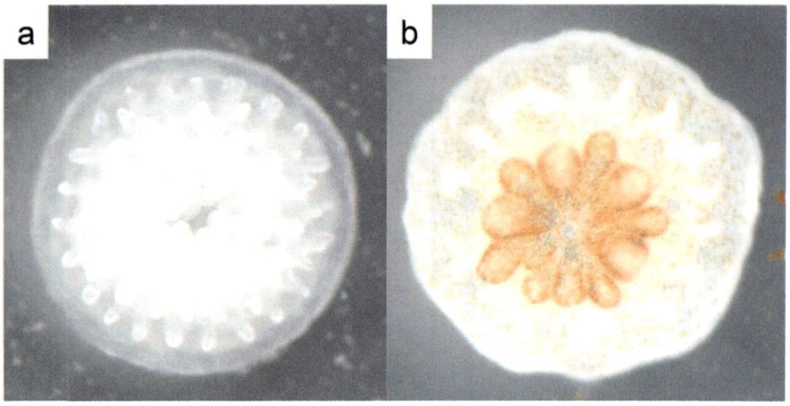

Fig. 6.8 Aposymbiotic (**a**) and symbiotic (**b**) primary polyps of *Acropora digitifera*. Brown granular material are symbiotic algae. After settlement, the polyps are infected with the symbiotic algae

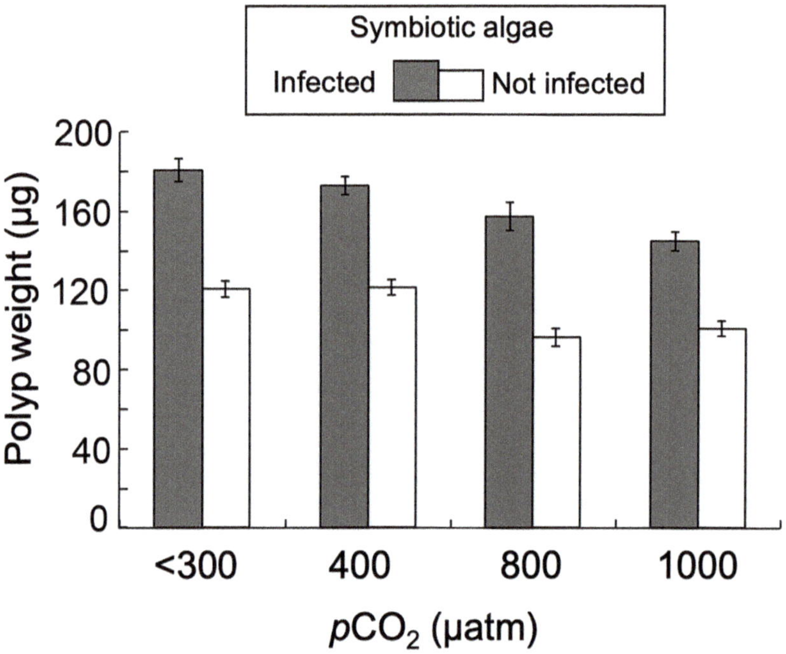

Fig. 6.9 Results of an experiment to evaluate the effects of ocean acidification on the primary polyps of *Acropora digitifera* using the first-generation AICAL system (Ohki et al. 2013). Polyp weight decreased with increasing pCO_2 in both polyps with and without symbiotic algal infection; however, symbiosis increased coral tolerance to ocean acidification

system; in both treatments, water temperature was ~27 °C. Overall calcification rates tended to decrease in the high-pCO_2 treatment in both species, but the differences were not statistically significant. However, within each coral species, the response to ocean acidification differed markedly. This result suggests that some local populations may be more tolerant of ocean acidification than others owing to intraspecies variation of tolerance. Kavousi et al. (2016) conducted a similar experiment in 15 colonies of *Montipora digitata* corals (water temperature 26.4–26.7 °C; control treatment, pCO_2 ~ 425 µatm; high-pCO_2 treatment, 786 µatm) in an outdoor AICAL system (Fig. 6.7). Their results differed from those of Sekizawa et al. (2017) in that there was a significant overall trend toward lower calcification rates in the high-pCO_2 treatment, but they were similar in that they also observed large intraspecific variability in the ocean acidification response among the colonies. Such intraspecific variation in the ocean acidification response of corals may support the adaptive potential of coral populations in future high CO_2 ocean (e.g., Kurihara et al. 2021).

Nakamura et al. (2017) evaluated the calcification and photosynthetic efficiency in the coral *Stylophora pistillata* to investigate the effect of ocean acidification on recovery from damage caused by intense light exposure. They used a first-generation AICAL system (~27 °C and five pCO_2 levels: 304, 413, 521, 668, and 800 µatm). In all three experimental groups, they found no decrease in the amount of calcification as the seawater pCO_2 increased. Therefore, at least some colonies of this species can tolerate ocean acidification with pCO_2 up to about 800 µatm. In addition, ocean acidification was not observed to affect the recovery of the coral from damage caused by high light exposure.

Sekizawa et al. (2017) and Nakamura et al. (2017) found variation in the response to progressively more acidic

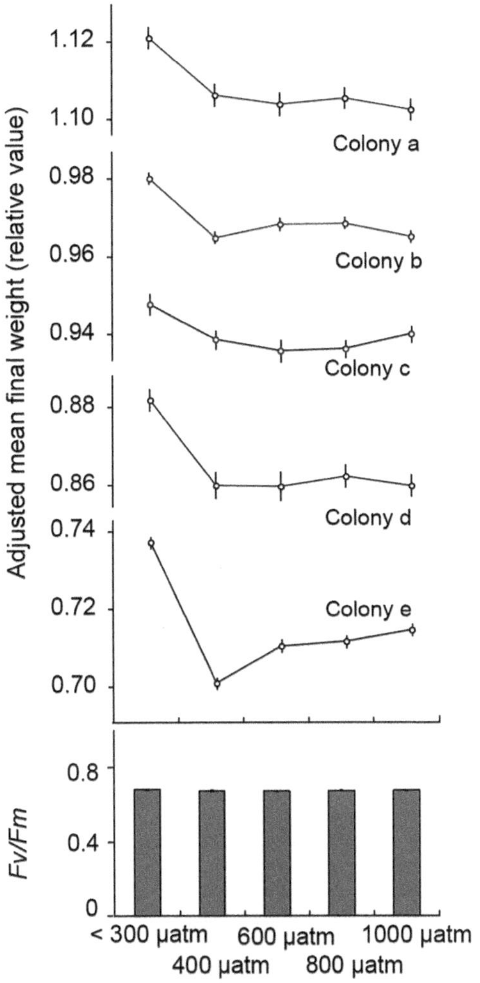

Fig. 6.10 Responses of adult corals to ocean acidification (Ohki et al. 2013). Adjusted mean final weights of coral fragments and mean *Fv/Fm* values from five colonies (colonies **a**–**e**) of *Acropora digitifera* subjected to five pCO_2 treatments: (1) preindustrial pCO_2, <300 µatm; (2) present-day pCO_2, 400 µatm, and three near-future pCO_2 conditions, (3) 600 µatm, (4) 800 µatm, and (5) 1000 µatm. Bars show ±S.E

seawater among coral species, and they also showed that different genotypes exist within coral species differed in their tolerance to ocean acidification. These results suggest that future ocean acidification may lead to changes in the species composition and intraspecific phylogeny of coral reef communities. This finding is expected to have applications in predicting changes to coral communities in response to future ocean acidification and in coral transplantation, a coral community restoration method.

Past and future changes in the ocean according to the latest prediction (IPCC 2019) and summary results of coral calcification from AICAL experiments are compared in Fig. 6.12. Although there are non-negligible interspecific variations in tolerance to ocean acidification, our results suggest that a reduction of coral calcification may become evident after the mid-twenty-first century.

6.7 Ocean Acidification, Coral Skeletal Composition, and Calcification Mechanisms

To evaluate how calcifying organisms will respond to future ocean acidification, interdisciplinary and comprehensive studies must be performed on the mechanisms of calcium carbonate production by individual calcifying organisms. In corals, skeletal formation takes place in a calcifying matrix sandwiched between the osteoblasts and the skeleton. During skeletal formation, hydrogen ions, which inhibit skeletal formation, must be removed. Ohno et al. (2017a, 2017b) developed a non-destructive fluorescence method for observing the calcification process of early coral polyps that uses confocal microscopy and low-toxicity fluorescent indicators. Upon adding a pH-sensitive fluorescent dye to the seawater, they could visualize and measure detailed pH changes at the skeletal formation site. In addition, they visualized the arrival of ambient seawater at the skeletal formation site. Accordingly, the group suggested that the active pH adjustment mechanism and the diversity of seawater inflow processes to the skeletal formation site may determine how resistant calcifying organisms are to ocean acidification.

Inoue et al. (2015) conducted culture experiments in *Porites australiensis* and measured $^{44}Ca/^{40}Ca$ abundance ratios in their skeletons among three different treatments. They postulated that calcium was actively transported via scleroblasts (bone-forming cells) to the calcification medium. They also inferred that the calcification rate depended on the carbonate species. Calcium and boron isotope ratios can be measured by thermal ionization mass spectrometry or by multi-collector inductively coupled plasma mass spectrometry, which is a novel analytical method in the field of geochemistry. It is also important to investigate whether boron isotope ratios and U/Ca ratios (Inoue et al. 2011) of skeletal samples obtained from ocean acidification experiments can be effective pH indicators for skeletal formation sites.

Pioneering studies on ocean acidification, such as those of Kleypas et al. (1999) and Orr et al. (2005), have considered Ω to be the primary parameter controlling the biological response to ocean acidification. However, more recent studies have suggested that when $\Omega > 1$, Ω may not be an appropriate indicator of the calcifying environment (Bach 2015; Cyronak et al. 2016a, 2016b; Thomsen et al. 2015; Waldbusser et al. 2016; Fassbender et al. 2016). Instead, calcification in some species may be regulated by the ratio of bicarbonate to hydrogen ions, referred to as the substrate-to-inhibitor (SIR) ratio:

$$SIR = [HCO_3^{\,-}]/[H^+], \qquad (6.2)$$

where $[HCO_3^-]$ and $[H^+]$ are the bicarbonate ion and hydrogen ion concentrations. Interestingly, SIR does not show the

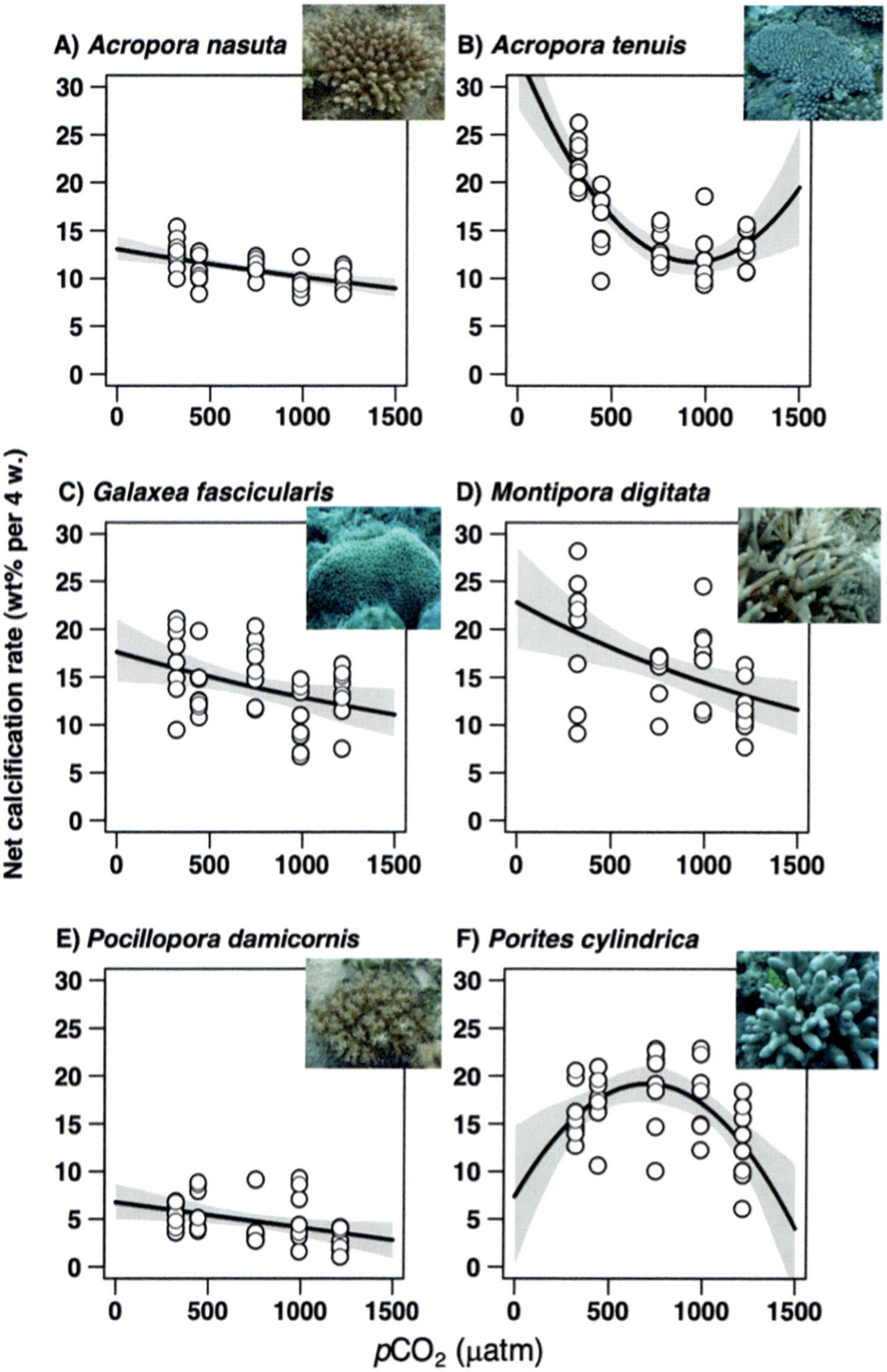

Fig. 6.11 The effect of five levels of pCO_2-adjusted seawater on calcification rates in six common coral species in the subtropical Ryukyu Islands: *Acropora nasuta*, *A. tenuis*, *Montipora digitata*, *Pocillopora damicornis*, *Porites cylindrica*, and *Galaxea fascicularis* (Bell et al. 2022). With permission from Journal of the Japanese Coral Reef Society: Bell et al. (2022) Galaxea J Coral Reef Stud 24: 63–68, Fig. 2

same temperature dependency as Ω and $[CO_3^{2-}]$, both of which show a strong poleward decline (Bach 2015). Therefore, Bach (2015) suggested that ocean acidification is a more globally uniform problem for marine calcifiers than was previously thought. Confirmation of the SIR hypothesis would support the view that ocean acidification is a global

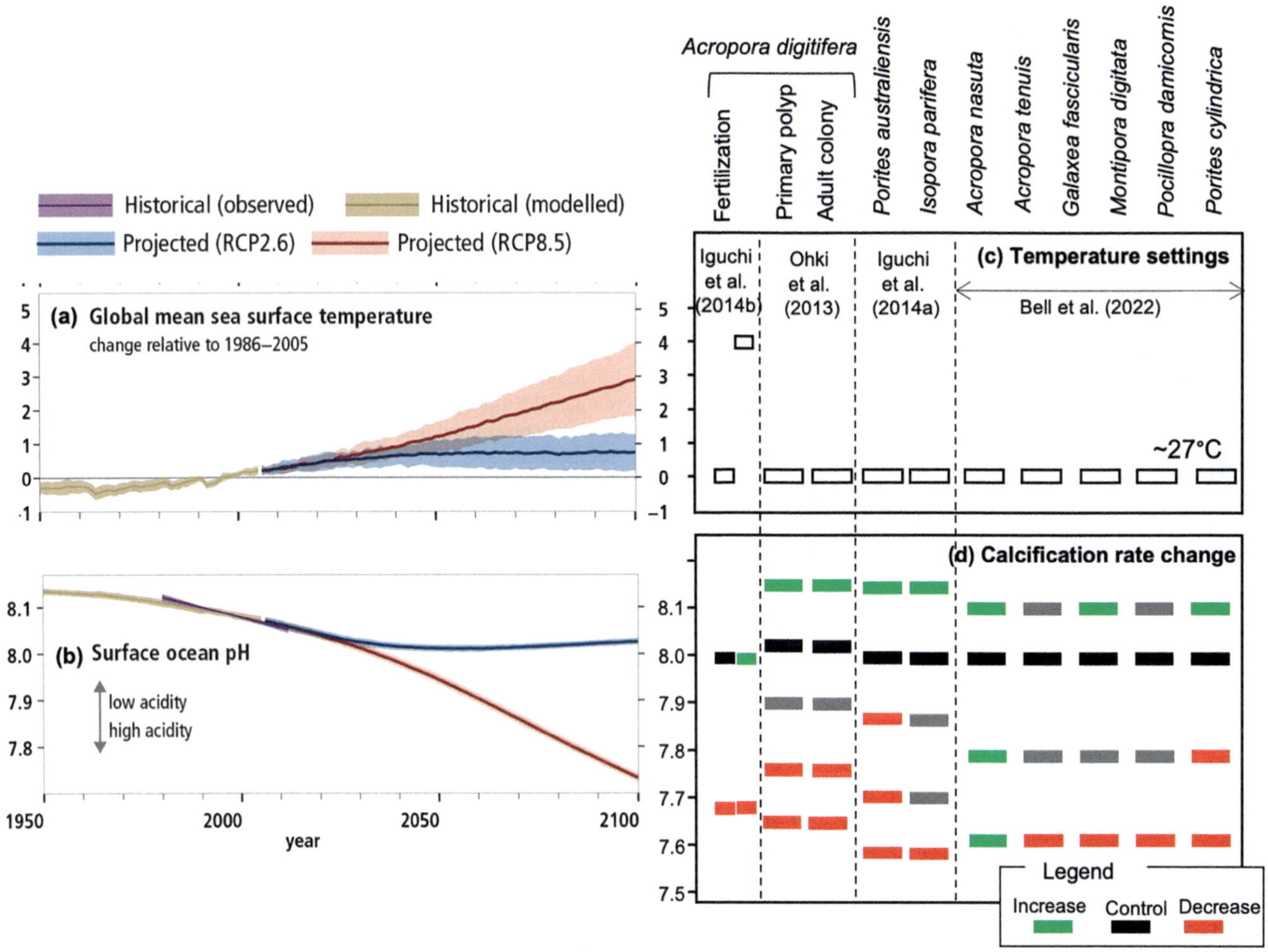

Fig. 6.12 Past and future changes in the ocean and summary results of coral calcification from AICAL experiments. Observed and modeled historical changes in the global mean sea surface temperature (**a**) and surface ocean pH in the total hydrogen ion scale (**b**) since 1950 (IPCC 2019). RCP2.6 and RCP8.5 represent Representative concentration pathways scenarios examined in IPCC reports (IPCC 2019, 2021). (**c**) Temperature settings of coral experiments with the AICAL device. For *Acropora digitifera* polyp experiments, results of uninfected with symbiotic algae are shown (Ohki et al. 2013). (**d**) Calcification rate changes at different seawater pH (the total hydrogen ion scale) detected by coral experiments using the AICAL device. More than 10% increase or decrease compared to the control values are shown in green and red bar, respectively

problem that affects both low-latitude and high-latitude regions. The SIR hypothesis should be examined further in diverse marine calcifying species, including both reef-building corals and reef-dwelling foraminifera. In this effort, experiments conducted with the various AICAL systems should generate useful datasets accompanied by precise seawater CO_2 system parameters.

Recently, micro-computed tomography (microCT) has been introduced as a method for analyzing the skeletal growth of corals reared under controlled conditions. For example, three-dimensional morphological measurements by microCT have revealed the role of symbiotic algae in the formation of coral polyp skeletons (Iwasaki et al. 2016; Fig. 6.13). In addition, Iwasaki et al. (2018) employed microCT to evaluate the influence of seawater turbulence on the formation of coral primary polyp skeletons and observed that seawater turbulence increases the thickness of the skeletal branches. This method has also been applied to reef-dwelling foraminifera (Kuroyanagi et al. 2021). Therefore, microCT data can provide new opportunities to test hypotheses regarding the effects of ocean acidification on biological calcification.

6.8 Future Prospects

In this review, we have introduced some of the studies on corals and other calcifying organisms being conducted by members of the AICAL project. In addition, another group of AICAL project members is actively evaluating the effects of ocean acidification on important fishery species such as abalone, horned turban, and sea urchins (Kimura et al. 2011;

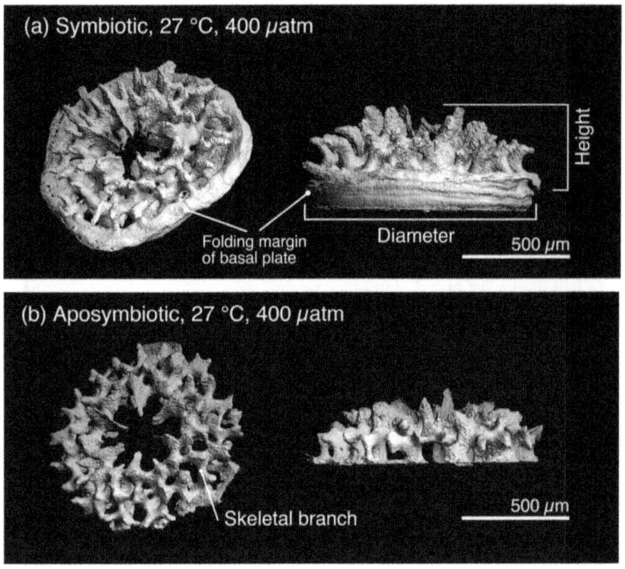

Fig. 6.13 Skeletal image of an initial *Acropora digitifera* polyp obtained by microCT (Iwasaki et al. 2016, 2018). Polyps (**a**) infected and (**b**) uninfected with symbiotic algae grown at a water temperature of 27 °C and a pCO_2 of 400 μatm

Onitsuka et al. 2014, 2018; Nishida et al. 2015, 2020). In the summer of 2016, the first large-scale bleaching event since 1998 occurred in coral reef systems around the Ryukyu Islands, and the decline of coral reef ecosystems is again attracting a great deal of attention (Afzal et al. 2023). In the future, we plan to promote research on assessing the effects of global warming and ocean acidification at species, community, and ecosystem levels using various types of AICAL devices. Our immediate task is to develop biological recruitment experiments as ecosystem experiments using outdoor AICAL devices. The AICAL device can also be used to simultaneously study ocean acidification and a variety of other anthropogenic effects. The combined effects of eutrophication and ocean acidification phenomena on corals have already been examined (Tanaka et al. 2014a), but it is also important to study the combined effects of herbicides and other hazardous chemicals (e.g., Flores et al. 2021; Kitada et al. 2008). Multi-interdisciplinary studies and research must be promoted by applying new techniques of molecular biology, cell physiology, genetic ecology, and advanced isotope ratio analysis methods.

Acknowledgments We thank Prof. Ichiro Takeuchi for providing us with this chance to introduce our work. Special thanks go to Professors Hodaka Kawahata and Takahiro Irie of the University of Tokyo, Takashi Nakamura, and Kazuhiko Fujita of the University of the Ryukyus, Mayuri Inoue of Okayama University, and Yasuaki Tanaka of Universiti Brunei Darussalam. This review discusses the outcomes of the following projects: "Experimental study on the effects of ocean acidification on calcifying organisms and their ecosystems" (A-0804, PI: Yukihiro Nojiri, National Institute for Environmental Studies; FY2008-2010) and "Experimental study on the combined effects of global warming and ocean acidification on marine organisms" (A-1203, PI: Yukihiro Nojiri, National Institute for Environmental Studies; FY2012-2014) of the Global Environment Research Fund of the Ministry of the Environment of Japan. This review was also supported by Grant-in-Aid for Scientific Research (KAKENHI) Program S for "Experiments to evaluate the effects of ocean acidification on coastal organisms and ecosystems" (PI: Yukihiro Nojiri, FY 2014-2018, 26220102), Program B for "Application of X-ray computed tomography on coral skeletal climatology" (PI: Atsushi Suzuki, FY 2018-2022, 18H03366), Program B for "Global environmental change reconstruction based on coral skeletal records around Japanese Islands" (PI: Hideki Takami, FY 2019-2022, 19H04288), Program A for "Unveiling the mechanism of phase-shift in coral reef ecosystem focusing on extracellular degradation system in corals" (PI: Akira Iguchi, FY 2020-2023, 20H00653), and Environment Research and Technology Development Fund of the Environmental Restoration and Conservation Agency of Japan for "Technical development and adaptation strategy to define local environmental threshold for coral reef conservation in an era of high CO_2 (PI: Akira Iguchi, FY 2019-2021, JPMEERF20194007)." Special thanks go to Kimoto Electric Co. Ltd. for technical assistance regarding the AICAL apparatus. I would like to acknowledge all of the people involved in the AICAL project. M. Iijima, K. Nishida, M. Hikami, T. Bell, K. Shimada, S. Ozaki, K. Shinmen, K. Ishikawa, C. Mori, and S. Kim conducted rearing experiments as part of their research for their doctoral or master degrees research works. We also thank Ryota Suwa, Hiromi Kinjyo, and the staff of the Sesoko Station, Tropical Biosphere Research Center, University of the Ryukyus (Motobu Town, Okinawa Prefecture), and the Demonstration Laboratory of the Marine Ecology Research Institute (Kashiwazaki City, Niigata Prefecture) for assisting with the culture experiments. This review was also supported by the Research Laboratory on Environmentally-conscious Developments and Technologies (E-code) at the National Institute of Advanced Industrial Science and Technology (AIST).

References

Afzal MS, Ikeda K, Ueno M, Nakamura T (2023) Dynamics of coral reef communities of Southern Japan under climate change: Sekisei Lagoon, Japan following the severe mass bleaching event of 2016. In: Takeuchi I, Yamashiro H (eds) Coral reefs of Eastern Asia under anthropogenic impacts. Coral reefs of the world, vol 17. Springer Nature Switzerland AG, Cham, pp 35–50

Bach LT (2015) Reconsidering the role of carbonate ion concentration in calcification by marine organisms. Biogeosciences 12:4939–4951

Bell T, Nishida K, Ishikawa K, Suzuki A, Nakamura T, Sakai K, Ohno Y, Iguchi A, Yokoyama Y (2017) Temperature-controlled culture experiments with primary polyps of coral *Acropora digitifera*: calcification rate variations and skeletal Sr/Ca, Mg/Ca, and Na/Ca ratios. Palaeogeogr Palaeoclimatol Palaeoecol 484:129–135

Bell T, Manullang C, Kumagai NH, Sakai K, Suzuki A, Iguchi A (2022) Calcification responses of subtropical corals to acidified seawater: a case study of Sesoko Island, Okinawa, Japan. Galaxea J Coral Reef Stud 24:63–68

Cyronak T, Schulz KG, Jokiel PL (2016a) Response to Waldbusser et al. (2016): "Calcium carbonate saturation state: on myths and this or that stories". ICES J Mar Sci 73:569–571

Cyronak T, Schulz KG, Jokiel PL (2016b) The omega myth: what really drives lower calcification rates in an acidifying ocean. ICES J Mar Sci 73:558–562

Fassbender AJ, Sabine CL, Feifel KM (2016) Consideration of coastal carbonate chemistry in understanding biological calcification. Geophys Res Lett 43:4467–4476

Flores F, Marques JA, Uthicke S, Fisher R, Patel F, Kaserzon S, Negri AP (2021) Combined effects of climate change and the herbicide diuron on the coral *Acropora millepora*. Mar Pollut Bull 169:112582

Fujita K, Hikami M, Suzuki A, Kuroyanagi A, Sakai K, Kawahata H, Nojiri Y (2011) Effects of ocean acidification on calcification of symbiont-bearing reef foraminifers. Biogeosciences 8:2089–2098

Gruber N, Clement D, Carter BR, Feely RA, van Heuven S, Hoppema M, Ishii M, Key RM, Kozyr A, Lauvset SK, Lo Monaco C, Mathis JT, Murata A, Olsen A, Perez FF, Sabine CL, Tanhua T, Wanninkhof R (2019) The oceanic sink for anthropogenic CO_2 from 1994 to 2007. Science 363:1193–1199

Hikami M, Ushie H, Irie T, Fujita K, Kuroyanagi A, Sakai K, Nojiri Y, Suzuki A, Kawahata H (2011) Contrasting calcification responses to ocean acidification between two reef foraminifers harboring different algal symbionts. Geophys Res Lett 38:19. https://doi.org/10.1029/2011GL048501

Hoegh-Guldberg O, Mumby PJ, Hooten AJ, Steneck RS, Greenfield P, Gomez E, Harvell CD, Sale PF, Edwards AJ, Caldeira K, Knowlton N, Eakin CM, Iglesias-Prieto R, Muthiga N, Bradbury RH, Dubi A, Hatziolos ME (2007) Coral reefs under rapid climate change and ocean acidification. Science 318:1737–1742

Iguchi A, Ozaki S, Nakamura T, Inoue M, Tanaka Y, Suzuki A, Kawahata H, Sakai K (2012) Effects of acidified seawater on coral calcification and symbiotic algae on the massive coral *Porites australiensis*. Mar Environ Res 73:32–36

Iguchi A, Kumagai NH, Nakamura T, Suzuki A, Sakai K, Nojiri Y (2014) Responses of calcification of massive and encrusting corals to past, present, and near-future ocean carbon dioxide concentrations. Mar Pollut Bull 89:348–355

Iguchi A, Suzuki A, Sakai K, Nojiri Y (2015) Comparison of the effects of thermal stress and CO_2-driven acidified seawater on fertilization in coral *Acropora digitifera*. Zygote 23:631–635

Iijima M, Yasumoto J, Iguchi A, Koiso K, Ushigome S, Nakajima N, Kunieda Y, Nakamura T, Sakai K, Yasumoto-Hirose M, Mori-Yasumoto K, Mizusawa N, Amano H, Suzuki A, Jimbo M, Watabe S, Yasumoto K (2021) Phosphate bound to calcareous sediments hampers skeletal development of juvenile coral. R Soc Open Sci 8:201214. https://doi.org/10.1098/rsos.201214

Inoue M, Suwa R, Suzuki A, Sakai K, Kawahata H (2011) Effects of seawater pH on growth and skeletal U/Ca ratios of *Acropora digitifera* coral polyps. Geophys Res Lett 38:L12809. https://doi.org/10.1029/201GL04786

Inoue M, Shinmen K, Kawahata H, Kawahata H, Nakamura T, Tanaka Y, Kato A, Shinzato C, Iguchi A, Kan H, Suzuki A, Sakai K (2012) Estimate of calcification responses to thermal and freshening stresses based on culture experiments with symbiotic and aposymbiotic primary polyps of a coral *Acropora digitifera*. Glob Planet Change 92–93:1–7

Inoue M, Gussone N, Koga Y, Iwase A, Suzuki A, Sakai K, Kawahata H (2015) Controlling factors of Ca isotope fractionation in scleractinian corals evaluated by temperature, pH and light controlled culture experiments. Geochim Cosmochim Acta 167:80–92

IPCC (2019) Summary for policymakers. In: Pörtner H-O, Roberts DC, Masson-Delmotte V, Zhai P, Tignor M, Poloczanska E, Mintenbeck K, Alegría A, Nicolai M, Okem A, Petzold J, Rama B, Weyer NM (eds) IPCC special report on the ocean and cryosphere in a changing climate. Cambridge University Press, Cambridge, New York, pp 3–35

IPCC (2021) Summary for policymakers. In: Masson-Delmotte V, Zhai P, Pirani A, Connors SL, Péan C, Berger S, Caud N, Chen Y, Goldfarb L, Gomis MI, Huang M, Leitzell K, Lonnoy E, Matthews JBR, Maycock TK, Waterfield T, YelekçI O, Yu R, Zhou B (eds) Climate change 2021: the physical science basis. Contribution of Working Group I to the sixth assessment report of the intergovernmental panel on climate change. Cambridge University Press

Ishii M, Kosugi N, Sasano D, Saito S, Midorikawa T, Inoue HY (2011) Ocean acidification off the south coast of Japan: a result from time series observations of CO_2 parameters from 1994 to 2008. J Geophys Res 116:C06022. https://doi.org/10.1029/2010JC006831

Iwasaki S, Inoue M, Suzuki A, Sasaki O, Kano H, Iguchi A, Sakai K, Kawahata H (2016) The role of symbiotic algae in the formation of the coral polyp skeleton: 3-D morphological study based on X-ray microcomputed tomography. Geochem Geophys Geosyst 17:3629–3637

Iwasaki S, Suzuki A, Iguchi A, Sasaki O, Kano H, Ohno Y, Enomoto K (2018) Effect of seawater turbulence on formation of coral primary polyp skeletons. Coral Reefs 37:939–944

Japan Meteorological Agency (2021) Annual changes in oceanic and atmospheric carbon dioxide (CO_2) in the western North Pacific at https://www.data.jma.go.jp/gmd/kaiyou/english/co2_trend/co2_trend_en.html. Accessed 28 Sept 2021

Kato A, Hikami M, Kumagai NH, Suzuki A, Nojiri Y, Sakai K (2014) Negative effects of ocean acidification on two crustose coralline species using genetically homogeneous samples. Mar Environ Res 94:1–6

Kavousi J, Tanaka Y, Nishida K, Suzuki A, Nojiri Y, Nakamura T (2016) Colony-specific calcification and mortality under ocean acidification in the branching coral *Montipora digitata*. Mar Environ Res 119:161–165

Kawahata H, Fujita K, Iguchi A, Inoue M, Iwasaki S, Kuroyanagi A, Maeda A, Manaka T, Moriya K, Takagi H, Toyofuku T, Yoshimura T, Suzuki A (2019) Perspective on the response of marine calcifiers to global warming and ocean acidification—behavior of corals and foraminifera in a high CO_2 world "hot house". Prog Earth Planet Sci 6:5. https://doi.org/10.1186/s40645-018-0239-9

Kayanne H, Suzuki A, Saito H (1995) Diurnal changes in the partial pressure of carbon dioxide in coral reef water. Science 269:214–216

Kimura RYO, Takami H, Ono T, Onitsuka T, Nojiri Y (2011) Effects of elevated pCO_2 on the early development of the commercially important gastropod, Ezo abalone *Haliotis discus hannai*. Fish Oceanogr 20:357–366

Kitada Y, Kawahata H, Suzuki A, Oomori T (2008) Distribution of pesticides and bisphenol A in sediments collected from rivers adjacent to coral reefs. Chemosphere 71:2082–2090

Kleypas JA, Buddemeier RW, Archer A, Gattuso J-P, Langdon C, Opdyke BN (1999) Geochemical consequences of increased atmospheric carbon dioxide on coral reefs. Science 284:118–120

Kubota K, Yokoyama Y, Ishikawa T, Suzuki A, Ishii M (2017) Rapid decline in pH of coral calcification fluid due to incorporation of anthropogenic CO_2. Sci Rep 7:7694. https://doi.org/10.1038/s41598-017-07680-0

Kurihara H (2008) Effects of CO_2-driven ocean acidification on the early developmental stages of invertebrates. Mar Ecol Prog Ser 373:275–284

Kurihara H, Wouters J, Yasuda N (2019) Seasonal calcification of the coral *Acropora digitifera* from a subtropical marginal Okinawa reef under ocean acidification. Coral Reefs 38:443–454

Kurihara H, Watanabe A, Tsugi A, Mimura I, Hongo C, Kawai T, Reimer JD, Kimoto K, Golbuu Y (2021) Potential local adaptation of corals at acidified and warmed Nikko Bay, Palau. Sci Rep 11:11192. https://doi.org/10.1038/s41598-021-90614-8

Kuroyanagi A, Kawahata H, Suzuki A, Fujita K, Irie T (2009) Impacts of ocean acidification on large benthic foraminifers: results from laboratory experiments. Mar Micropaleontol 73:190–195

Kuroyanagi A, Irie T, Kinoshita S, Kawahata H, Suzuki A, Nishi H, Sasaki O, Takashima R, Fujita K (2021) Decrease in volume and density of foraminiferal shells with progressing ocean acidification. Sci Rep 11:19988. https://doi.org/10.1038/s41598-021-99427-1

Lui HK, Chen CTA, Lee J, Wang SL, Gong GC, Bai Y, He X (2015) Acidifying intermediate water accelerates the acidification of seawater on shelves: an example of the East China Sea. Cont Shelf Res 111:223–233

Morita M, Suwa R, Iguchi A, Nakamura M, Shimada K, Sakai K, Suzuki A (2010) Ocean acidification reduces sperm flagellar motility in broadcast spawning reef invertebrates. Zygote 18:103–107

Nakamura M, Morita M (2012) Sperm motility of the scleractinian coral *Acropora digitifera* under preindustrial, current, and predicted ocean acidification regimes. Aquat Biol 15:299–302

Nakamura M, Ohki S, Suzuki A, Sakai K (2011) Coral larvae under ocean acidification: survival, metabolism, and metamorphosis. PLoS One 6:e14521. https://doi.org/10.1371/journal.pone.0014521

Nakamura T, Iguchi A, Suzuki A, Sakai K, Nojiri Y (2017) Effects of acidified seawater on calcification, photosynthetic efficiencies and the recovery processes from strong light exposure in the coral *Stylophora pistillata*. Mar Ecol 38:e12444. https://doi.org/10.1111/maec.12444

Nishida K, Iguchi A, Ishimura T, Sakai K, Suzuki A (2014a) Skeletal isotopic responses of the Scleractinian coral *Isopora palifera* to experimentally controlled water temperatures. Geochem J 48:e9–e14. https://doi.org/10.2343/geochemj.2.0317

Nishida K, Ishikawa K, Iguchi A, Tanaka Y, Sato M, Ishimura T, Inoue M, Nakamura T, Sakai K, Suzuki A (2014b) Skeletal oxygen and carbon isotope compositions of *Acropora* coral primary polyps experimentally cultured at different temperatures. Geochem Geophys Geosyst 15:2840–2849

Nishida K, Suzuki A, Isono R, Hayashi M, Watanabe Y, Yamamoto Y, Irie T, Nojiri Y, Mori C, Sato K, Sato M, Sasaki T (2015) Thermal dependency of shell growth, microstructure, and stable isotopes in laboratory-reared *Scapharca broughtonii* (Mollusca: Bivalvia). Geochem Geophys Geosyst 16:2395–2408

Nishida K, Chew YC, Miyairi Y, Hirabayashi S, Suzuki A, Hayashi M, Yamamoto Y, Sato M, Nojiri Y, Yokoyama Y (2020) Novel reverse radioisotope labelling experiment reveals carbon assimilation of marine calcifiers under ocean acidification conditions. Methods Ecol Evol 11:739–750

Ohki S, Irie T, Inoue M, Shinmen K, Kawahata H, Nakamura T, Kato A, Nojiri Y, Suzuki A, Sakai K, van Woesik R (2013) Calcification responses of symbiotic and aposymbiotic corals to near-future levels of ocean acidification. Biogeosciences 10:6807–6814

Ohno Y, Iguchi A, Shinzato C, Gushi M, Inoue M, Suzuki A, Sakai K, Nakamura T (2017a) Calcification process dynamics in coral primary polyps as observed using a calcein incubation method. Biochem Biophys Rep 9:289–294

Ohno Y, Iguchi A, Shinzato C, Inoue M, Suzuki A, Sakai K, Nakamura T (2017b) An aposymbiotic primary coral polyp counteracts acidification by active pH regulation. Sci Rep 7:40324. https://doi.org/10.1038/srep40324

Onitsuka T, Kimura R, Ono T, Takami H, Nojiri Y (2014) Effects of ocean acidification on the early developmental stages of the horned turban, *Turbo cornutus*. Mar Biol 161:1127–1138

Onitsuka T, Takami H, Muraoka D, Matsumoto Y, Nakatsubo A, Kimura R, Ono T, Nojiri Y (2018) Effects of ocean acidification with pCO_2 diurnal fluctuations on survival and larval shell formation of Ezo abalone, *Haliotis discus hannai*. Mar Environ Res 134:28–36

Ono H, Kosugi N, Toyama K, Tsujino H, Kojima A, Enyo K, Iida Y, Nakano T, Ishii M (2019) Acceleration of ocean acidification in the Western North Pacific. Geophys Res Lett 46: 13:161–13, 169

Orr JC, Fabry VJ, Aumont O, Bopp L, Doney SC, Feely RA, Gnanadesikan A, Gruber N, Ishida A, Joos F, Key RM, Lindsay K, Maier-Reimer E, Matear R, Monfray P, Mouchet A, Najjar RG, Plattner GK, Rodgers KB, Sabine CL, Sarmiento JL, Schlitzer R, Slater RD, Totterdell IJ, Weirig MF, Yamanaka Y, Yool A (2005) Anthropogenic ocean acidification over the twenty-first century and its impact on calcifying organisms. Nature 437:681–686

Riebesell U, Fabry VJ, Hansson L, Gattuso J-P (2011) Guide to best practices for ocean acidification research and data reporting. Publications Office for the European Union, Luxembourg, pp 1–258

Saito H, Tamura N, Kitano H, Mito A, Takahashi C, Suzuki A, Kayanne H (1995) A compact seawater pCO_2 measurement system with membrane equilibrator and nondispersive infrared gas analyzer. Deep-Sea Res I Oceanogr Res Pap 42:2025–2033

Sekizawa A, Uechi H, Iguchi A, Nakamura T, Kumagai NH, Suzuki A, Sakai K, Nojiri Y (2017) Intraspecific variations in responses to ocean acidification in two branching coral species. Mar Pollut Bull 122:282–287

Suwa R, Nakamura T, Iguchi A, Nakamura M, Morita M, Kato A, Fujita K, Inoue M, Sakai K, Suzuki A, Koike I, Shirayama Y, Nojiri Y (2010a) A review of the influence of ocean acidification on marine organisms in coral reefs. Oceanogr Jpn 19:21–40. (In Japanese with English abstract)

Suwa R, Nakamura M, Morita M, Shimada K, Iguchi A, Sakai K, Suzuki A (2010b) Effects of acidified seawater on early life stages of scleractinian corals (genus *Acropora*). Fish Sci 76:93–99

Suzuki A (2018) Coral-based approaches to paleoclimate studies, future ocean environment assessment, and disaster research. In: Endo K, Kogure T, Nagasawa H (eds) Biomineralization: from molecular and nano-structural analyses to environmental science. Springer, Singapore, pp 275–282

Takahashi A, Kurihara H (2013) Ocean acidification does not affect the physiology of the tropical coral *Acropora digitifera* during a 5-week experiment. Coral Reefs 32:305–314

Tanaka Y, Iguchi A, Inoue M, Mori C, Sakai K, Suzuki A, Kawahata H, Nakamura T (2013) Microscopic observation of symbiotic and aposymbiotic juvenile corals in nutrient-enriched seawater. Mar Pollut Bull 68:93–98

Tanaka Y, Iguchi A, Nishida K, Inoue M, Nakamura T, Suzuki A, Sakai K (2014a) Nutrient availability affects the response of juvenile corals and the endosymbionts to ocean acidification. Limnol Oceanogr 59: 1468–1476

Tanaka Y, Inoue M, Nakamura T, Suzuki A, Sakai K (2014b) Loss of zooxanthellae in a coral under high seawater temperature and nutrient enrichment. J Exp Mar Biol Ecol 457:220–225

Tanaka Y, Grottoli AG, Matsui Y, Suzuki A, Sakai K (2015) Partitioning of nitrogen sources to algal endosymbionts of corals with long-term ^{15}N-labelling and a mixing model. Ecol Model 309–310:163–169

Tanaka Y, Suzuki A, Sakai K (2016) Effects of elevated seawater temperature and phosphate enrichment on the crustose coralline alga *Porolithon onkodes* (Rhodophyta). Phycol Res 65:51–57

Tanaka Y, Grottoli AG, Matsui Y, Suzuki A, Sakai K (2017) Effects of nitrate and phosphate availability on the tissues and carbonate skeleton of scleractinian corals. Mar Ecol Prog Ser 570:101–112

Thomsen J, Haynert K, Wegner KM, Melzner F (2015) Impact of seawater carbonate chemistry on the calcification of marine bivalves. Biogeosciences 12:4209–4220

Waldbusser GG, Hales B, Haley BA (2016) Calcium carbonate saturation state: on myths and this or that stories. ICES J Mar Sci 73:563–568

Anthropogenic Stresses in Coral Reefs and Adjacent Ecosystems of the East China Sea 7

Ichiro Takeuchi

Abstract

In this chapter, anthropogenic stressors in the East China Sea (ECS) impacting the coral reefs of the Ryukyu Archipelago are reviewed. While seawater temperature of the ECS has increased at +1.20–1.25 °C/100 years, the increase has been lower, +0.85 °C/100 years, in the Sakishima Islands, western Ryukyus Archipelago. The ocean acidification rate near the Ryukyu Archipelago is slower than that in the middle ECS. The inputs of nutrients and pesticides into the ECS from the continent have recently increased. Recent monitoring suggests that persistent organic pollutants (POPs) and marine debris from the continent and Taiwan reach the Ryukyu Archipelago. More than 60% of the coastline of Okinawa Island has been altered by landfills and armouring. The island-based run-off is promptly discharged into coastal waters via short rivers after high rainfall. Red soils and herbicides from agricultural land have been closely monitored. Pharmaceutical and personal care products were detected in relatively high concentrations in rivers in Okinawa Island. The high concentration of linear alkylbenzene sulfonate in treated wastewater in Okinawa Prefecture was estimated based on the pollutant release and transfer register (PRTR) system. Additionally, POPs and additives in antifouling paints have been detected in seawaters around the Ryukyu Archipelago. Military facilities are another likely significant source of anthropogenic chemicals. Thus, detailed monitoring of anthropogenic activities, including concentration levels of various chemicals, is urgently required to protect the coral reefs of the Ryukyu Archipelago.

I. Takeuchi (✉)
Graduate School of Agriculture, Ehime University, Matsuyama, Ehime, Japan

Center of Advanced Technology for the Environment, Graduate School of Agriculture, Ehime University, Matsuyama, Ehime, Japan
e-mail: takeuchi@agr.ehime-u.ac.jp

Keywords

Anthropogenic chemicals · Climate changes · Coastal-based impacts · Continental origin · Island-based run-off · Ryukyu Archipelago

7.1 Introduction

The East China Sea (ECS), located off the East Asian mainland, is a marginal sea of the northwest Pacific Ocean that connects to the South China Sea (SCS) through the Taiwan Strait in the south, the Sea of Japan through the Tsushima Strait in the northeast, and the Yellow Sea in the north (Fig. 7.1).

The ECS consists of two parts, a wide, shallow continental shelf (depth, <200 m) and a deep Okinawa Trough (depth, >2000 m), and is separated from the Philippine Sea in the northwest Pacific Ocean by a submarine ridge associated with the Ryukyu Island chain (Matsuno 2020; Nakamura 2020). The shelf region is strongly affected by the surrounding land through the massive inflow of freshwater with nutrients and suspended organic matter from the Yangtze and Qiantang Rivers (J Chen et al. 2020b; Matsuno 2020). The Pearl River (Zhujiang), which is supersaturated with pCO_2, CH_4, and N_2O, ranks as the second largest river after the Yangtze River in China in terms of annual water discharge entering the northern coast of the SCS, close to the ECS (Chen et al. 2008). The Kuroshio Current flows in the ECS through the channel east of Taiwan (depth, 775 m) and out through the Tokara Strait (depth, 690 m) between Kyushu Island and the Amami Islands (Nakamura 2020). Thus, the Kuroshio Current contributes to the exchange of seawater in the ECS section shallower than ~700 m between the Okinawa Trough and the Philippine Sea (Nakamura 2020). The seawater of the northern South China Sea flows into the ECS through the Taiwan Strait channel, which is closely linked with the shoreward current across the slope off the ECS and the northward current through the Tsushima Strait (Gan et al.

I. Takeuchi, H. Yamashiro (eds.), *Coral Reefs of Eastern Asia under Anthropogenic Impacts*, Coral Reefs of the World 17,
https://doi.org/10.1007/978-3-031-27560-9_7

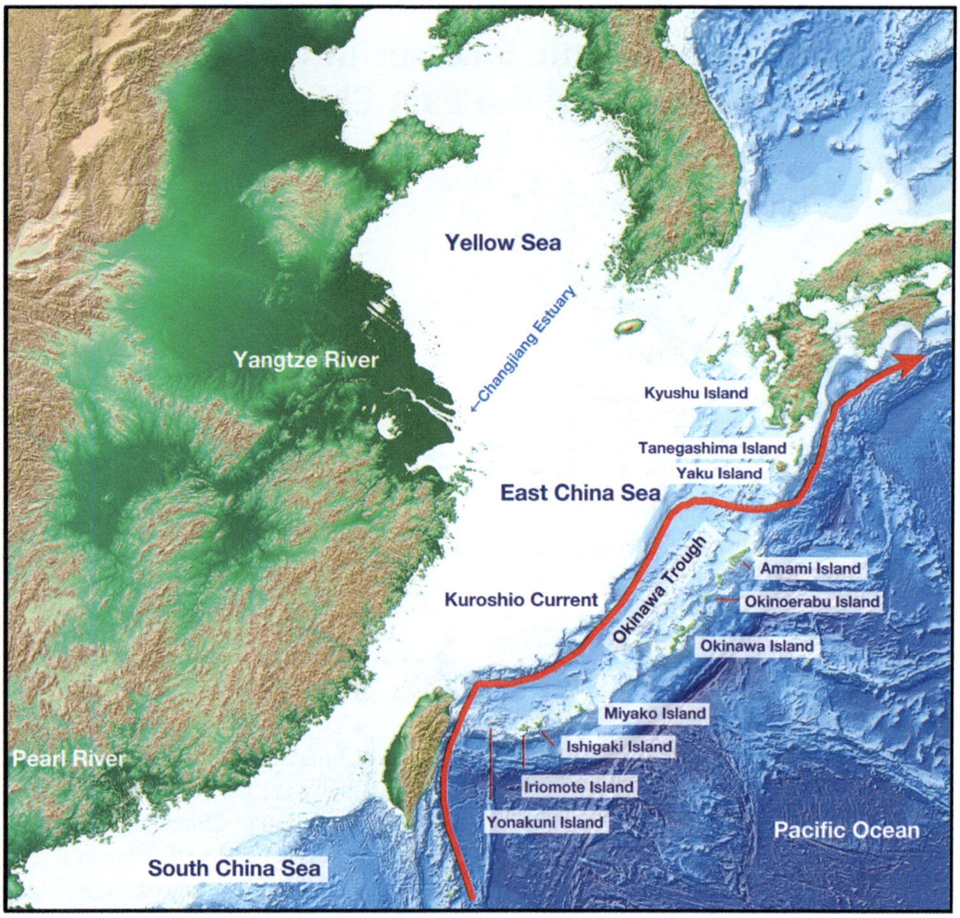

Fig. 7.1 Schematic view of the East China Sea. The red line indicates the Kuroshio warm current. Map was provided by the Marine Information Research Center, Japan Hydrographic Association

2020). Wind direction and stress over the ECS differ between winter (December to February) and summer (June to August) (Liu et al. 2021). In winter, extensive northerly winds prevail over the continental shelf, increasing in magnitude from the inshore area to the east near the Ryukyu Archipelago (Liu et al. 2021). Southerly winds, which are much weaker than those in winter, appear over the continental shelf to the northeast of the Taiwan Strait in summer (Liu et al. 2021).

As mentioned in the Introduction—The relevance of anthropogenic factors to coral reef conservation in the coastal areas of the East China Sea—(Takeuchi 2023), the ECS is surrounded by countries with intensive economic development. At the same time, the coral reefs of the ECS, distributed along the coasts of the Ryukyu Archipelago, have been drastically degraded over the past several decades (see Omori 2011; Hongo and Yamano 2013). Moreover, coral diseases have been intensifying in this region, as well as in the Atlantic and other Pacific regions, at least since 2000 (Yamashiro et al. 2023). Among the major coral faunal provinces in the world, the Japan-Vietnam province, including the coral reefs around the ECS, is estimated to possess the least resistance to thermal stress (McClanahan et al. 2020). Moreover, as the climate changes, the northern distribution limit is shifting towards the north of the ECS (see Yamano et al. 2011). In this section, I review various anthropogenic stressors that are considered to affect the health of coral reefs and adjacent ecosystems along the coast of the ECS, especially the Ryukyu Archipelago.

7.2 Climate Change

The Japan Meteorological Agency presented a long-term trend of sea surface temperature (SST) in the ECS from around 1910 to 2020 (Japan Meteorological Agency 2021). The annual increase rate in SST around Japan until 2020 was +1.16 °C/100 years, which is larger than the global average rate of SST (+0.56 °C/100 years) and similar to the air temperature increase rate in Japan (+1.26 °C/100 years). In seas around Japan, the increase rate of SST in the ECS and

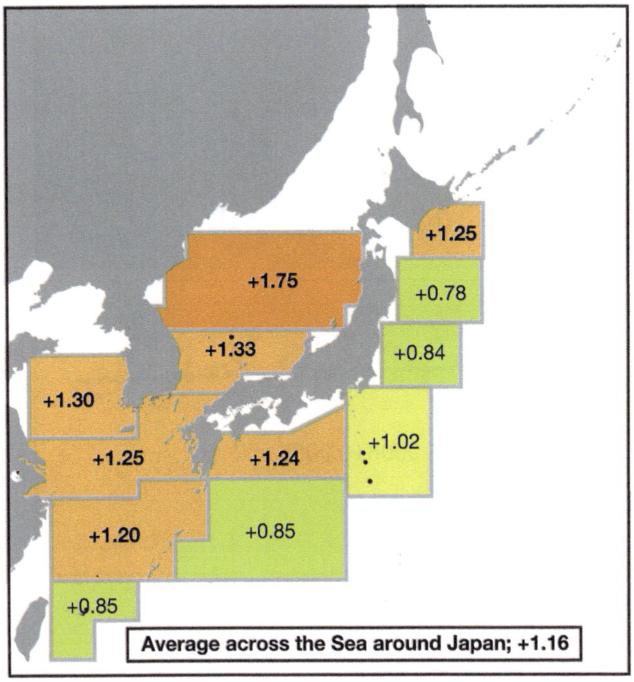

Fig. 7.2 Increase in seawater temperature around the Japanese Archipelago over approximately 100 years until 2020. Data are obtained from Japan Meteorological Agency (2021) (https://www.data.jma.go.jp/kaiyou/data/shindan/a_1/japan_warm/japan_warm.html)

the central part of the Sea of Japan was above the average, +1.20 °C/100 years in the southern ECS and + 1.25 °C/100 years in the northern ECS. In contrast, the SST in the Sakishima Islands (Yonaguni Island to Miyako Islands in the Ryukyu Archipelago) was +0.85 °C/100 years, lower than the average (Japan Meteorological Agency 2021) (Fig. 7.2). Decade-long fluctuations in SST have been monitored in the seas around Japan; after the last maximum around 2000, SST in this region reached a minimum around 2010 and has been increasing since then (Japan Meteorological Agency, 2021). Severe thermal stress events driving coral bleaching (temperature exceeding 30.0 °C) occurred along the coasts of Okinawa Island in the summer of 1998 and 2001 and have been recorded again in the summer of 2016 (Hongo and Yamano 2013; Nishiguchi et al. 2018; Sakai et al. 2019).

From 1998 to 2016, SST in the SCS increased by 0.012 °C/yr., which is less than the rate of warming in high-latitude oceans (C-TA Chen et al. 2020a). The decrease in SST after 2000 observed in the ECS by the Japan Meteorological Agency (2021) has also been widely recorded along the Chinese coast, including the north of the SCS and the Taiwan Strait (Belkin and Lee 2014; Y Li et al. 2021b).

While ocean acidification in the western North Pacific Ocean is progressing, the surface seawater of the western part, including the ECS, tends to show lower acidification rates (Lui et al. 2020). Nagasaki Marine Observatory (after 2013, Nagasaki Local Meteorological Observatory), Japan Meteorological Agency, has been conducting oceanographic observations along the Pollution Nagasaki (PN) Line between Yangtze River mouth and near Okinoerabu Island in the middle of the ECS since the 1970s (Takayanagi et al. 2006). Many scientists have analysed various oceanographic observation data available at the Observatory's website. For example, the acidification rate of the surface water at the shelf bank in the middle of the PN Line is −0.0019 pH_{25}/yr (1982–2010), higher than that of the open oceans at the same depths (Lui et al. 2015) (Fig. 7.3). This higher acidification rate is likely a consequence of enhanced productivity caused by increased anthropogenic nutrient inputs from the East Asian mainland (Lui et al. 2020).

In contrast, the acidification rate of the surface water at St. 1, the end site of the PN line located ~30 km northwest of Okinoerabu Island, is −0.0006 pH_{25}/yr (1982–2010) (Lui et al. 2015). These relatively low acidification rates extending to a depth of approximately 300 m at St. 1 (Fig. 7.3) were attributed to the enhanced biological uptake and mixing of seawater between the euphotic zone and the zones below (Lui et al. 2015).

The acidification rate at the South East Asia Time Series Study (SEATS) station (18°N, 116°E) located north of the SCS was −0.0015 ± 0.0009 pH unit/yr, which is nearly equivalent to −0.0017 pH unit/yr assuming air-sea CO_2 equilibrium (Lui et al. 2020) and − 0.0019 pH_{25}/yr on the bank of the ECS (Lui et al. 2015). A wide range of seasonal and diurnal variation of $pH_{total\ scale}$ was reported from Sesoko Island located northwest of Okinawa Island (Kurihara et al. 2019).

The mean increase in sea level of the ECS from 1993 to 2010 was estimated at ~3.5 mm/yr, which is slightly higher than the global mean (Moon and Song 2017). Mesoscale westward-propagating eddies around the main island of the Ryukyu Archipelago depress the increase in sea level around these islands (Konda et al. 2005; Moon and Song 2017).

The rising maximum wind speed of the mean tropical cyclone under climate change increases the risk of coastal damage by wave set-up via increasing wave heights at the outer ocean (Hongo and Kiguchi 2021). Hongo and Kiguchi (2021) presented several scenarios for the reef flat of Ishigaki Island facing the ECS, including (1) a healthy reef will grow at the same rate as the rising sea levels (+0.98 m by 2100 according to the RCP 8.5 scenario), and (2) the reef will degrade without reef growth. The wave height (defined as the mean wave height of the highest 33% of waves on the reef flat) at the reef flat of Ishigaki Island facing the ECS is estimated to increase from 1.69 m at present to a maximum of 2.63 m for healthy reefs and 3.00 m for degraded reefs under the assumption that the maximum wind speeds of typhoons will be intensified by 2100 (Hongo and Kiguchi 2021).

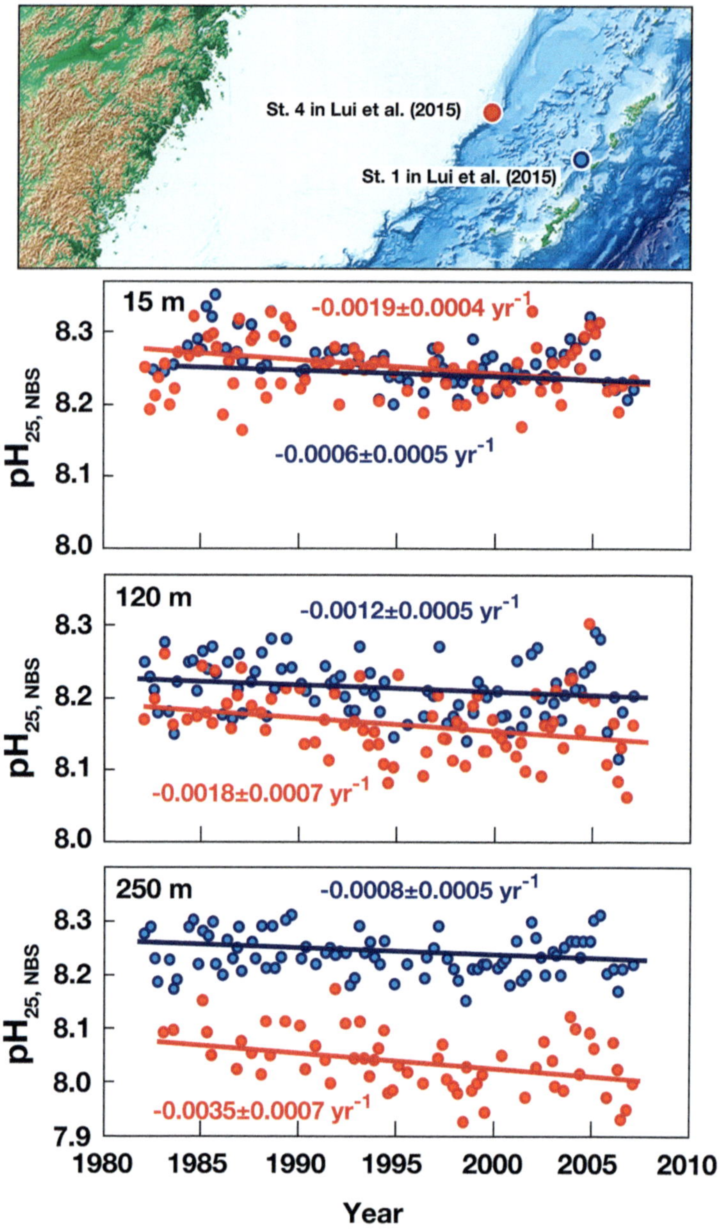

Fig. 7.3 Succession of CO_2 concentration in seawater at the middle of the East China Sea and near Okinoerabu Island, Ryukyu Archipelago, between 1982 and 2007. pH values are expressed on the NBS scale at 25 °C. Data are from Lui et al. (2015, Cont Shelf Res 111, 223–233) with permission from Elsevier. Copyright (2015) Elsevier Ltd. The map was provided by Marine Information Research Center, Japan Hydrographic Association

7.3 Continental Origin Anthropogenic Impacts

7.3.1 Eutrophication

In the ECS, offshore waters are major sources of phosphate and silicate to the continental shelf, whereas river waters, especially the Yangtze River, are significant contributors of nitrogen to this region (J Chen et al. 2020b). Liu et al. (2018) simulated the flow of nitrogen (N) and phosphorus (P) from land to sea in the Yangtze River basin from 1900 to 2010 using combined models for hydrology, nutrient input to surface water, and instream retention. Their simulation indicated a dramatic increase in nutrient delivery to the ECS over 110 years: 17.5-fold increase in N (from 337 Gg N/yr to 5896 Gg N/yr), 6.5-fold increase in P (from 58 Gg P/yr to 381 Gg P/yr), and the change in N:P molar ratio from 13 to 35, indicating river waters to the ECS switched to P limitation. Wang et al. (2020) similarly estimated a rapid increase in nitrogen inflow to the ECS (from 1.0 in 1970 to 4.6 Tg/yr in 2010) and the SCS (from 1.4 in 1970 to 3.8 Tg/yr in 2010) from surrounding rivers using the Integrated Model to Assess the Global Environment–Global Nutrient Model (IMAGE-GNM). In the ECS, the dominance of nitrogen supply from rivers, including the Yangtze River from China, increased from 89% in 1970 to 97% in 2010 (Wang et al. 2020). In addition, Wang et al. (2018) estimated that the input of dissolved inorganic nitrogen, dissolved inorganic phosphorus, and dissolved inorganic silicate by submarine groundwater discharge of Yangtze River into the continental shelf of the ECS was respectively 0.7-, 22-, and 1.4-fold that of the riverine inputs. With the progress of eutrophication, phytoplankton fauna changed in the Changjiang Estuary, off the Yangtze River. The dominant species in the summer phytoplankton were chain-forming diatoms (*Skeletonema*, *Chaetoceros*, and *Pseudo-nitzschia*), but after 2005 the proportion of large-celled dinoflagellates (*Ceratium*) and filamentous cyanobacteria increased with increasing total phytoplankton density (J Chen et al. 2020b). Marine-sourced organic matter, formed by eutrophication-induced marine primary production, is the dominant oxygen consumer in the multiple hypoxic zone in the ECS off the Changjiang Estuary (Wang et al. 2016; Zhou et al. 2020).

Similar to other areas in the world, the average dissolved oxygen in the ECS has decreased to −0.448 (surface samples), −0.608 (water column samples), and −0.736 μmol/kg/yr (bottom samples) along a 32°N transect (122°–127°E) (Ning et al. 2011; J Chen et al. 2020b). The recent increase in eutrophication affects the seasonality of ocean acidification; the average CO_2 in the inner part of the Changjiang Estuary decreased notably during summer, while the average CO_2 increased in autumn and winter from the 1990s to the 2000s (Chou et al. 2012).

As a result, large-scale plumes are formed at the front, where nutrient-rich river water meets with relatively low-nutrient oceanic water on the continental shelf off the large rivers (J Chen et al. 2020b). During flood years, such as 1998 and 2010, the surface area affected by the Yangtze River discharge in the ECS more than doubled compared to that during non-flooding years, reaching the middle of the ECS, approximately100 km west of Okinawa Island, in July 2010 (Fig. 7.4) (Gong et al. 2011). The maximum DIN (sum of nitrate, nitrite, and ammonium) at the plume edge increased from ~16 μmol/L in the 1980s to ~25 μmol/L in the 2000s (J Chen et al. 2020b). The average and extreme precipitation intensities and the frequency of extremely heavy precipitation in the Yangtze River basin have significantly increased according to the analysis of extreme precipitation indices over 60 years, from 1960 to 2019, conducted by X Li et al. (2021a). The extreme precipitation in the Yangtze River is estimated to push discharge water close to the Ryukyu Archipelago and beyond in the near future.

7.3.2 Anthropogenic Chemicals

Persistent organic pollutants (POPs) constitute a diverse group of persistent organic chemicals with toxic and bioaccumulative characteristics that are prone to long-range transport on Earth (Lohmann et al. 2007; Fiedler et al. 2019). The UNEP Stockholm Convention currently regulates the so-called 'dirty dozen' or 'legacy' POPs; initially, those included chlorinated chemicals such as pesticides or industrial chemicals, and since 2009, 16 new POPs, mostly integrated into articles or products such as flame retardants and water repellent chemicals have been added to this group (Lohmann et al. 2007; Fiedler et al. 2019).

Pesticide production and usage in China have been rapidly increasing since 1990, with China becoming one of the largest producers and consumers of pesticides in the 2010s (Grung et al. 2015; Takeuchi 2023). Most of China's freshwater resources are distributed in large rivers, such as the Yangtze River, the Yellow River and the Pearl River, and in lakes (Bao et al. 2012). The levels of POPs in water in China are generally at the high end of the worldwide range (Bao et al. 2012). A comparison of China's regulatory limits indicates that polychlorinated biphenyls (PCBs) in rivers and coastal waters might pose a potential human health risk (Bao et al. 2012). Similarly, the levels of dichlorodiphenyltrichloroethane and its metabolites (DDTs) in several rivers of China are a potential health hazard to humans based on the regulatory limits of DDT recommended by the European Union (Bao et al. 2012). The water supply per capita in China is one-fourth of the global average, whereas the country possesses 1/20 of the global renewable water resources (Bao et al. 2012). Ma et al. (2020) reported that the inadequate freshwater quality of China exacerbates water scarcity, with uneven distribution in the country. The combined water scarcity index including quality and quantity has deteriorated in the middle and lower reaches of the Yangtze River facing the ECS (Ma et al. 2020).

The large-scaled 'hotspot' of hexachlorobenzene (HCB), hexachlorocyclohexanes (HCHs), and DDTs is reported to be present in the southwest areas of the northern SCS and the Yangtze River estuary in the ECS (Ya et al. 2019). The seasonal and spatial distributions of HCB, HCHs, and DDTs in the northern SCS (18–23°N) and ESC (26–32°N) were based on 68–115 seawater samples (~1–2 m depth) collected during 2009–2011 (Ya et al. 2017, 2019). While the residual levels of HCB in the northern SCS were significantly higher than those in the ECS from spring to autumn, significantly higher levels of HCB were detected in the ECS in winter due to increased inputs of ultrafine particles by air movement from northern China (Fig. 7.5A) (Ya et al. 2019). DDTs flowing into the northern SCS showed significantly higher residual levels than those in the ECS during summer and autumn, which is attributed to the increased usage and higher environmental abundance of organochlorine pesticides in southern China (Fig. 7.5A) (Ya et al. 2019). Thus, anthropogenic chemicals that originate in southern China are transported into the ECS by the northward-flowing Kuroshio Current via the northern SCS. The POPs in air and surface seawater from the ECS to the North Pacific were monitored in the spring of 2015 (Wu et al. 2020). The concentrations of α-HCH and *p,p*-DDT levels remained

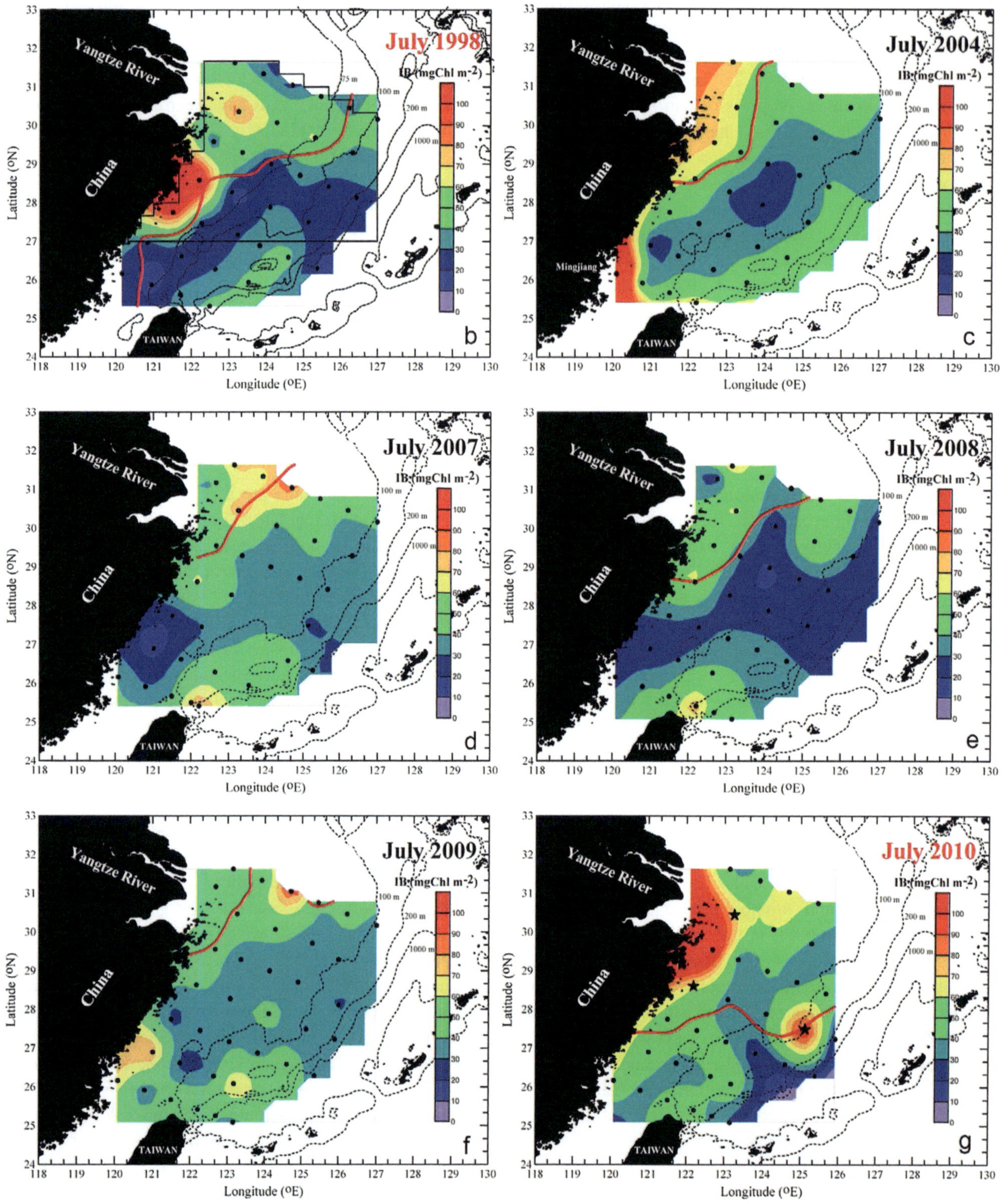

Fig. 7.4 Distribution of vertically integrated algal chlorophyll *a* inventory (IB) in the East China Sea between 1998 and 2010 from Gong et al. (2011, Geophys Res Lett 38, L13603) with permission from John Wiley & Sons, Inc. Copyright (2011) the American Geophysical Union. Observations in July 1998 and July 2010 were conducted during periods of intense flooding in the Yangtze River drainage basin. Red line indicates 31 psu salinity (Gong et al. 2011; Fig. 1) and represents the outer boundary of Changjiang (Yangtze River) Diluted Water (Gong et al. 2011)

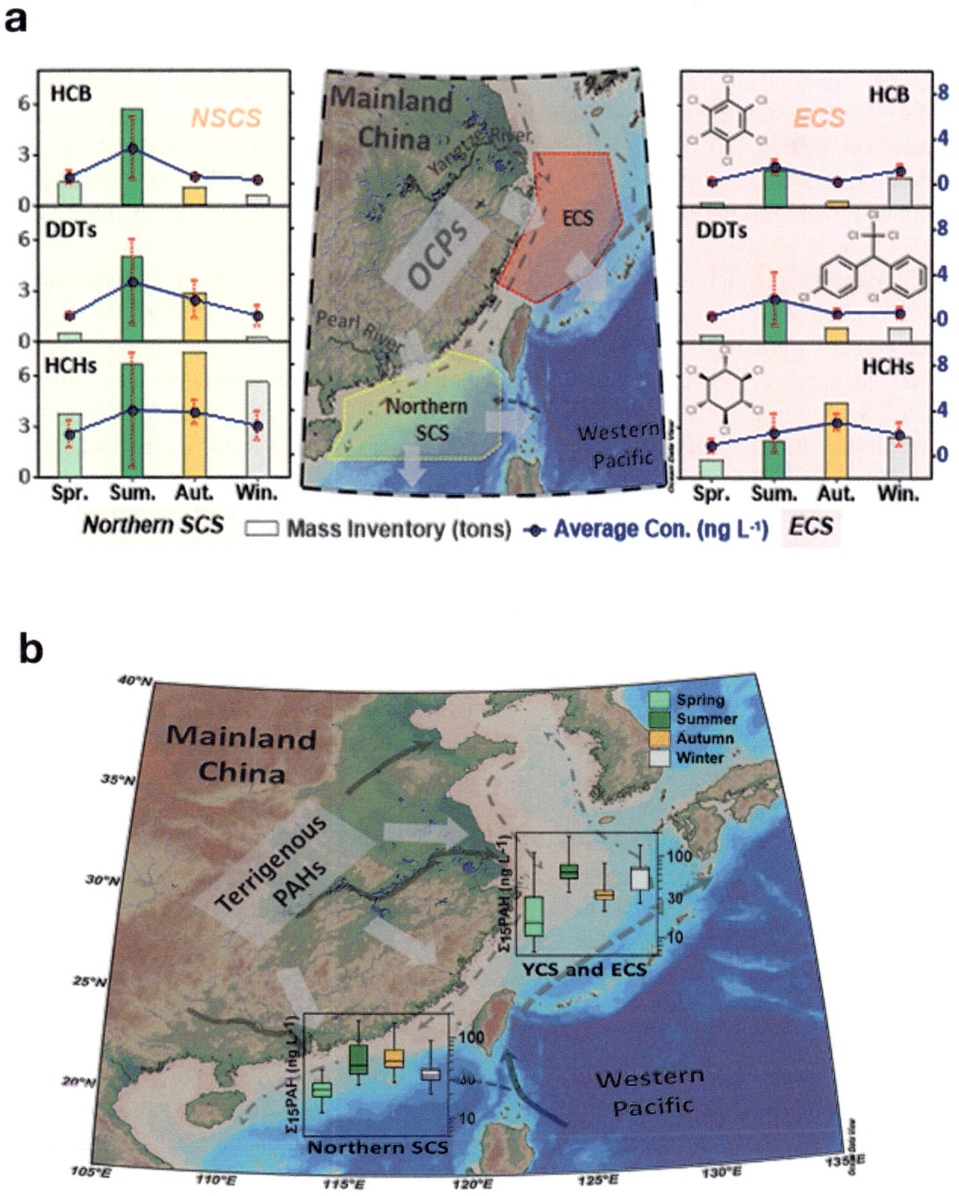

Fig. 7.5 Distribution of anthropogenic chemicals along the marginal seas of China. A, Hexachlorobenzene (HCB), hexachlorocyclohexanes (HCHs), and dichlorodiphenyltrichloroethane and its metabolites (DDTs) in the northern South China Sea and East China Sea. Reprinted Ya et al. (2019, Environ Sci Technol 53, 13088–13097) with permission from the American Chemical Society. Copyright (2019) American Chemical Society. B, Polycyclic aromatic hydrocarbons (PAHs) in surface water of the Yellow Sea, Each China Sea, and northern South China Sea. Reprinted from Ya et al. (2017, Environ Sci Technol 51, 9072–9079) with permission from the American Chemical Society. Copyright (2017) American Chemical Society

high in the west North Pacific across the Ryukyu Archipelago long after the regulation of the use of technical HCHs and DDTs in East Asia was enforced (Wu et al. 2020).

Polycyclic aromatic hydrocarbons (PAHs) are composed only of carbon and hydrogen with two or more fused aromatic rings (Takeuchi et al. 2009; Ramesh et al. 2011). Of the more than 100 compounds belonging to PAHs, the U.-S. Environmental Protection Agency has designated 16 compounds as priority pollutants for environmental monitoring purposes (Ramesh et al. 2011). The seasonal distribution of PAHs also indicates different trends between the northern SCS and ECS; the highest season of the total PAHs (Σ15PAHs) is autumn (October and November; 82 ± 78 ng/L), followed by summer (July and August: 77 ± 50 ng/L) in the northern SCS, and summer (August: 80 ± 32 ng/L), followed by winter (December: 67 ± 25 ng/L) in the ECS and the Yellow Sea located north of the ECS (Ya et al. 2017) (Fig. 7.5B). PAHs originating from industrial coal burning and firewood combustion during winter in the northern parts of the continent could be transported and deposited on the surface water of the ECS and Yellow Sea by the prevailing northwest wind (Ya et al. 2017).

Yang et al. (2021) monitored the concentrations of 22 commonly used herbicides in China in Bohai Bay and the Yellow Sea. They detected herbicides in all sampling locations within 100 km of the shoreline. The maximum quantum efficiency of photosystem II (PSII), *Fv*/*Fm*, in phytoplankton significantly decreases when exposed to in situ concentrations of triazines (Yang et al. 2021).

7.3.3 Plastic Wastes

The global riverine plastic outflows were considered the main pathways for the transport of land-based plastic waste to the oceans (Mai et al. 2020a). It was estimated at 57,000–265,000 metric tonnes (MT) in 2018, and it is expected to continue growing in the next 10 years due to the increasing use of plastic products (Mai et al. 2020b). The Yangtze River exports the largest amount of plastics globally (6750–33,600 MT/yr) to the ECS (Mai et al. 2020b). Microplastics were accumulated in the surface micro layer (SML; the upper 1–1000 μm of the ocean) relative to deeper water (Song et al. 2014; Anderson et al. 2018). The microplastics in the surface seawater of the northern ECS were five times higher than those in the southern ECS due to the Yangtze River estuary and Hangzhou Bay (Huang et al. 2022). The distribution of MPs in the seas around Japan was surveyed from July to September 2014 at sampling sites located mainly in the ECS, the Sea of Japan, and off the Tohoku District in the North Pacific along the Tsushima Current (Isobe et al. 2015). The total particle count of microplastics in the east Asian seas was 1,720,000 pieces/km^2, 20 times greater than that in the Seto Inland Sea (a sea surrounded by Honshu, Kyushu, and Shikoku Islands), 16 times greater than that in the North Pacific, and 27 times greater than that in oceans worldwide (Isobe et al. 2015). Thus, the east Asian seas are thought to be one of the hotspots of pelagic microplastics globally (Isobe et al. 2015). Moreover, Isobe et al. (2015) reported that small plastic fragments are transported northward along Japan by the north-eastward Tsushima and Kuroshio currents. Thus, these small plastic fragments likely originate southwest of their study area, in the Yellow Sea and ECS.

Polystyrene constitutes approximately 70% of the stranded marine debris in Japan; drifting and stranded polystyrene is considered to generate styrene oligomers (SOs) in the marine environment (Saido et al. 2012). Saido et al. (2012) detected SOs (styrene monomer phenylethylene, SM; styrene dimer, 2,4-diphenyl-1-butene, SD; styrenetrimer, 2,4,6-triphenyl-1-hexene, ST) in seawater and sand from the coasts of Iriomote and Ishigaki Islands, both located near the west end of the Ryukyu Archipelago close to Taiwan, and from Okinawa Island. On average, 116.9 ng/L SM, 238.3 ng/L SD, and 479.3 ng/L ST were detected in seawater, and 246.9 ng/g SM, 270.8 ng/g SD, and 1221.9 ng/g ST in sand on all three islands (Saido et al. 2012). A relatively higher concentration of ST (>1800 ng/g) was measured in the sand from the northern coasts of Iriomote and Ishigaki Islands, close to the Kuroshio Current flowing from the area near Taiwan to the middle of the ECS (Saido et al. 2012).

Fujieda et al. (2006) estimated the origin and dispersal of marine debris along the coasts of Japan by monitoring the dispersal of disposable cigarette lighters. They identified the locality of manufacture from the symbols engraved on the bottom of the lighters and the discharge location from the lettering, including advertisements, printed on the lighter's surface (Fujieda et al. 2006). Over half of 6609 lighters collected from 92 coastal sites across Japan that washed up on the coastline from Yonaguni Island in the westernmost part of the Ryukyu Archipelago to Yaku Shima Island close to southern Kyushu originated from continental China. Fujieda et al. (2014) repeated the analysis and surveyed in greater detail the origin of >60,000 lighters collected between 2003 and 2014. The lighters collected from the coastal areas from Yonaguni Island to Amami Island along the Ryukyu Archipelago originated from Guangdong Province to Shanghai and the west coast of Taiwan. The ratio of lighters of Taiwanese origin increased westward from the Ryukyu Archipelago (Fig. 7.6) (Fujieda et al. 2014). The results of both Saido et al. (2012) and Fujieda et al. (2014) suggested that marine debris in the ECS that washes up on the islands of the Ryukyu Archipelago south of the Kuroshio Current is transported by the northerly wind.

After October 2021, pumice from the volcano located near the Ogasawara Islands south of mainland Japan was dispersed to Kita-Daito Island and Kikai Island, which are located west and north of Okinawa Island, respectively (Yoshida et al. 2022). A large amount of pumice stones were blown to the northeast coast of Okinawa Island by strong seasonal north-easterly winds (Ohno et al. 2022). The stranding of a large amount of pumice stones on the coral reef coast northwest of Okinawa Island was confirmed (Fig. 7.7). The southward dispersal of pumice to Okinawa Islands suggests a difference in the movements of currents (including the Kuroshio Current) and surface seawater.

Taken together, my observations and the above studies (Fujieda et al. 2006, 2014; Saido et al. 2012; Ohno et al. 2022; Yoshida et al. 2022) indicate that contamination and waste, including plastics, in the ECS and the Pacific north of the Ryukyu Archipelago drift towards the coral reefs of the Ryukyu Archipelago as a result of the extensive northerly winds during winter (Liu et al. 2021). These results contrast with Isobe et al. (2015) which reported that small plastic fragments are transported north along Japan by the north-eastward Tsushima and Kuroshio currents.

7.4 Island-Based Anthropogenic Impacts

7.4.1 Urbanisation

Coral reefs around Okinawa Islands from 24° to 27°N are categorised as fringing-type reefs; the width of the reef flat is ca. 1 km and its depth is 1 to 3 m (Kayanne et al. 2004). Thus, reefs adjacent to islands are often characterised by relatively

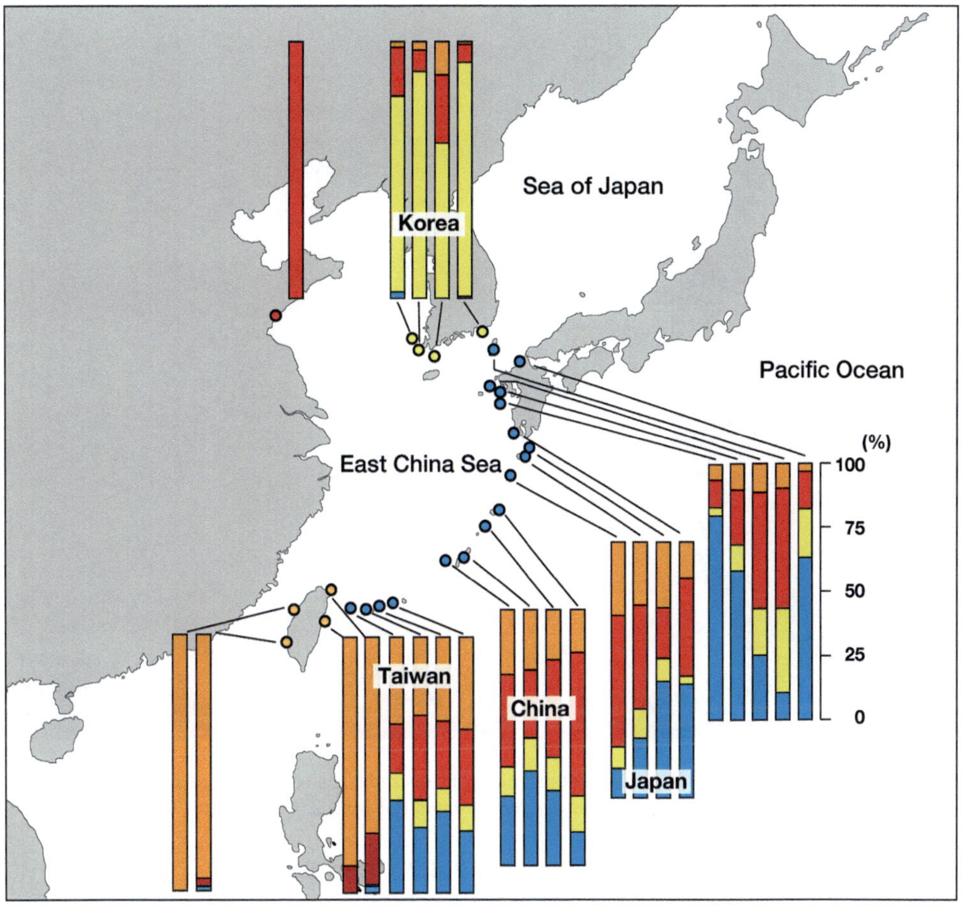

Fig. 7.6 Distribution of portable cigarette lighter stranded on the coasts along the coasts of the East China Sea. Reprinted from Fujieda et al. (2014). The percentage distribution categorised by the origin of the cigarette lighter

semi-enclosed environments (Kinjo 2017). Of the 682.8 km total coastline around Okinawa Island, 63.2% has been altered by land-filling (309.2 km, 45.3%), hard armouring (98.9 km, 14.5%), and soft armouring (23.7 km, 3.4%) (Masucci and Reimer 2019). Corals and coral reefs near coastal cities are subjected to various stressors (Heery et al. 2018). Tetrapod breakwater structures in northern Okinawa Island reduced the survival rate of the branching hermatypic corals by affecting the physical environment and benthic community composition around the tetrapods (Masucci et al. 2020). Population density is highest in the southern half of Okinawa Island, where Naha, the capital city of Okinawa Prefecture, is located (Heery et al. 2018). The coastline in this area is degraded or destroyed by landfills, construction of shore protection and roads, dredging and harbours, and outflow from adjacent terrestrial areas (Reimer et al. 2015).

Okinawa Islands are located on the north end of the Asian tropical forest zone with high rainfall of 2106 mm/yr (average for Ishigaki Island for the period 2003–2012; Terada et al. 2017). There were 55 rainy days per year, occurring at approximately 7 d intervals (Terada et al. 2017). The monitoring of a small, 2.2 km long river on Ishigaki Island revealed that total nitrogen (TN) and total phosphorus (TP) flux from upstream to the mouthpart on a rainy day were 9.1 times and 3.4 times of a clear day, respectively (Terada et al. 2017). Higa et al. (2001) reported that the concentrations of TN and TP during rainfall show an increasing trend with suspended solids (SS) at river estuaries along the Okinawa Island coast. Both TN and TP concentrations are higher in rivers in southern Okinawa Island than in rivers in the northern part of the island, which was contributed to the differences in soil and soil texture, land usage pattern, and the inflow of domestic wastewater. However, the sewerage population rate of Okinawa Prefecture is 72.0%, lower than the average rate of 79.7% in Japan; the sewerage population rate is calculated as [the population in the area with sewerage system]/[the administrative population] (Okinawa Prefecture 2021). The biological oxygen demand (BOD) and suspended particulate matter (SPM) at the river mouthparts is strongly positively related with human population density in the catchment area along the Okinawa Island coast (West and van Woesik 2001). The human population density within the catchment is also significantly correlated with the inorganic

Fig. 7.7 Dense, moraine-like aggregation of stranded pumice stones along the intertidal zone close to coral reefs of Nakijin-son located north-west of Okinawa Island on 25 November 2022. The image was photographed using a DJI Mini 2 (DJI Co. Ltd., Guangdong, China)

nutrient concentrations at the river mouth around Okinawa Island, and the increase in inorganic nutrient concentrations leads to a decrease in coral cover at the reef edge around the southern part of the Island (Shilla et al. 2013).

Reimer et al. (2015) pointed out the weakness of the Japanese environmental assessment system. The Environmental Impact Statements are performed by the project proponents themselves, not by an external independent agency, while environmental rules and regulations for large-scale developments have become more assertive in Japan over the past four decades (Reimer et al. 2015). Reimer et al. (2015) reported the apparent impacts of the 4.75-km-long causeway constructed 40 years ago near the coral reef off the west coast of Okinawa Island. Most sites along the causeway dividing a large tidal flat showed a changing environment, likely resulting from a reduced water flow (Reimer et al. 2015). In particular, the site closest to Okinawa Island, heavily impacted by poor water quality, low diversity, and disturbed biotic communities, had an unusual microbial community highly dominated by the enterobacterial genus *Serratia*, an opportunistic human pathogen and agent of coral diseases (Reimer et al. 2015; Soliman et al. 2017). Tashiro et al. (2004) reported a high concentration of PCBs in the sediment, 5.07 μg/g, detected near the bridge of the Manko Flat near Naha. They attributed this high concentration of PCBs to the residue of paints used on the bridge in the 1970s.

7.4.2 Red Soil

After the return of Okinawa Prefecture to Japan in 1972, extensive urban and agricultural development led to a significant increase in sediment discharge to coastal waters along the coast of the prefecture (Omija 2004; Yamano et al. 2015). These sediments are called ‘red soil’ because of their colour (see Yamano et al. 2015). Pineapple and sugarcane fields, especially immediately after planting, were the main sources of red soil along the coast of the Okinawa Islands (Higa et al. 1995). Nakandakari et al. (2007) pointed out that particulate nutrients (particulate nitrogen and phosphorus) are closely linked with the run-off of suspended soil from an upland watershed in the middle of Okinawa Island, where sugarcane is the dominant crop. The inflow of red soil into the coral reef from adjacent land leads to high increase in turbidity and sedimentation in the coral reef for an extended period (Kinjo 2017). Red soil pollution tends to occur when natural factors facilitate the flow of red soil to reefs together with anthropogenic factors. In Okinawa, there are three natural factors: (1) soils that are easily dispersed due to low organic matter, (2) steep and short river topography, and (3) heavy rainfall, three times stronger than the average rainfall of Japan (Omija 2004). In addition to these natural factors, Omija (2004) referred to three anthropogenic factors: run-off from agricultural land, various development projects, and U.S. military bases. In Okinawa, the ratio of farmland to total land area is

18% (Hashimoto 2005). A simplified measurement method for suspended particles in sea sediment (SSPS) has been proposed (Omija 2004; Kinjo 2017). Of eight ranks from 1 (< 0.4 kg/m^3) to 8 (≥ 400 kg/m^3), maintaining rank 3 to 5a (1 to <30 kg/m^3) is required to conserve coral reef ecosystems; ranking above 6 indicates the presence of pollution due to red soil run-off and other anthropogenic factors (Kinjo 2017).

The total run-off of red soils along the coast of Okinawa Prefecture in 2011 was 298,300 tonnes, which corresponded to approximately 80% of that in 2001; the most significant source of red soils was farmland, recording 85.5% of total red soils, far from the other two major sectors, development projects and U.S. military bases (Okinawa Prefecture 2013) (Fig. 7.8). Therefore, Okinawa Prefecture established the 'Okinawa Prefecture Basic Plan for Prevention of Red Soil Discharge' in September 2013 to comprehensively prevent the discharge of red soil (Okinawa Prefecture 2013; Kinjo 2017). The main parts of the Basic Plan are as follows:

1. Selecting priority areas for implementation (priority monitoring sites)
2. Deciding environmental protection targets for red soil
3. Estimating targeted reduction amounts of red soil run-off

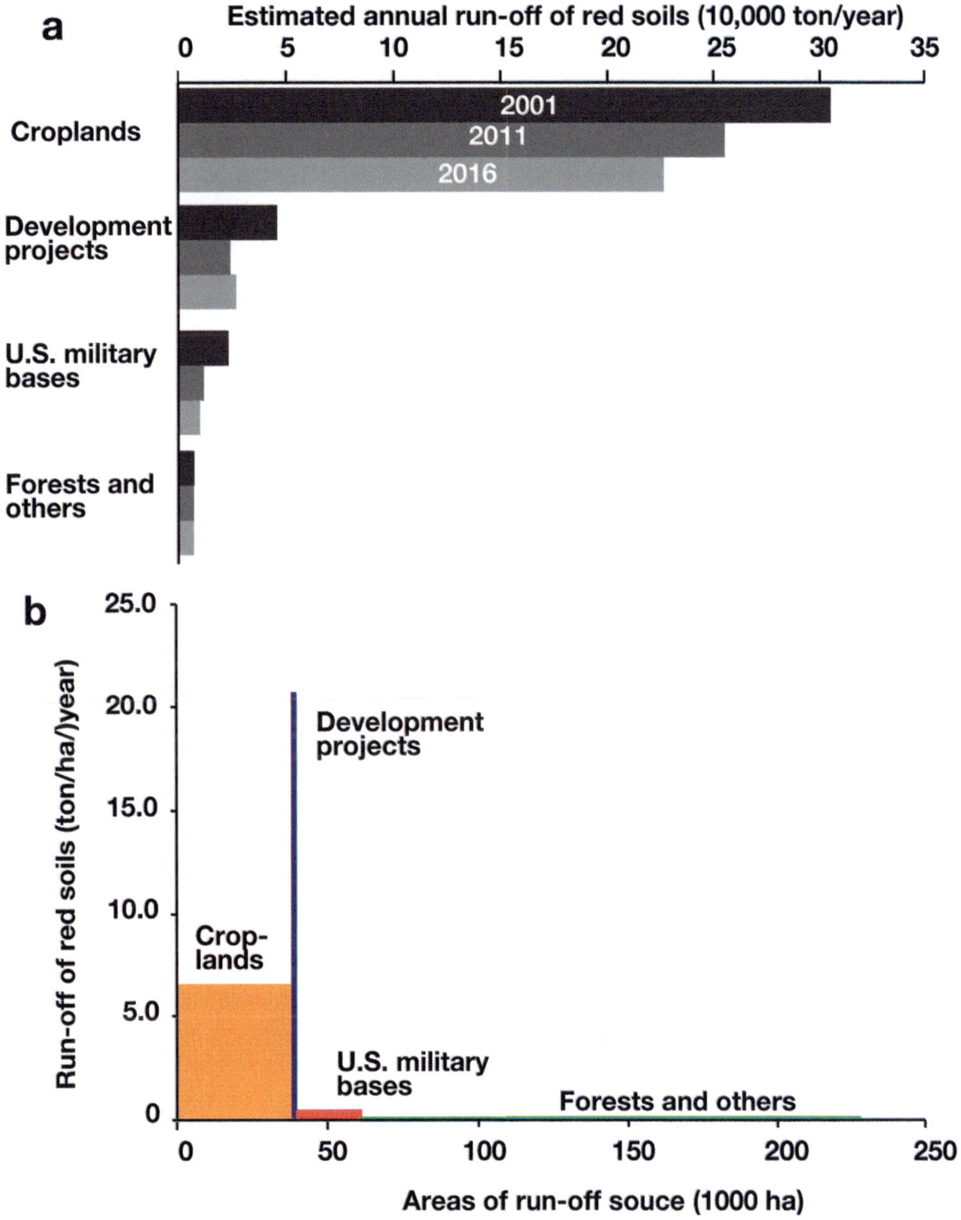

Fig. 7.8 Estimated red soil run-off in Okinawa Prefecture. A, Annual amounts from four major sectors in 2001, 2011, and 2018 (Fig. 2-8 on p. 9 in Okinawa Prefecture 2013; Fig. 3.2-1 on p. 7 in Okinawa Prefecture 2019). B, Comparison of red soil annual run-off from four major sectors in 2011 (Fig. 2-9 on p. 9 in Okinawa Prefecture 2013)

The total run-off of red soils in 2016 was 270,900 tonnes, with farmland contributing 83.6% (Okinawa Prefecture 2019). In 2016, the red soil run-off from the U.S. military base was reduced by <50% of that in 2001; this reduction was more significant than that of the other two dominant sectors, farmland and development projects (Fig. 7.8). The red soils run-off per hectare from the U.S. military base is 0.51 t/ha/yr, which corresponds to 8.5% and 2.3% of the run-off from farmlands and development project areas, respectively (Okinawa Prefecture 2019).

7.4.3 Nutrients

Nutrient monitoring along the coast of Okinawa Prefecture has been conducted since 2009 (Kinjo 2017). To preserve >50% of coral coverage, Kinjo (2017) recommended maintaining the following water qualities: <0.08 mg/L TN, <0.01 mg/L TP, <0.01 mg/L nitric acid + nitrous acid, <0.006 mg/L phosphoric acid, <0.11 degrees turbidity, and >14 m horizontal transparency. He established the relationship between TN and coral coverage. However, according to Kinjo (2017), coral coverage >50% was present only in approximately 10% of monitored sites with <0.08 mg/L TN, whereas approximately 60% of monitored sites with <0.08 mg/L TN had <10% coral coverage (Fig. 7.9). Kinjo et al. (2006) reported the occurrence of *Acropora* with a 20 cm diameter at a tidal pool off the southern coast of Sesoko Island; the pool had no inflow of red soil and the concentration of TN was recorded at 0.2 mg/L. Thus, the lower coral coverage in oligotrophic areas and the observations reported by Kinjo et al. (2006) suggest that lower nutrient levels are one of the required factors for maintaining high coral coverage, whereas various other factor(s) contribute to reduced coral coverage. While Kinjo (2017) described the above quality guideline, he also noted that, in addition to red soil and nutrients, chemicals such as pesticides, chemical fertilisers, surfactants, sunscreens, heavy metals, and microplastics would be stress factors for the coral reef ecosystem. Hashimoto (2005) suggested that, in addition to red soil, eutrophication and/or pollution caused by agricultural sectors should be tracked to monitor the effects of

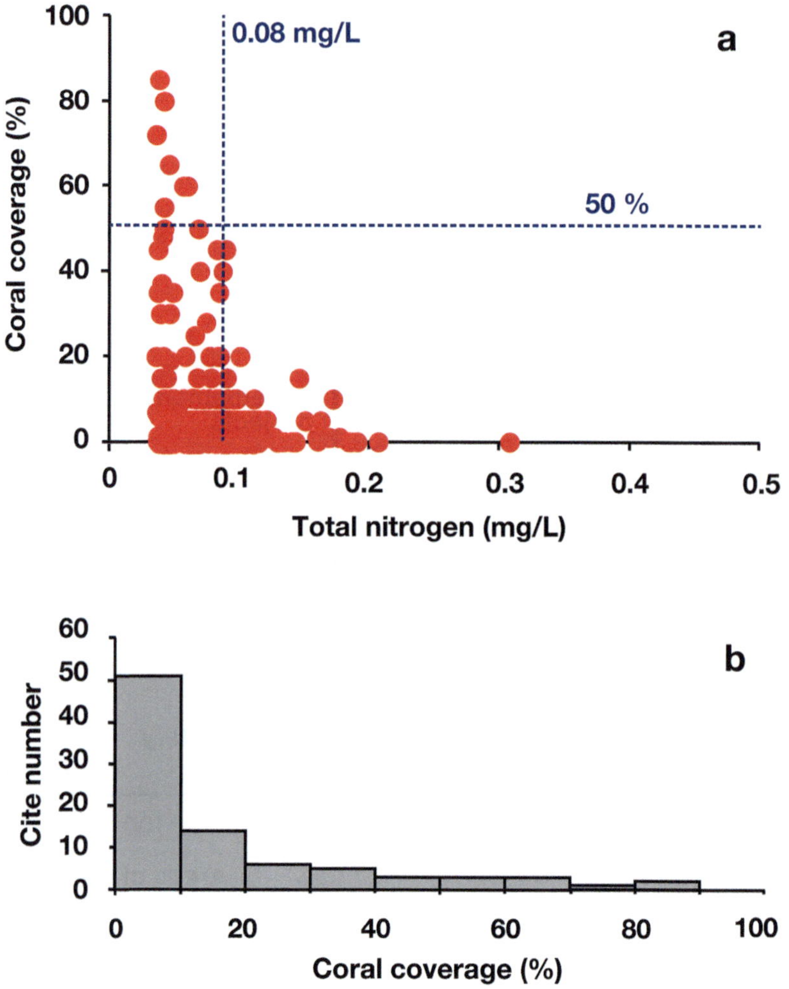

Fig. 7.9 Coral coverage (%) along the coasts of Okinawa Prefecture, Japan after 2009. A, The relationship between coral coverage (%) and total nitrogen concentration (Modified from: Kinjo, K., 2017, J Japan Coral Reef Soc 19, 87—94, Fig. 5, with permission from © Japanese Coral Reef Society); B, the proportion of coral coverage (%) at <0.08 mg/L total nitrogen. These data are based on visual count of dots presented in Fig. 5 by Kinjo (2017)

agriculture on coral reefs. Yamano et al. (2015) stated that the release of sediment has caused the degradation of freshwater and coastal ecosystems and biodiversity along the island coastlines in the Okinawa Prefecture. Hongo and Yamano (2013) reported a steady decline in coral cover of anthropogenic turbid reefs around Okinawa Island from 1995 to 2009, with the possibility that these reefs, especially the branching *Acropora* spp., may be less resilient to thermal stress than non-turbid reefs. The decline of branching *Acropora* spp. were also reported from the coral communities at Ishigaki Island (Harii et al. 2014).

7.4.4 Submarine Groundwater Discharge

Along with limestone areas, such as the Ryukyu Archipelago (Shinjo 2014), submarine groundwater discharge is one of the major transportation routes for nutrients released by human activities (Saito et al. 2020). The freshwater lens is a thin layer of underground freshwater overlaying seawater in aquifers of islands and peninsulas including the Ryukyu Archipelago (Yoshimoto et al. 2016). Subsurface dams were created for providing underground freshwater to farmlands in southern Okinawa Island (Nakano et al. 2013). The 75% probability of nitrate nitrogen from the underground water there was slightly higher than 10 mg/L, the upper limit of environmental criteria in Japan (Nakano et al. 2013). Yoshimoto et al. (2016) reported that nitrate in the groundwater at Tarama Island, located between Miyako and Ishigaki Islands, originated from multiple sources such as chemical fertilisers, livestock manure, and domestic wastewater. Higuchi et al. (2014) reported relatively high concentrations of nitrate and nitrite ions (NO_x) brought by coastal groundwater inflow in the reef moat off Bise, northwest Okinawa Island, where coral *Montipora digitata* and seagrass *Thalassia hemprichii* co-inhabit.

7.4.5 Herbicides and Pesticides

Sugarcane farmers in Okinawa use fertilisers and herbicides (including Diuron), similar to the farming practices in Australia that are affecting the Great Barrier Reef (GBR), the world's largest coral reefs located off Queensland, Australia (Hashimoto 2005). In Okinawa, 53% of farmlands are cultivated with sugarcane (Hashimoto 2005). In 2004, approximately 17,000 kg of Diuron was applied on farmed land in Okinawa Prefecture, which was the highest amount among Japanese prefectures (Kitada et al. 2009). Kitada et al. (2008) surveyed the concentration of Diuron and Irgarol 1051 (hereafter Irgarol) in the natural sediment of river mouthparts adjacent to coral reefs on Okinawa Island in July 2004 and September 2005. Both herbicides are of PSII type and have been used as 'booster' agents in copper-based antifouling paints (Dafforn et al. 2011). Diuron was detected in river sediments from both rural and urban areas of Okinawa Island with a maximum concentration of 0.22 μg/kg (Kitada et al. 2008). Their monitoring revealed a higher concentration of Diuron near Naha, suggesting that its consumption is related to home gardens in the area (Kitada et al. 2008). Irgarol in these river sediment samples was likely transported from the river mouths during flood tides (Kitada et al. 2008).

Diuron concentrations were monitored during 2007 and 2009 at the coral reef off Shiraho, Ishigaki Island, brought by the Todoroki River, which runs through sugarcane farms (Sheikh et al. 2009a). This herbicide was detected in all stations from the river at concentrations up to 754 ng/L (Fig. 7.10), and its concentration at the coral reef off Shiraho ranged from not detected (ND) to 90 ng/L; the concentration increased more than tenfold from May to August compared to its levels in November (7.8 ng/L) (Sheikh et al. 2009a). In their personal communication with local farmers in 2007, Sheikh et al. (2009a) confirmed that Diuron is extensively used for sugarcane cultivation during the summer season. The authors presented the following scenario for the fate of Diuron after its application: Diuron sprayed on a sugarcane field is retained in the soil during summer, but it is leached into the Todoroki River during rainfalls after the summer season and transported to the reef. A relatively high concentration of Diuron (14 ng/L) was detected in the estuary of the Arakawa River located west of Ishigaki Island in September 2019 (Tanita and Onduka 2021).

Glyphosates are widely used in Okinawa Prefecture. According to the Japan Plant Protection Association (JPPA) (2021), 1600.3 tonnes of pesticides were shipped to Okinawa Prefecture, including 986.1 tonnes of insecticides and 360.4 tonnes of herbicides, during 2019 pesticide year (Oct 2018 to Set 2019). The shipment of glyphosate and related herbicides in Okinawa Prefecture was approximately 160 tonnes (including 115.8 tonnes of glyphosate isopropylamine salt and related salts), which corresponds to nearly 45% of total herbicide shipment and is approximately 10 times the shipment of Diuron and related herbicides (JPPA 2021). Due to the extensive worldwide usage of glyphosate, relatively high concentrations, up to 1377 μg/L, of this compound have been detected from seawaters in the western Pacific (see Matozzo et al. 2020). Most toxicity experiments on glyphosates have focused on molluscs (Matozzo et al. 2020). A 21-day exposure of the mussel *Mytilus galloprovincialis* to 10, 100, and 1000 μg/L glyphosate induced 111, 124, and 211 differentially regulated transcripts, respectively, and disrupted several critical biological processes, including energy metabolism and Ca^{2+} homeostasis, as indicated by the functional analysis of differentially expressed genes (Milan et al. 2018). A combination

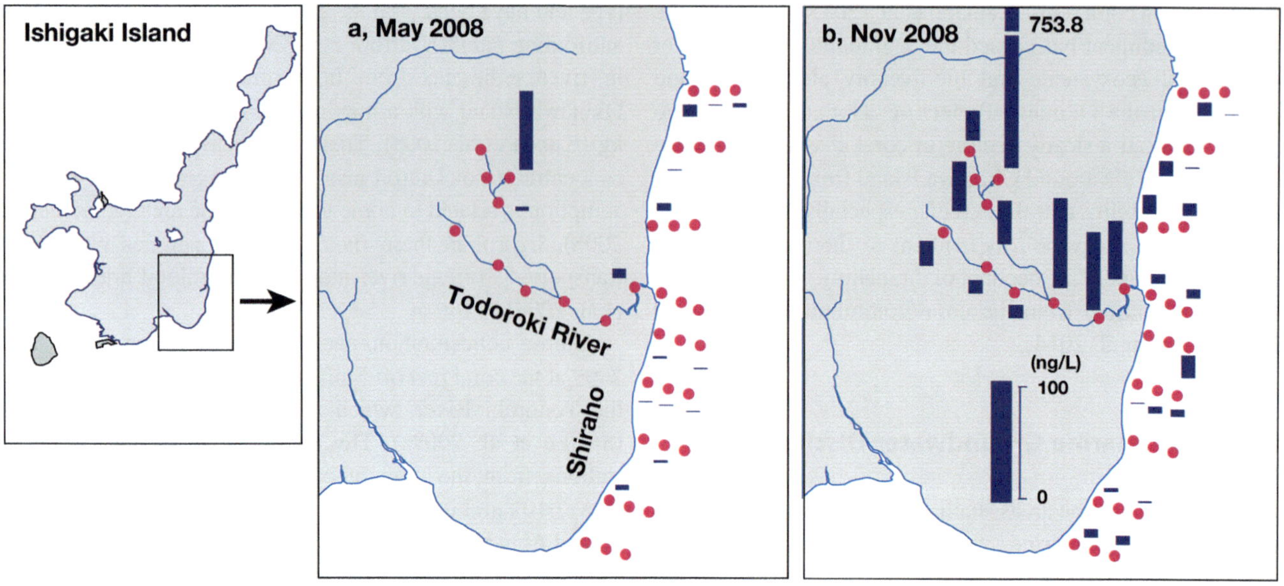

Fig. 7.10 Concentration of Diuron in seawater of coral reefs off Shiraho and Todoroki River, Ishigaki Island, Japan in May (A) and August (B) 2006. Sampling sites and concentration data are from Sheikh et al. (2009a, Mar Pollut Bull 58, 1922–1926) with permission from Elsevier. Copyright (2009) Elsevier Ltd.

of elevated water temperature (31 °C) and glyphosate exposure (12.0 mg/L) significantly decreases chlorophyll *a* and induces colour loss of *Acropora formosa* (Amid et al. 2018).

In addition to Diuron and Irgarol, chlorpyrifos, an organophosphate pesticide, has been widely detected in Okinawa Island (Kitada et al. 2008, 2009). Moreover, a passive sampler continuously deployed for 2 weeks in 2013 recorded 19 species of pesticides in rivers from Okinawa Island, of which EPN, an organophosphate pesticide, was the most abundant at 260 ng/day per sampler (Tashiro and Kameda 2015). The occurrence of seven species of neonicotinoids and fipronil (phenylpyrazole insecticides) was reported in the estuary of the Seto Inland Sea, which are most likely due to the downstream flow of agricultural effluents (Hano et al. 2019). The high concentrations of fipronil detected in June and July were estimated to have a significant effect on benthic and planktonic crustaceans (Hano et al. 2019). Yamamuro et al. (2019) reported that the use of neonicotinoids in watersheds since 1993 led to the collapse of smelt harvest by approximately 90% and a nearly 80% decrease in zooplankton biomass in spring in Lake Shinji, an oligohaline lake located in western Japan near the Sea of Japan.

The above studies on various herbicides and insecticides in Okinawa suggest that the red soils flowing into coral reefs, especially that originating from farmlands, contain some amount of Diuron and one or more additional species of herbicides and/or insecticides. Moreover, rivers are polluted with several species of insecticides. Despite Australia's Reef Water Quality Protection Plan 2009 for the GBR, which includes guidelines for herbicide concentrations (The State of Queensland 2009), herbicide residues from river discharge were widespread in the GBR ecosystem (Lewis et al. 2009). These effects may cause a change in the community structure of mangrove, seagrass, and coral reef ecosystems, which are already affected by other pressures such as climate change (Lewis et al. 2009). In contrast to the water quality management of the GBR, pesticide measurements were not included in the monitoring of red soil run-off conducted to enable the protection of coral reefs in Japan (Okinawa Prefecture 2013, 2019) despite the recommendation by Yamano et al. (2015) to monitor other pollutants, in addition to red soil, in the Ryukyu Archipelago.

7.4.6 Pharmaceutical and Personal Care Products (PPCPs)

Pharmaceutical and personal care products (PPCPs), widely used for human health, cosmetic, agriculture, and veterinary purposes, are considered as emerging contaminants (ECs) in the environment due to their known or suspected adverse ecological effects and human health risks (Keerthanan et al. 2021).

Of PPCPs, nonylphenol (NP) and bisphenol A (BPA) have been detected in river and coral reef sediments in Ishigaki and Okinawa Islands (Kawahata et al. 2004; Kitada et al. 2008). NP is a degradation product of nonylphenol polyethoxylates used worldwide in various industrial processes, whereas BPA as part of polycarbonate plastics is released through their degradation (Kawahata et al. 2004). Unlike PCBs, NPs and octylphenol do not bioaccumulate through the coastal food web of Tokyo Bay, that is, concentrations of NPs and octylphenol in molluscs occupying lower trophic levels were higher than in their predators (Takeuchi et al. 2009). Both chemicals are known to act as endocrine disruptors in various organisms including fish (Fraser et al. 2017). Kawahata et al. (2004) detected up to 0.17 μg/L of NP and 0.058 μg/L of BPA in Aja River, Naha, on Okinawa Island and up to 0.12 μg/L of NP and 0.08 μg/L on Ishigaki Island. These concentrations of NPs were positively related to human population density, not to the red soil from farmland (Kawahata et al. 2004). NP and BPA pollution on Okinawa Island is considered a consequence of human waste discharge, both domestic and industrial (Kawahata et al. 2004). Since the river mouths with high NP and BPA concentrations are located close to coral reefs, Kawahata et al. (2004) considered both chemicals are likely already affecting the coral reefs off Okinawa and Ishigaki Islands.

Recently, the effects of 4-NP, a common isomer of NP, on endocrine and metabolic homeostasis in the hermatypic coral *Pocillopora damicornis* have been reported (Rougée et al. 2021). Colonies of *P. damicornis* exposed to sublethal concentration of 1.0 μg/L 4-NP for 6 weeks, a period longer than the lunar cycle, exhibited a significant increase in the activity of the steroid-clearing enzyme UDP-glycosyltransferase (UGT) (Rougée et al. 2021). Exposure to 4-NP altered the natural fluctuation in UGT activity with the lunar cycle to consistently high UGT activity throughout the reproductive cycle. Rougée et al. (2021) concluded that environmentally relevant levels of 4-NP may affect the reproductive health of corals and threaten the persistence of coral reefs. Nonylphenol concentration of 0.17 μg/L, which corresponds to approximately 1/6 of the concentration used by Rougée et al. (2021), was reported from the river mouth in Naha City in Jun 2001 (Kawahata et al. 2004). Thus, the results of Rougée et al. (2021) support the consideration of Kawahata et al. (2004) regarding the impact of NP and BPA on coral reefs.

Mano et al. (2017) analysed the concentration of representative nine compounds of PPCPs from six rivers in southern Okinawa Island. High concentrations of caffeine (86–2200 ng/L) were detected in these rivers (Fig. 7.11); the median and 90th percentile of caffeine in 109 major Japanese rivers were 54 ng/L and 310 ng/L, respectively (Komori et al. 2013; Mano et al. 2017). The sampling site in Mano et al. (2017) was located downstream of the discharge area of grey water (domestic wastewater without faecal contamination), industrial effluent, or village drainage facilities. We consume caffeine daily in a variety of beverages (coffee, tea, and caffeinated soft drinks) and numerous food products (chocolate, pastries, and dairy desserts) (Ferreira 2005). Vieira et al. (2022) reported the occurrence of caffeine in the coastline worldwide. The highest concentration of 11,000 ng/L was recorded from the sewage outfall site at the tropical estuary of Darwin Harbour, NT, Australia (French et al. 2015; Vieira et al. 2022). Darwin Harbour is a rapidly expanding city in northern Australia, with a population of above 120,000 in its catchment (French et al. 2015). Aguirre-Martínez et al. (2015) conducted embryo-larval development experiments of the sea urchin *Paracentrotus lividus* exposed to four PPCPs including caffeine. The larval development of *P. lividus* significantly decreased at 0.01 μg/L (10 ng/L) concentration for all four PPCPs (Aguirre-Martínez et al. 2015). The sea urchin showed high sensitivity to zinc pyrithione (ZnPT), which is a widely used anti-fungal ingredient in shampoos and antifouling biocides (Dafforn et al. 2011; Soon et al. 2019; Mangion et al. 2021). In the 1970s, more than 2000 tonnes of sea urchin *Tripneustes gratilla* were caught annually in Okinawa Prefecture, but that amount declined drastically to less than 100 tonnes by around 2010 (Ohta et al. 2017).

In addition to caffeine, benzafibrate, a lipid-regulating drug for hyperlipidaemia, was detected on Okinawa Island at concentrations of 13–320 ng/L, except for one case of 0.27 ng/L (Mano et al. 2017). The maximum detected benzafibrate concentration is approximately 10 times the 90th percentile concentrations in the 109 major rivers in Japan (Komori et al. 2013; Mano et al. 2017). The maximum detected concentration of sulfamethoxazole (140 ng/L) was >15 times higher than the corresponding 90th percentile concentration in major Japanese rivers (Komori et al. 2013; Mano et al. 2017). Sulfamethoxazole, which is the antibiotic agent used for humans and livestock in Japan, is likely discharged through livestock effluents into these rivers (Mano et al. 2017). Tashiro et al. (2003) detected a high concentration of estrone (E1) and equivalent 17β-estradiol (E2) concentration from seawater and sediments at the tidal flat near Naha. They thought that small-scale pig farming located upstream of the rivers that flows into the tidal flat is the major source of these estrogens.

Most sewage treatment waters in Okinawa Prefecture are directly discharged into the ocean (Okinawa Prefecture 2021). Sewage is one of the major land-derived sources of pollution driving the degradation of coral reefs (Wear and Thurber 2015). Figure 7.11 indicates that a high concentration of caffeine (12–858 ng/L) was detected from the secondary effluent waters in a wastewater treatment plant in Naha (Mano et al. 2017). Takeda et al. (2018) conducted an initial health and ecological risk assessment of anthropogenic chemicals contained in treated wastewater in Okinawa Prefecture based on the pollutant release and transfer register (PRTR) system. Of the 66 chemicals estimated, eight substances, including hydrazine, were with high ecological hazard quotient (HQ $\geq$ 1) and listed as compounds that require monitoring (Takeda et al. 2018). Hydrazine is one of the dominant reagents in the fine chemical and

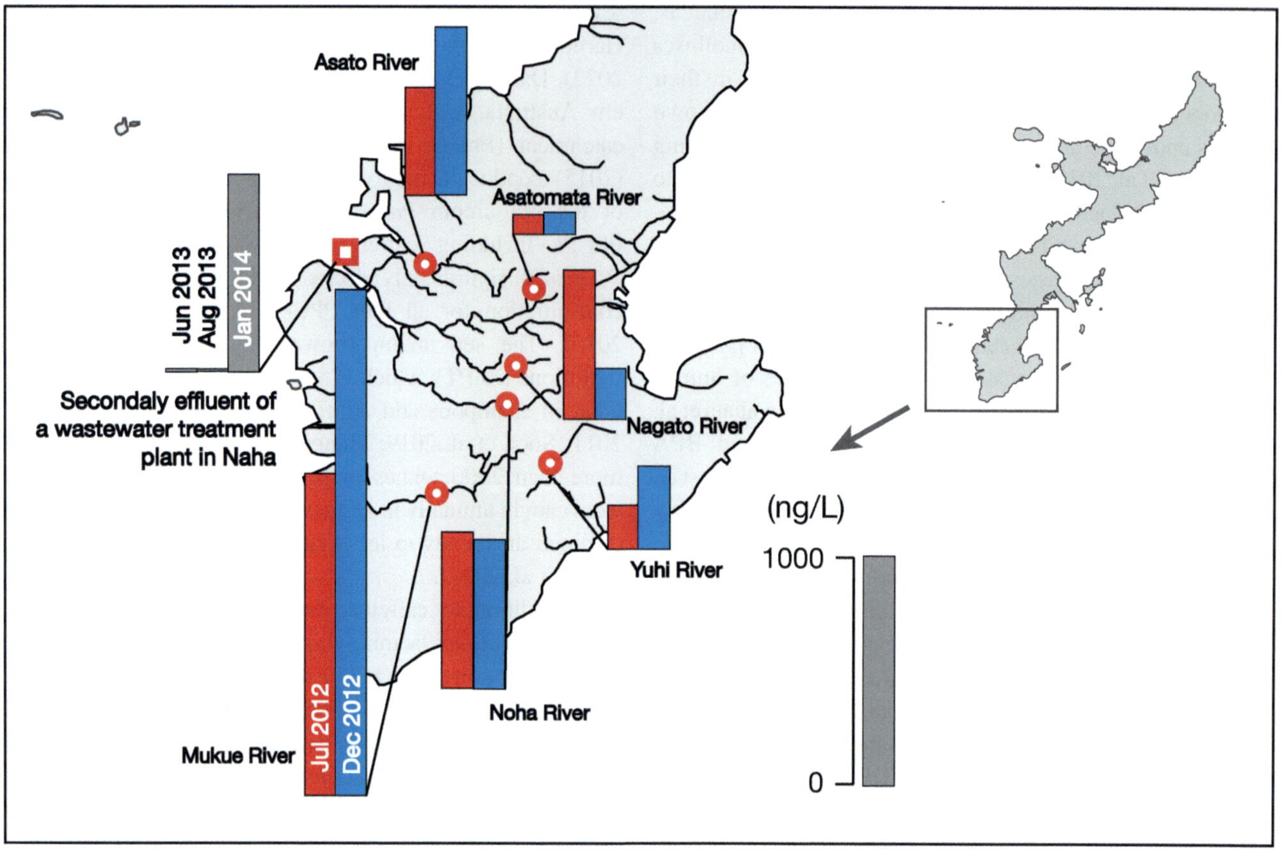

Fig. 7.11 Concentration of caffeine, one of the representatives of PPCPs, in river waters on July 2012 and December 2012 and in secondary the secondary effluent of a wastewater treatment plant from June 2013 to January 2014 in southern Okinawa Island. Sampling sites and numerical data are from Mano et al. (2017, Environ Monitor Assess 189, 442) with permission from Springer. Copyright (2017) Springer International Publishing AG

pharmaceutical industries and is extensively used as a rocket propellant (Niemeier and Kjell 2013). In addition to these high HQ chemicals, the concentrations of five other chemicals were estimated at >10 μg/L in treated wastewater in Okinawa Prefecture (Takeda et al. 2018). Of these five chemicals, linear alkylbenzene sulfonate (LAS) was estimated at maximum levels of 190 μg/L. LAS is one of the most widely used anionic surfactants in the world. The maximum concentrations reported by Takeda et al. (2018) induce severe tissue loss in the hermatypic coral *Pocillopora verrucosa* after 84 h of exposure (Kegler et al. 2015).

Various organisms, including humans, are exposed to novel and interactive combinations of UV radiation and other environmental factors owing to rapid climate change (Barnes et al. 2019). A number of sun-blocking chemicals are used in personal care products as UV filters and UV light stabilisers. In 2011, Tashiro and Kameda (2013) monitored these chemicals at four beaches, two coral reefs, and water from a river mouth around Okinawa Island. They detected 14 species of these chemicals, with the highest level of 1.258 μg/L recorded for 2-hydroxy-4-methoxybenzophenone (BP-3) at a beach in summer (Tashiro and Kameda 2013). Danovaro et al. (2008) reported the effects of UV filters, including BP-3, at low concentrations on hermatypic corals; *Acropora* spp. bleached 24 to 48 h after exposure to 33–50 μL (42.9–65.0 μg)/L of BP-3. The maximum concentrations of 1395 μg/L and 19.2 μg/L of BP-3 were detected from St. John, U.S. Virgin Island and Oahu Island, Hawaii (Downs et al. 2016). Consequently, BP-3 with octinoxate was banned from use in sunscreen in Hawaii on January 1, 2021 (Raffa et al. 2019; Conway et al. 2021). Recently, however, BP-3 was reported not to pose an acute mortality risk to the coral *Galaxea fascicularis* under environmentally relevant concentrations based on three independent acute toxicity tests using 0.31–10 mg/L nominal concentrations (Conway et al. 2021).

Thus, the prompt run-off containing PPCPs reaching rivers, treatment waters, and other terrestrial waters likely influences the health of adjacent reefs. However, the acute risk assessment of BP-3 contained in the sun-blocking chemicals would not induce coral bleaching at environmentally relevant concentrations (Conway et al. 2021). Therefore, the monitoring of various anthropogenic chemicals in these terrestrial waters and eco-toxicological experiments of various compounds are required to elucidate the effects of PPCPs and other anthropogenic chemicals on corals.

7.4.7 Persistent Organic Pollutants (POPs)

Along the coasts of Okinawa Island, the concentration of POPs in the small rock oyster *Saccostrea mordax* was recently reported (Mukai et al. 2020). Small rock oysters were collected from 22 locations across Okinawa Island in 2017 and 2018 (Fig. 7.12), and representative POPs were detected in all samples at median concentrations of 470 ng/g lipid weight (lw) (chlordane compounds; CHLs), 200 ng/g lw (DDTs), 260 ng/g lw (PCBs), and 500 ng/g lw (hexabromocyclododecanes, HBCDs) (Mukai et al. 2020). The highest concentrations of CHLs, PCBs, and DDTs were detected in oysters from Naha and Kadena in the middle to the southern part of the island, whereas the highest concentration of HBCDs was found near Cape Hedo, at the northern end of the island. Expanded polystyrene (EPS) buoys floating on seawater along the coast of Okinawa have been considered potential sources of HBCDs (Mukai et al. 2020). Similar high concentrations of HBCDs (>800 ng/g lw) were detected from the middle to the northern part of the island. This distribution tendency (Fig. 7.12) suggests the possibility: a relatively large number of EPS buoys has landed on the coasts of Okinawa Island from a significantly larger source (i.e., the Yangtze River) via ECS as discussed in Sect. 7.3.; in addition to that EPS buoys are distributed almost uniformly along the coast of Okinawa Island, as discussed by Mukai et al. (2020).

7.4.8 Outbreak of the Crown-of-Thorns Starfish

The outbreak of the crown-of-thorns starfish, *Acanthaster planci* sensu lato, is a significant threshold for coral reefs worldwide. The significant reduction in hermatypic corals reported along the coast of the Ryukyu Islands after the 1970s has been contributed to several outbreaks of the crown-of-thorns starfish (Yasuda 2018). Terrestrial run-off after heavy rainfall is suggested to increase the level of phytoplankton required for the growth and survival of crown-of-thorns starfish larvae (Okaji et al. 2019). Okaji et al. (2019) noted the importance of improving water quality to reduce the recruitment of crown-of-thorns starfish larvae and protect coral reefs.

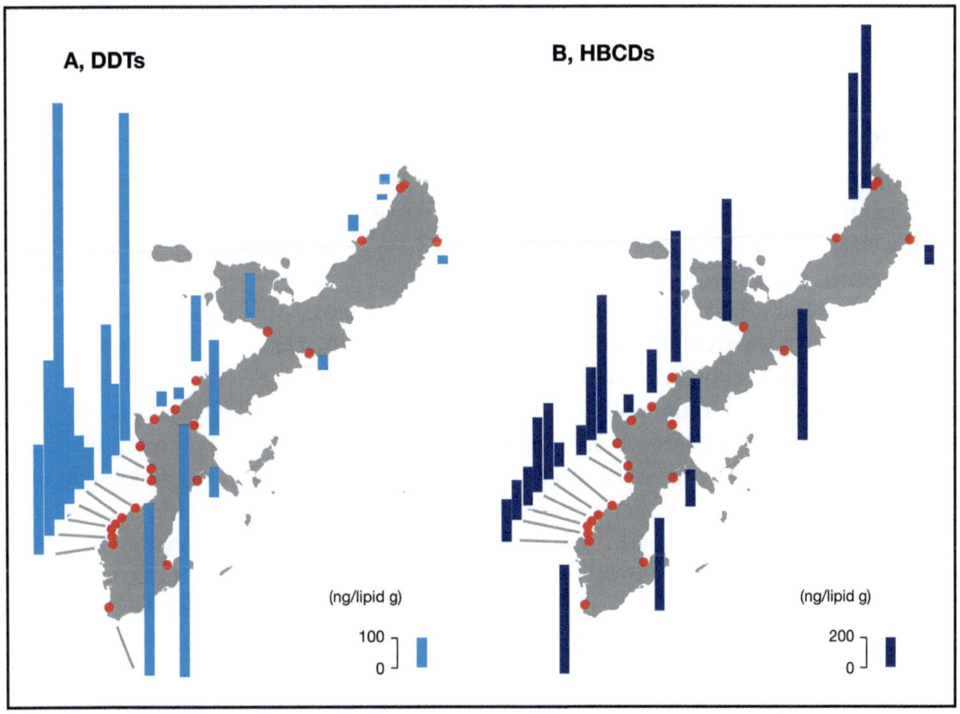

Fig. 7.12 Concentration of dichlorodiphenyltrichloroethanes and its metabolites (DDTs) and hexabromocyclododecanes (HBCDs) in oysters *Saccostrea mordax* collected along the coasts of Okinawa Island from August 2017 to July 2019. Sampling sites and numerical data are from Mukai et al. (2020, Sci Total Environ 739, 140049) with permission from Elsevier. Copyright (2020) Elsevier B.V.

7.4.9 Military Base-Related Anthropogenic Effects

Relatively high concentrations of PCBs, DDTs, and polychlorinated dibenzo-*p*-dioxins (PCDDs) and dibenzofurans (PCDFs) were detected in habu snake, *Protobothrops flavoviridis*, collected within 1 km of the boundary of U.S. military facilities. Recently, Tashiro et al. (2021) reported elevated concentrations of POPs (CHLs, DDTs, and PCBs), and chlorinated dioxins and related compounds in habus collected from urban areas in the southern part of the island in comparison to the rural areas of the northern part of the island. The population density of the southern area is 560 people/km^2, and the area is covered by housing, commercial facilities, and several U.S. military facilities (Tashiro et al. 2021). In contrast, the density of the northern part is quite low at 1.7 people/km^2, and the area is dominated by forest and farmland (Tashiro et al. 2021). The higher median concentrations of CHLs (4400 ng/g lw), DDTs (610 ng/g lw), and PCBs (1600 ng/g lw) in the adipose tissue of habus collected from an urban area suggest the existence of environmental contamination by these organochlorine chemicals in the southern part of Okinawa Island (Tashiro et al. 2021). The concentrations of CHLs in oysters from Kadena-cho and Chatan-cho, both located in the middle of Okinawa Island close to one of the largest U.S. air military bases in far-east Asia, tended to be higher than those from southern locations around Naha, whereas PCB concentrations were higher in oysters from Naha (Mukai et al. 2020). These differences indicate the different POP contamination sources between the area of Kadena and Chatan and the area of Naha. Nearly double PCB concentrations (7.6 ng/g) compared to non-military sites (3.9 ng/g) were reported in mussels collected from formerly used defence sites on Unalaska Island in the Aleutian Archipelago (Adams et al. 2019). High concentrations of PCBs were also detected in salmons from Adak Island, the Aleutian Archipelago, where military bases were located (Hardell et al. 2010). These reports suggest that higher concentrations of PCBs and other POPs accumulate in various biota of coral ecosystems near military facilities. Tashiro et al. (2021) highlighted the need to survey these chemicals inside or near military facilities to elucidate whether pollution sources are present on military bases on Okinawa Island.

Per- and polyfluoroalkyl substances (PFASs) are a group of persistent anthropogenic contaminants that include perfluoroalkane sulfonic acids (PFSAs; such as perfluorooctane sulfonate, PFOS) and perfluoroalkyl carboxylic acids (PFCAs; which include perfluorooctanoic acid, PFOA) (Catherine et al. 2019; Yukioka et al. 2020). These chemicals have been used in various kinds of products including firefighting foam (Yukioka et al. 2020). Similar to POPs, PFASs exhibit persistence, bioaccumulation, and toxicity and are globally distributed in the environment. Therefore, PFOS and its related compounds were included in the Stockholm Convention list of POPs in 2009 (Catherine et al. 2019). Long-chain perfluoroalkyl acids and some of their precursors were specifically detected around the firefighting training area of the U.S. Air Military Base at Kadena (Yukioka et al. 2020). Higher concentrations of PFASs were also detected in three rivers, Hija, Nagata, and Tengan Rivers, of Kadena and Uruma City located in the central part of Okinawa Island (Shiokawa and Tamaki 2017). The high concentration of lead in the hermatypic coral *Porites* in Hija River mouth indicates anthropogenic inputs of the river run-off from upstream areas (Ramos et al. 2004).

Mitchell (2020) noted that PFAS contamination in Okinawa Prefecture originates from three main sources: firefighter training, accidental leaks of aqueous film-forming foam (AFFF), and AFFF disposal. In 2016, most residents of Okinawa Prefecture first learnt of PFAS contamination when the Prefecture government announced elevated levels of PFOS detected in rivers near the drinking water facility (Mitchell 2020). The main source of drinking water for Okinawa Prefecture is Chatan Water Treatment Plant adjacent to the U.S. Air Military Base (Mitchell 2020). In June 2020, the Ministry of the Environment, Japan, revealed that the concentration of PFOS in the river running through the base was 1462 ng/L, the highest in Japan (Mitchell 2020). Thus, although the red soil run-off from the U.S. military base was reduced significantly, the monitoring of anthropogenic chemicals including POPs and PFASs around the bases and adjacent waters is urgently required. Tokuda and Barnett (2017) pointed out that the people of Okinawa were pointed out to have already experienced several adverse health and environmental effects from the military bases locating Okinawa Island.

7.5 Coastal Waters-Based Anthropogenic Impacts

Organotin compounds (OTs), including tributyltin (TBT) and triphenyltin (TPhT), have been used as fungicides and insecticides in agricultural activities, for industrial impregnation of textiles, and as biocides in antifouling paints on ship hulls and fish cages at mariculture sites (Dafforn et al. 2011; Sham et al. 2020). With the discovery of the antifouling efficacy of trialkyltins, TBT has been used in antifouling paints since the 1960s (Dafforn et al. 2011; Sham et al. 2020). However, the extremely high toxicity of TBT to marine invertebrates at concentrations of several nanograms per litre has been reported. Alzieu (2000) mentioned that low levels of TBT ($<$2 ng/L) disturbed the masculinisation of female gastropods known as imposex, affected phyto- and zooplankton growth, and inhibited calcification of the pacific oyster *Crassostrea gigas*, which resulted in gelatinous pockets inside the shell. In Japan, legislation restricting TBT usage as an antifouling agent for coastal boats and aquaculture construction was implemented in 1990 (Takeuchi et al. 2004). In 2008, the International Convention on the Control of Harmful Antifouling Systems on Ships (AFS Convention) prohibited OT-based antifouling paints on ship hulls (Dafforn et al. 2011; Sham et al. 2020).

The OTs dibutyltin (DBT), monobutyltin (MBT), TBT, TPhT, diphenyltin (DPhT), and monophenyltin (MPhT) were reported along the coast of Okinawa in the first decade of the twenty-first century (Takeuchi et al. 2004; Sheikh et al. 2007; Imo et al. 2008). Four di- and mono-OTs are metabolised deliveries of TBT or TPhT. High concentrations of TBT (160 ng/L) were detected in the seawater of Naha Port (Takeuchi et al. 2004). The concentrations of TBT in seawater ranged from not detectable to 26.1 ng/L in the island-wide survey conducted by Imo et al. (2008) along Okinawa Island during June–September 2002. While the highest concentration of TBT, 26.1 ng/L, was detected from Itoman Port in the southern part, increased concentrations >2.4 ng/L were measured across the island including its northern, less populated part (5.47 ng/L at Kunigami-son and 7.65 ng/L at Motobu-cho on the northwest coast). A relatively high concentration of TBT has also been detected in the less populated areas of western Japan (Takeuchi et al. 2004).

Chen et al. (2019) reported the concentrations of OTs (TBT, DBT, MBT, TPhT, DPhT, and MPhT) in oysters collected along the Chinese coast during 2014–2015. The ΣOTs in commercial oysters from seafood markets in Shanghai ranged from 648 to 5705 ng/g dry weight (dw); maximum concentrations of TBT and TPhT were 166 ng/g dw and 2202 ng/g dw, respectively (Chen et al. 2019). A higher concentration (425–4376 ng/g dw) of TPhT was detected in wild oysters from south of the Yangtze River than north of the river in 13–83 ng/g dw owing to the wider usage of TPhT-based biocides in local mariculture and agriculture (Chen et al. 2019). Furthermore, a high concentration of TPhT, ranging from 3.8 to 11.7 ng/L in seawater, 71.8 to 91.7 ng/g dw in sediment, and 9.6 to 1079.9 ng/g wet weight in biotic samples (molluscs, crustaceans, and fish) based on 2015–2016 sampling, has also been reported in marine environments around Hong Kong (Sham et al. 2020). This high concentration of TPhT in Hong Kong poses a potential risk to the subtidal marine ecosystem and health risks to humans via the consumption of highly contaminated commercial fish (Sham et al. 2020).

After the global ban on the use of TBT, butyltin-containing antifouling paints were replaced with copper compounds, which were predominantly used before the commercialisation of TBT (Dafforn et al. 2011). Several 'booster' agents including Irgarol and Diuron are mixed in these copper-based antifouling paints to prevent the attachment of algae that are tolerant to copper (Dafforn et al. 2011). These herbicides have been reported to inhibit the photosynthetic activity of hermatypic corals. The EC_{50} for the maximum effective quantum yield ($\Delta F/F_m'$) of hermatypic coral *Seriatopora hystrix* after 10 h exposure to Irgarol and Diuron was 0.7 ± 0.07 μg/L and 2.3 ± 0.04 μg/L, respectively (Jones and Kerswell 2003).

Higher concentrations of Irgarol (>1.0 μg/L) were recorded in areas adjacent to the ECS, namely Hong Kong, Peninsular Malaysia, and Singapore (Basheer et al. 2002; Lam et al. 2005; Ali et al. 2013). Most of the sites with >1.0 μg/L were located in international ports or adjacent areas, although 1.37 μg/L was recorded from the local port for leisure and high-speed traffic boats at Redang Island, a coral reef resort island on the east coast of Peninsular Malaysia (Ali et al. 2013). The exposure of cultured *Acropora tenuis*, which was initially collected from the west coast of Okinawa Island, to 1.0 μg/L Irgarol led to coral bleaching after 10–12 d (Hirayama et al. 2017; Takeuchi et al. 2020).

The concentration of Irgarol was investigated in commercial fishing ports in central and southern areas of Okinawa Island and Naha Bay from 2007 to 2008 (Sheikh et al. 2009b) (Fig. 7.13). The average concentration of Irgarol detected at commercial fishery ports was 24.70 ± 9.88 ng/L, and its concentration in Naha Bay was 10.00 ± 12.98 ng/L (Sheikh et al. 2009b). The relatively constant mean concentration of Irgarol in commercial and fishery ports suggests that Irgarol-containing antifouling paints were used widely along the coast of Okinawa Island at least around 2007–2008. Furthermore, the mean Irgarol concentration detected in commercial and fishery ports corresponds to the concentration that reduces $\Delta F/F_m'$ of *Acropora tenuis* reported by Kamei et al. (2020). Recently, *Acropora tenuis*, one of the representative species of *Acropora* from the Ryukyu Archipelago, showed a significant 8% reduction in $\Delta F/F_m'$ after 7 days of exposure to 20 ng (0.02 μg)/L of Irgarol (Kamei et al. 2020).

Tanita and Onduka (2021) registered Diuron at concentrations from <0.6 to 32 ng/L along the coasts of Ishigaki and the Iriomote Islands in September 2019. Relatively high concentrations of Diuron of 19 and 32 ng/L were detected at a boat mooring on Ishigaki Island and in fishing ports of Iriomote Island, respectively, which have been attributed to antifouling paints applied to boats (Tanita and Onduka 2021). Diuron was also detected in Naha Port and its adjacent areas, as reviewed in Sect. 7.4.5 Island-Based Anthropogenic Impacts. Herbicides and Pesticides.

Recently, Zhang et al. (2021) reported that an extremely high concentration of Diuron up to 3056 ng/L was detected from mariculture areas in nearshore waters of the north Yellow Sea, China. The probabilistic risk assessment by Zhang et al. (2021) indicated an adverse effect of Diuron on up to 8% of the aquatic species in nearshore areas.

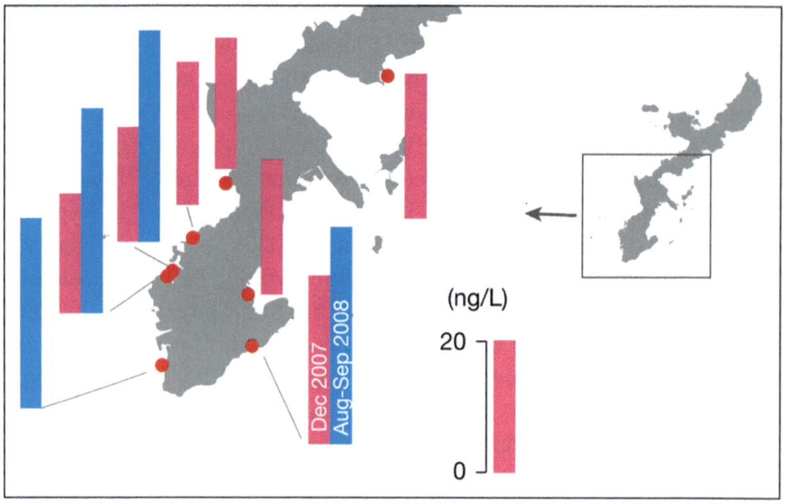

Fig. 7.13 Concentration of Irgarol 1051 in seawater along the coasts of southern Okinawa Island, Japan from December 2007 to August 2008. Sampling sites and concentration data are obtained from Sheikh et al. (2009b, Int J Environ Sci Tech 6, 353–358) with permission from Springer Nature. Copyright (2009) Iranian Society of Environmentalists (IRSEN), Center for Environment and Energy Research and Studies (CEERS), Islamic Azad University (IAU)

7.6 Tourism

The economic ripple effect of travel and tourism in Okinawa Prefecture in ficancial year 2015 was estimated at 1.01 trillion yen (Okinawa Prefecture 2018a). Of that, 0.49 trillion yen was the value-added effect, corresponding to approximately 12% of the gross regional product (4.14 trillion yen) in Okinawa Prefecture (Okinawa Prefecture 2018a). Sea bathing and marine leisure account for 24.8–28.4% and diving for 6.4–7.8% of tourist activities to Okinawa Prefecture (Okinawa Prefecture 2018b).

Although the conservation of the coral reef ecosystem at the Ryukyu Archipelago is of value as a tourism resource for the local community, excessive and inappropriate utilisation of these marine areas is a matter of concern (The Ministry of the Environment, 2017[2016]). The coasts of Okinawa Prefecture have been estimated to attract >1500 scuba divers, >2000 snorkelers, and >120 boats per day (The Ministry of the Environment 2017[2016]). The Ministry of the Environment (2017[2016]) conducted a basic survey of the impact of diving tourism in surrounding areas of the headland of Onnason, on the west coast of Okinawa Island. The survey revealed that coral coverage has decreased from approximately 65% to <20% with increasing diving activity, while the density of small colonies of *Acropora* spp. remained stable (The Ministry of the Environment, 2017[2016]). During the monitoring of the underwater behaviour of SCUBA divers in Okinawa, Japan, 7% of divers contacted live coral cover causing skeletal breakage, while 91% made contact at least once but did not cause visible damage (Toyoshima and Nadaoka 2015). Toyoshima and Nadaoka (2015) also reported that pre-dive briefings and buoyancy control skills during SCUBA diving could reduce physical contacts with reefs.

7.7 Conclusion

Several anthropogenic activities are present around the coral reefs of the Ryukyu Islands facing the ECS. The increased seawater temperature and acidification rate in the ECS compared to those in the other oceans are mitigated along the Ryukyu Archipelago by the complex topography. The inputs of anthropogenic origin, including nutrients and chemicals, to the ECS from the Yangtze River have increased lately owing to recent economic development. Recent monitoring suggests that anthropogenic impacts from China arrived to the coastal regions of the Ryukyu Archipelago. These inputs flow along the coastlines of the Ryukyu Archipelago, affecting the ecosystems. The coastlines along the archipelago have already been largely altered. The high rainfall of 2000 mm/yr and the shorter length of the rivers on the Ryukyu Islands generate a prompt discharge of red soil run-off, river waters, and other terrestrial waters into the coastal region. The input of red soils is considered one of the main factors damaging coral reefs, and reducing nutrients are numerically targeted for maintaining healthy coral reefs; however, other factors should be considered when establishing water quality guidelines for coral reef protection. The treatment and river waters in southern Okinawa Island may contain relatively high concentrations of PPCPs. Various anthropogenic chemicals, including POPs, herbicides (such as Diuron), and anthropogenic chemicals contained in antifouling paints (TBT and Irgarol), were detected in coastal waters around the Ryukyu Archipelago. Although previous studies have mentioned the importance of monitoring these anthropogenic chemicals to assess their potential health risk to coral reefs, anthropogenic chemicals were not included in the list of monitored items in action plans to protect coral reefs from red soil run-off in Japan. The red soil run-off from military facilities has been reduced, and yet the possibility that military facilities on Okinawa Island are the source of anthropogenic chemicals in coastal areas has not been scrutinised. Thus, the detailed monitoring of various anthropogenic activities is an urgent research topic for protecting coral reefs around the ECS.

Acknowledgments The concept of the present review was mainly formed by several research grants from the Japan Society for the Promotion of Science (JSPS) Grants-in-Aid for Scientific Research (KAKENHI; Grant nos. 26281048, 15K12253, 18H03387, and 21H03619). I would like to thank Editage (www.editage.com) for English language editing of this manuscript.

References

Adams EM, von Hippel FA, Hungate BA, Buck CL (2019) Polychlorinated biphenyl (PCB) contamination of subsistence species on Unalaska Island in the Aleutian Archipelago. Heliyon 5: e02989. https://doi.org/10.1016/j.heliyon.2019.e02989

Aguirre-Martínez GV, Owuor MA, Garrido-Pérez C, Salamanca MJ, Del Valls TA, Martín-Díaz ML (2015) Are standard tests sensitive enough to evaluate effects of human pharmaceuticals in aquatic biota? Facing changes in research approaches when performing risk assessment of drugs. Chemosphere 120:75–85

Ali HR, Arifin MM, Sheikh MA, Shazili NAM, Bachok Z (2013) Occurrence and distribution of antifouling biocide Irgarol-1051 in coastal waters of Peninsular Malaysia. Mar Pollut Bull 70:253–257

Alzieu C (2000) Environmental impact of TBT: the French experience. Sci Total Environ 258:99–102

Amid C, Olstedt M, Gunnarsson JS, Le Lan H, Tran Thi Minh H, Van den Brink PJ, Hellström M, Tedengren M (2018) Additive effects of the herbicide glyphosate and elevated temperature on the branched coral *Acropora formosa* in Nha Trang, Vietnam. Environ Sci Pollut Res 25:13360–13372

Anderson ZT, Cundy AB, Croudace IW, Warwick PE, Celis-Hernandez O, Stead JL (2018) A rapid method for assessing the accumulation of microplastics in the sea surface microlayer (SML) of estuarine systems. Sci Rep 8:9428. https://doi.org/10.1038/s41598-018-27612-w

Bao L-J, Maruya KA, Snyder SA, Zeng EY (2012) China's water pollution by persistent organic pollutants. Environ Pollut 163:100–108

Barnes PW, Williamson CE, Lucas RM, Robinson SA, Madronich S, Paul ND, Bornman JF, Bais AF, Sulzberger B, Wilson SR, Andrady AL, McKenzie RL, Neale PJ, Austin AT, Bernhard GH, Solomon KR, Neale RE, Young PJ, Norval M, Rhodes LE, Hylander S, Rose KC, Longstreth J, Aucamp PJ, Ballare CL, Cory RM, Flint SD, de Gruijl FR, Hader DP, Heikkila AM, Jansen MAK, Pandey KK, Matthew RT, Sinclair CA, Wangberg S, Worrest RC, Yazar S, Young AR, Zepp RG (2019) Ozone depletion, ultraviolet radiation, climate change and prospects for a sustainable future. Nat Sustain 2: 569–579

Basheer C, Tan KS, Lee HK (2002) Organotin and Irgarol-1051 contamination in Singapore coastal waters. Mar Pollut Bull 44:697–703

Belkin IM, Lee MA (2014) Long-term variability of sea surface temperature in Taiwan Strait. Clim Chang 124:821–834

Catherine M, Nadège B, Charles P, Yann A (2019) Perfluoroalkyl substances (PFASs) in the marine environment: spatial distribution and temporal profile shifts in shellfish from French coasts. Chemosphere 228:640–648

Chen C, Chen L, Huang Q, Chen Z, Zhang W (2019) Organotin contamination in commercial and wild oysters from China: increasing occurrence of triphenyltin. Sci Total Environ 650:2527–2534

Chen C-TA, Wang S-L, Lu X-X, Zhang S-R, Lui H-K, Tseng H-C, Wang B-J, Huang H-I (2008) Hydrogeochemistry and greenhouse gases of the Pearl River, its estuary and beyond. Quatern Int 186:79–90

Chen C-TA, Yu S, Huang T-H, Lui H-K, Bau Y, He X (2020a) Changing biogeochemistry in the South China Sea. In: Chen C-TA, Guo X (eds) Changing Asia-Pacific marginal seas. Springer, Singapore, pp 203–206

Chen J, Li D, Jin H, Jiang Z, Wang B, Wu B, Hao Q, Sun X (2020b) Changing nutrients, oxygen and phytoplankton in the East China Sea. In: Chen C-TA, Guo X (eds) Changing Asia-Pacific marginal seas. Springer, Singapore, pp 155–178

Chou W-C, Gong G-C, Cai W-J (2012) Seasonality of CO_2 in coastal oceans altered by increasing anthropogenic nutrient delivery from large rivers: evidence from the Changjiang-East China Sea system. Biogeosci Discuss 9:18993–19017

Conway AJ, Gonsior M, Clark C, Heyes A, Mitchelmore CL (2021) Acute toxicity of the UV filter oxybenzone to the coral *Galaxea fascicularis*. Sci Total Environ 796:148666

Dafforn KA, Lewis JA, Johnston EL (2011) Antifouling strategies: history and regulation, ecological impacts and mitigation. Mar Pollut Bull 62:453–465

Danovaro R, Bongiorni L, Corinaldesi C, Giovannelli D, Damiani E, Astolfi P, Greci L, Pusceddu A (2008) Sunscreens cause coral bleaching by promoting viral infections. Environ Health Perspect 116:441–447

Downs CA, Kramarsky-Winter E, Segal R, Fauth J, Knutson S, Bronstein O, Ciner FR, Jeger R, Lichtenfeld Y, Woodley CM, Pennington P, Cadenas K, Kushmaro A, Loya Y (2016) Toxicopathological effects of the sunscreen UV filter, oxybenzone (benzophenone-3), on coral planulae and cultured primary cells and its environmental contamination in Hawaii and the U.S. Virgin Islands. Arch Environ Contam Toxicol 70:265–288

Ferreira AP (2005) Caffeine as an environmental indicator for assessing urban aquatic ecosystems. Cad Saúde Pública 21:1884–1892

Fiedler H, Kallenborn R, de Boer J, Sydnes LK (2019) The Stockholm convention: a tool for the global regulation of persistent organic pollutants. Chem Int 41:4–11

Fraser TW, Khezri A, Lewandowska-Sabat AM, Henry T, Ropstad E (2017) Endocrine disruptors affect larval zebrafish behavior: testing potential mechanisms and comparisons of behavioral sensitivity to alternative biomarkers. Aqua Toxicol 193:128–135

French VA, King SC, Kumar A, Northcott G, McGuinness K, Parry D (2015) Characterisation of microcontaminants in Darwin harbour, a tropical estuary of northern Australia undergoing rapid development. Sci Total Environ 536:639–647

Fujieda S, Kojima A, Kanehiro H (2006) Monitoring marine debris using disposable lighters as indicator. J Japan Soc Waste Manage Experts 17:117–124. (In Japanese with English abstract)

Fujieda S, Ohkura Y, Kojima A (2014) Flow of marine litter around Japan visualized by the relationship of outflow locations and collected sites of the drifted disposable lighters. J Japan Drift Soc 12:29–42. (In Japanese with English abstract)

Gan J, Liu Z, Hui CR, Tang Y, Cai Z, Li J (2020) The changing circulation of Asia-Pacific marginal seas in the South China Sea. In: Chen C-TA, Guo X (eds) Changing Asia-Pacific marginal seas. Springer, Singapore, pp 179–201

Gong G-C, Liu K-K, Chiang K-P, Hsiung T-M, Chang J, Chen C-C, Hung C-C, Chou W-C, Chung C-C, Chen H-Y, Shiah F-K, Tsai A-Y, Hsieh C-H, Shiao J-C, Tseng C-M, Hsu S-C, Lee H-J, Lee M-A, Lin I-I, Tsai F (2011) Yangtze River floods enhance coastal ocean phytoplankton biomass and potential fish production.

Geophys Res Lett 38:L13603. https://doi.org/10.1029/2011GL047519
Grung M, Lin Y, Zhang H, Steen AO, Huang J, Zhang G, Larssen T (2015) Pesticide levels and environmental risk in aquatic environments in China—a review. Environ Int 81:87–97
Hano T, Ito K, Ohkubo N, Sakaji H, Watanabe A, Takashima K, Sato T, Sugaya T, Matsuki K, Onduka T, Ito M, Mochida K (2019) Occurrence of neonicotinoids and fipronil in estuaries and their potential risks to aquatic invertebrates. Environ Pollut 252:205–215
Hardell S, Tilander H, Welfinger-Smith G, Burger J, Carpenter DO (2010) Levels of polychlorinated biphenyls (PCBs) and three organochlorine pesticides in fish from the Aleutian Islands of Alaska. PLoS One 5:e12396. https://doi.org/10.1371/journal.pone.0012396
Harii S, Hongo C, Ishihara M, Ide Y, Kayanne H (2014) Impacts of multiple disturbances on coral communities at Ishigaki Island, Okinawa, Japan, during a 15 year survey. Mar Ecol Prog Ser 509: 171–180
Hashimoto K (2005) Characteristics of coral reef islands and impacts of agriculture on coral reefs. In: JIRCAS 2005 International Symposium in Ishigaki, pp 33–38
Heery EC, Hoeksema BW, Browne NK, Reimer JD, Ang PO, Huang D, Friess DA, Chou LM, Loke LHL, Saksena-Taylor P, Alsagoff N, Yeemin T, Sutthacheep M, Vo ST, Bos AR, Gumanao GS, Hussein MAS, Waheed Z, Lane DJW, Johan O, Kunzmann A, Jompa J, Suharsono TD, Bauman AG, Todd PA (2018) Urban coral reefs: degradation and resilience of hard coral assemblages in coastal cities of East and Southeast Asia. Mar Pollut Bull 135:654–681
Higa E, Omija T, Hanashiro K, Mitsumoto H (1995) Quantity of soil run-off on Okinawa Island. Ann Rep Okinawa Pref Inst Health Environ 29:83–88. (In Japanese)
Higa E, Nakasone K, Omija T, Mitumoto H (2001) Pollutants run-off from mouth of rivers to sea in Okinawa Island. Ann Rep Okinawa Pref Inst Health Environ 40:107–113. (In Japanese)
Higuchi T, Takagi KK, Matoba K, Kobayashi S, Tsurumi R, Arakaki S, Nakano Y, Fujimura H, Oomori T, Tsuchiya M (2014) The nutrient and carbon dynamics that mutually benefit coral and seagrass in mixed habitats under the influence of groundwater at Bise coral reef, Okinawa, Japan. Int J Mar Sci 4:1–15
Hirayama K, Takayama K, Haruta S, Ishibashi H, Takeuchi I (2017) Effect of low concentrations of Irgarol 1051 on RGB (R, red; G, green; B, blue) colour values of the hard-coral *Acropora tenuis*. Mar Pollut Bull 124:678–686
Hongo C, Kiguchi M (2021) Assessment to 2100 of the effects of reef formation on increased wave heights due to intensified tropical cyclones and sea level rise at Ishigaki Island, Okinawa, Japan. Coast Eng J 63:216–226
Hongo C, Yamano H (2013) Species-specific responses of corals to bleaching events on anthropogenically turbid reefs on Okinawa Island, Japan, over a 15-year period (1995–2009). PLoS One 8: e60952. https://doi.org/10.1371/journal.pone.0060952
Huang Y, Chen M, Wang Z, Jiang L, Fan S, Zheng R, Yu X (2022) Impacts of terrestrial input on the distribution characteristics of microplastics in the East China Sea characterized by chromophoric dissolved organic matter (CDOM) analysis. Sci Total Environ 838: 156599
Imo ST, Sheikh MA, Sawano K, Fujimura H, Oomori T (2008) Distribution and possible impacts of toxic organic pollutants on coral reef ecosystems around Okinawa Island, Japan. Pacif Sci 62:317–326
Isobe A, Uchida K, Tokai T, Iwasaki S (2015) East Asian seas: a hot spot of pelagic microplastics. Mar Pollut Bull 101:618–623
Japan Meteorological Agency (2021) Long-term trend of sea surface temperature change in the sea around Japan published on 3 March 2021. https://www.data.jma.go.jp/kaiyou/data/shindan/a_1/japan_warm/japan_warm.html. Accessed 21 Aug 2021. (In Japanese)
Japan Plant Protection Association (JPPA) (2021) Pesticides Catalogue 2020. JPPA, Kita-ku, Tokyo, 774 pp. (In Japanese)
Jones RJ, Kerswell AP (2003) Phytotoxicity of photosystem II (PSII) herbicides to coral. Mar Ecol Prog Ser 261:149–159
Kamei M, Takayama K, Ishibashi H, Takeuchi I (2020) Effects of ecologically relevant concentrations of Irgarol 1051 in tropical to subtropical coastal seawater on hermatypic coral *Acropora tenuis* and its symbiotic dinoflagellates. Mar Pollut Bull 150:110734
Kawahata H, Ohta H, Inoue M, Suzuki A (2004) Endocrine disrupter nonylphenol and bisphenol a contamination in Okinawa and Ishigaki Islands, Japan—within coral reefs and adjacent river mouths. Chemosphere 55:1519–1527
Kayanne H, Hongo C, Yamano H (2004) Distribution of coral reefs. In: Tsuchiya M, Nadaoka K, Kayanne H, Yamano H (eds) Coral reefs of Japan. Ministry of the Environment, Tokyo and The Japanese Coral Reef Society, pp 64–68. (In Japanese)
Keerthanan S, Jayasinghe C, Biswas JK, Vithanage M (2021) Pharmaceutical and personal care products (PPCPs) in the environment: plant uptake, translocation, bioaccumulation, and human health risks. Crit Rev Environ Sci Technol 51:1221–1258
Kegler P, Baum G, Indriana LF, Wild C, Kunzmann A (2015) Physiological response of the hard coral *Pocillopora verrucosa* from Lombok, Indonesia, to two common pollutants in combination with high temperature. PLoS One 10:e0142744. https://doi.org/10.1371/journal.pone.0142744
Kinjo K (2017) Measures to reduce negative impacts of terrestrial impacts on coral reefs. J Japan Coral Reef Soc 19:15–21. (In Japanese with English abstract)
Kinjo K, Higa E, Oshiro Y (2006) Nutrient state in coral reef coastal sea around Okinawa. Ann Rep Okinawa Pref Inst Health Environ 40: 107–113. (In Japanese with English abstract)
Kitada Y, Kawahata H, Suzuki A, Oomori T (2008) Distribution of pesticides and bisphenol a in sediments collected from rivers adjacent to coral reefs. Chemosphere 71:2082–2090
Kitada Y, Kawahata H, Suzuki A, Oomori T (2009) Environmental pollution by toxic chemicals - contamination of pesticides within coral reefs and adjacent river mouths in the Ryukyu islands. Chishitsu News 656:39–44. (In Japanese)
Komori K, Suzuki Y, Minamiyama M, Harada A (2013) Occurrence of selected pharmaceuticals in river water in Japan and assessment of their environmental risk. Environ Monitor Assess 185:4529–4536
Konda M, Ichikawa H, Han I-S, Zhu X-H, Ichikawa K (2005) Variability of current structure due to meso-scale eddies on the bottom slope southeast of Okinawa Island. J Oceanogr 61:1089–1099
Kurihara H, Wounter J, Yasuda N (2019) Seasonal calcification of the coral Acropora digitifera from a subtropical marginal Okinawa reef under ocean acidification. Coral Reefs 38:443–454
Lam K-H, Cai Z, Wai H-Y, Tsang VW-H, Lam MH-W, Cheung RY-H, Yu H, Lam PK-S (2005) Identification of a new Irgarol-1051 related *s*-triazine species in coastal waters. Environ Pollut 136: 221–230
Lewis SE, Brodie JE, Bainbridge ZT, Rohde KW, Davis AM, Masters BL, Maughan M, Devlin MJ, Mueller JF, Schaffelke B (2009) Herbicides: a new threat to the great barrier reef. Environ Pollut 157:2470–2484
Li X, Zhang K, Gu P, Feng H, Yin Y, Chen W, Cheng B (2021a) Changes in precipitation extremes in the Yangtze River Basin during 1960–2019 and the association with global warming, ENSO, and local effects. Sci Total Environ 760:144244
Li Y, Wang Q, Li Q, Liu Y, Wang Y (2021b) An asymmetric variation of hot and cold SST extremes in the China seas during the recent warming hiatus period. Sci Rep 11:2014. https://doi.org/10.1038/s41598-020-79854-2
Liu X, Beusen AHW, Van Beek LPH, Mogollón JM, Ran X, Bouwman AF (2018) Exploring spatiotemporal changes of the Yangtze River (Changjiang) nitrogen and phosphorus sources, retention and export to the East China Sea and Yellow Sea. Water Res 142:246–255

Liu Z, Gan J, Hu J, Wu H, Cai Z, Deng Y (2021) Progress on circulation dynamics in the East China Sea and southern Yellow Sea: origination, pathways, and destinations of shelf currents. Prog Oceanogr 193:102553

Lohmann R, Breivik K, Dachs J, Muir D (2007) Global fate of POPs: current and future research directions. Environ Pollut 150:150–165

Lui HK, Chen C-TA, Lee J, Wang S-L, Gong G-C, Bai Y, He X (2015) Acidifying intermediate water accelerates the acidification of seawater on shelves: an example of the East China Sea. Cont Shelf Res 111:223–233

Lui HK, Chen CTA, Hou W-P, Yu S, Chan J-W, Bai Y, He Z (2020) Transient carbonate chemistry in the expanded Kuroshio Region. In: Chen C-TA, Guo X (eds) Changing Asia-Pacific marginal seas. Springer, Singapore, 307–320

Ma T, Sun S, Fu G, Hall JW, Ni Y, He L, Yi J, Zhao N, Du Y, Pei T, Cheng W, Song C, Fang C, Zhou C (2020) Pollution exacerbates China's water scarcity and its regional inequality. Nat Commun 11: 650. https://doi.org/10.1038/s41467-020-14532-5

Mai L, He H, Bao L-J, Liu L-Y, Zeng EY (2020a) Plastics are an insignificant carrier of riverine organic pollutants to the coastal oceans. Environ Sci Technol 54:15852–15860

Mai L, Sun X-F, Xia L-L, Bao L-J, Liu L-Y, Zeng EY (2020b) Global riverine plastic outflows. Environ Sci Technol 54:10049–10056

Mangion SE, Holmes AM, Roberts MS (2021) Targeted delivery of zinc pyrithione to skin epithelia. Int J Mol Sci 22:9730. https://doi.org/10.3390/ijms22189730

Mano H, Takeda F, Kitamura T, Okamoto S, Suzuki Y, Park C-B, Yasui N, Kobayashi K, Tanaka Y, Yamashita N, Minamiyama M (2017) Water quality comparison of secondary effluent and reclaimed water to ambient river water of southern Okinawa Island via biological evaluation. Environ Monitor Assess 189:442

Masucci GD, Reimer JD (2019) Expanding walls and shrinking beaches: loss of natural coastline in Okinawa Island, Japan. PeerJ 7:e7520. https://doi.org/10.7717/peerj.7520

Masucci GD, Acierno A, Reimer JD (2020) Eroding diversity away: impacts of a tetrapod breakwater on a subtropical coral reef. Aquat Conserv Mar Freshwater Ecosyst 30:290–302

Matozzo V, Fabrello J, Marin MG (2020) The effects of glyphosate and its commercial formulations to marine invertebrates: a review. J Mar Sci Engineer 8:399. https://doi.org/10.3390/jmse8060399

Matsuno T (2020) The changing East China sea—a physical view. In: Chen C-TA, Guo X (eds) Changing Asia-Pacific marginal seas. Springer, Singapore, pp 139–153

McClanahan TR, Maina JM, Darling ES, Guillaume MMM, Muthiga NA, D'agata S, Leblond J, Arthur R, Jupiter SD, Wilson SK, Mangubhai S, Ussi AM, Humphries AT, Patankar V, Shedrawi G, Julius P, Ndagala J, Grimsditch G, Grimsditch G (2020) Large geographic variability in the resistance of corals to thermal stress. Glob Ecol Biogeogr 29:2229–2247

Milan M, Dalla Rovere G, Smits M, Ferraresso S, Pastore P, Marin MG, Bogialli S, Patarnello T, Bargelloni L, Matozzo V (2018) Ecotoxicological efects of the herbicide glyphosate in non-target aquatic species: transcriptional responses in the mussel *Mytilus galloprovincialis*. Environ Pollut 237:442–451

Mitchell J (2020) PFAS Contamination from US military facilities in mainland Japan and Okinawa. Asia-Pac J Japan Focus 18:5457

Moon JH, Song YT (2017) Decadal Sea level variability in the East China Sea linked to the North Pacific gyre oscillation. Continet Shelf Res 143:278–285

Mukai Y, Goto A, Tashiro Y, Tanabe S, Kunisue T (2020) Coastal biomonitoring survey on persistent organic pollutants using oysters (*Saccostrea mordax*) from Okinawa, Japan: geographical distribution and polystyrene foam as a potential source of hexabromocyclododecanes. Sci Total Environ 739:140049

Nakamura H (2020) Changing Kuroshio and its affected shelf sea: a physical view. In: Chen C-TA, Guo X (eds) Changing Asia-Pacific marginal seas. Springer, Singapore, pp 265–304

Nakandakari T, Yoshinaga A, Sakai K, Akiyoshi Y, Osawa K (2007) Runoff of suspended soil and nutrients from an upland watershed Okinawa Island. J Rainwater Catchment Syst 13:37–43. (In Japanese with English abstract)

Nakano T, Yasumoto J, Terasawa H, Nawa N (2013) Distribution and influent factors of nitrate concentration in groundwater. Trans Jpn Soc Irrigat Drainage Rural Eng 286:1–9. (In Japanese with English abstract)

Niemeier JK, Kjell DP (2013) Hydrazine and aqueous hydrazine solutions: evaluating safety in chemical processes. Org Process Res Dev 17:1580–1590

Ning X, Lin C, Su J, Liu C, Hao Q, Le F (2011) Long-term changes of dissolved oxygen, hypoxia, and the responses of the ecosystems in the East China Sea from 1975 to 1995. J Oceanogr 67:59–75

Nishiguchi S, Wada N, Yamashiro H, Ishibashi H, Takeuchi I (2018) Continuous recordings of the coral bleaching process on Sesoko Island, Okinawa, Japan, over about 50 days using an underwater camera equipped with a lens wiper. Mar Pollut Bull 131: 422–427

Ohno Y, Iguchi A, Ijima M, Yasumoto K, Suzuki A (2022) Coastal ecological impacts from pumice rafts. Sci Rep 12:11187. https://doi.org/10.1038/s41598-022-14614-y

Ohta I, Kubo H, Watanabe T, Uehara M (2017) A review of fisheries and biology of the sea urchin *Tripneustes gratilla* in Okinawa: implication for fisheries management and stock recovery. Okinawa Pref Fish Ocean Technol Center Project Rep 77:135–143. (In Japanese)

Okaji M, Ogasawara K, Yamakawa E, Kitamura M, Kumagai N, Nakatomi N, Yamamoto S, Nakajima R, Kinjo K, Nakamura M, Yasuda N (2019) Comprehensive management program of crown-of-thorns starfish outbreaks in Okinawa. J Jpn Coral Reef Soc 21:91–110. (In Japanese with English abstract)

Okinawa Prefecture (2013) Okinawa prefecture basic plan for prevention of red soil runoff. Okinawa Prefecture, Naha, Okinawa, 63 pp. (In Japanese)

Okinawa Prefecture (2018a) Current stats and problems of Okinawa tourism. Okinawa Prefecture, Naha, Okinawa, 8 pp. (In Japanese)

Okinawa Prefecture (2018b) Fiscal Year 2017. Statistics survey of tourism [summary version]. Okinawa Prefecture, Naha, Okinawa, 8 pp. (In Japanese)

Okinawa Prefecture (2019) Okinawa prefecture basic plan for prevention of red soil runoff. Midterm review. Okinawa Prefecture, Naha, Okinawa, 28 pp. (In Japanese)

Okinawa Prefecture (2021) Status of Sewerage System Development in Okinawa Prefecture [with appendix of Introduction to the Sewerage System]. https://www.pref.okinawa.jp/site/doboku/gesui/kokyo/seibi.html (appendix; https://www.pref.okinawa.jp/site/doboku/gesui/kokyo/documents/r2aramashi_seibijokyo.pdf). Accessed 27 Apr 2022. (In Japanese)

Omija T (2004) Terrestrial inflow of soils and nutrients. In: Tsuchiya M, Nadaoka K, Kayanne H, Yamano H (eds) Coral reefs of Japan. Tokyo and The Japanese Coral Reef Society, Ministry of the environment, pp 66–70. (In Japanese)

Omori M (2011) Degradation and restoration of coral reefs: experience in Okinawa, Japan. Mar Biol Res 7:3–12

Raffa RB, Pergolizzi JV Jr, Taylor R Jr, Kitzen JM (2019) Sunscreen bans: coral reefs and skin cancer. J Clin Pharm Ther 44: 134–139

Ramesh A, Archibong AE, Hood DB, Guo Z, Loganathan BG (2011) Global environmental distribution and human health effects of polycyclic aromatic hydrocarbons. In: Loganathan BG, Lam PKS (eds) Global contamination trends of persistent organic chemicals, 63: 97–126. CRC Press, Taylor & Francis Group

Ramos AA, Inoue Y, Ohde S (2004) Metal contents in *Porites* corals: anthropogenic input of river run-off into a coral reef from an urbanized area, Okinawa. Mar Pollut Bull 48:281–294

Reimer JD, Yang S-Y, White KN, Asami R, Fujita K, Hongo C, Ito S, Kawamura I, Maeda I, Mizuyama M, Obuchi M, Sakamaki T, Tachihara K, Tamura M, Tanahara A, Yamaguchi A, Jenke-Kodama H (2015) Effects of causeway construction on environment and biota of subtropical tidal flats in Okinawa, Japan. Mar Pollut Bull 94:153–167

Rougée LR, Collier AC, Richmond RH (2021) Chronic exposure to 4-nonylphenol alters UDP-glycosyltransferase and sulfotransferase clearance of steroids in the hard coral, *Pocillopora damicornis*. Front Physiol 12:608056. https://doi.org/10.3389/fphys.2021.608056

Saido K, Amamiya K, Sato H, Okabe A, Ogawa N, Kamaya Y, Kogure K, Nishimura M, Okukawa K, Kusui T (2012) Analysis of styrene oligomer contaminants generated from marine debris polystyrene on the coast of Okinawa. Bunseki Kagaku 61:629–636. (In Japanese with English abstract)

Saito M, Yasumoto J, Sugiyama A (2020) Groundwater and ecosystems; previous research trend and future perspective. J Groundw Hydrol 62:525–545. (In Japanese with English absract)

Sakai K, Singh T, Iguchi A (2019) Bleaching and post-bleaching mortality of *Acropora* corals on a heat-susceptible reef in 2016. PeerJ 7: e8138. https://doi.org/10.7717/peerj.8138

Sham RC-T, Ho KKY, Zhou G-J, Li Y, Wang X, Leung KMY (2020) Occurrence, ecological and human health risks of phenyltin compounds in the marine environment of Hong Kong. Mar Pollut Bull 154:111093

Sheikh MA, Tsuha K, Wang X, Sawano K, Imo ST, Oomori T (2007) Spatial and seasonal behaviour of organotin compounds in protected subtropical estuarine ecosystems in Okinawa, Japan. Inter J Environ Anal Chem 87:847–861

Sheikh MA, Fujimura H, Miyagi T, Uechi Y, Yokota T (2009a) Detection and ecological threats of PSII herbicide diuron on coral reefs around the Ryukyu archipelago, Japan. Mar Pollut Bull 58:1922–1926

Sheikh MA, Higuchi T, Fujimura H, Imo TS, Miyagi T, Oomori T (2009b) Contamination and impacts of new antifouling biocide Irgarol-1051 on subtropical coral reef waters. Int J Environ Sci Tech 6:353–358

Shilla DJ, Mimura I, Takagi KK, Tsuchiya M (2013) Preliminary survey of the nutrient discharge characteristics of Okinawa Rivers, and their potential effects on inshore coral reefs. Galaxea J Coral Reef Stud (Special Issue) Proc 2nd APCRS: 172–181

Shinjo R (2014) Geology and its development of the Ryukyu Island Arc; An example of geology of Okinawa Island, Central Ryukyus. J Japan Soc Civil Engineer Ser A2 (Appl Mechanic (AM)) 70:I_3–I-11. (In Japanese)

Shiokawa A, Tamaki F (2017) The survey on environmental pollution of Perfluorinated compounds (PFCs) in rivers and the sea areas of Okinawa Island. Ann Report Okinawa Pref Inst Health Environ 51: 33–48. (In Japanese)

Soliman T, Reimer JD, Yang S-Y, Villar-Briones A, Roy MC, Jenke-Kodama H (2017) Diversity of microbial communities and quantitative chemodiversity in layers of marine sediment cores from a causeway (Kaichu-Doro) in Okinawa Island, Japan. Front Microbiol 8:2451. https://doi.org/10.3389/fmicb.2017.02451

Song YK, Hong SH, Jang M, Kang J-H, Kwon OY, Han GM, Shim WJ (2014) Large accumulation of micro-sized synthetic polymer particles in the sea surface microlayer. Environ Sci Technol 48: 9014–9021

Soon ZY, Jung J-H, Jang M, Kang J-H, Jang M-C, Lee J-S, Kim M (2019) Zinc pyrithione (ZnPT) as an antifouling biocide in the marine environment—a literature review of its toxicity, environmental fates, and analytical methods. Water Air Soil Pollut 230:310

Takayanagi K, Nishiguchi K, Yokouchi K, Miyagi K (2006) A possible collaboration with China on marine ecosystem research in the East China Sea. Japan Agr Res Quart JARQ 40:59–64

Takeda F, Mano H, Minamiyama M, Okamoto S (2018) Initial health and ecological risk assessment of chemicals in treated wastewater in Okinawa prefecture based on Japanese PRTR system. J Japan Soc Civil Eng Ser G Environ Res 74:25–34. (In Japanese with English abstract)

Takeuchi I (2023) Introduction–The relevance of anthropogenic factors to coral reef conservation in the coastal areas of the East China Sea. In: Takeuchi I, Yamashiro H (eds) Coral reefs of eastern Asia under anthropogenic impacts. Coral reefs of the world, vol 17. Springer Nature Switzerland AG, Cham, pp 111–135

Takeuchi I, Takahashi S, Tanabe S, Miyazaki N (2004) Butyltin concentrations along the Japanese coast from 1997 to 1999 monitored by *Caprella* spp. (Crustacea: Amphipoda). Mar Env Res 57: 397–414

Takeuchi I, Miyoshi N, Mizukawa K, Takata H, Ikemoto T, Omori K, Tsuchiya K (2009) Biomagnification profiles of polycyclic aromatic hydrocarbons, alkylphenols and polychlorinated biphenyls in Tokyo Bay elucidated by $\delta^{13}C$ and $\delta^{15}N$ isotope ratios as guides to trophic web structure. Mar Pollut Bull 58:663–671

Takeuchi I, Takaichi D, Katsumata M, Ishibashi H (2020) Succession of delayed fluorescence correlated with coral bleaching in the hermatypic coral *Acropora tenuis*. Mar Pollut Bull 154:111008

Tanita I, Onduka T (2021) Diuron concentrations along shorelines of Ishigaki and Iriomote islands, Japan. Galaxea J Coral Reef Stud 23: 11–15

Tashiro Y, Kameda Y (2013) Concentration of organic sun-blocking agents in seawater of beaches and coral reefs of Okinawa Island, Japan. Mar Pollut Bull 77:333–340

Tashiro Y, Kameda Y (2015) Pesticide contamination monitored by passive sampling in environmental water of Japanese coral Island. J Water Res Ocean Sci 4:39–43

Tashiro Y, Takemura A, Fujii H, Takahira K, Nakanishi Y (2003) Livestock wastes as a source of estrogens and their effects on wildlife of Manko tidal flat, Okinawa. Mar Pollut Bull 77: 143–147

Tashiro Y, Takahira K, Osada H, Fujii H, Tokuyama A (2004) Distribution of polychlorinated biphenyls (PCBs), lead, and cadmium in Manko tidal flat, Okinawa. Limnology 5:177–183

Tashiro Y, Goto A, Kunisue T, Tanabe S (2021) Contamination of habu (*Protobothrops flavoviridis*) in Okinawa, Japan by persistent organochlorine chemicals. Environ Sci Pollut Res 28:1018–1028

Terada K, Koibuchi Y, Isobe M (2017) Rainfall effect on sediment and nutrient fluxes in a small mangrove river, Okinawa, Japan. J Mar Res 75:1–17

The Ministry of the Environment, Japan (2017[2016]) The action plan to conserve coral reef ecosystem in Japan 2016–2020. To protect blessings from the coral sea. Nature Conservation Bureau, Ministry of the Environment, Japan, Chiyoda-ku, Tokyo, Japan. v + 48 pp

The State of Queensland 2009 Reef Water Quality Protection Plan (2009) For the Great Barrier Reef World Heritage Area and adjacent catchments. 30 p. Published by the Reef Water Quality Protection Plan Secretariat, Brisbane Qld, Australia

Tokuda Y, Barnett PB (2017) Constructing a new U.S. military base: a health threat to Okinawan people. Environ Just 10:23–25

Toyoshima J, Nadaoka K (2015) Importance of environmental briefing and buoyancy control on reducing negative impacts of SCUBA diving on coral reefs. Ocean Coast Manage 116:20–26

Vieira LR, Soares AMVM, Freitas R (2022) Caffeine as a contaminant of concern: a review on concentrations and impacts in marine coastal systems. Chemosphere 286:131675

Wang H, Dai M, Liu J, Kao S-J, Zhang C, Cai W-J, Wang G, Qian W, Zhao M, Sun Z (2016) Eutrophication-driven hypoxia in the East

China Sea off the Changjiang Estuary. Environ Sci Technol 50: 2255–2263

Wang J, Beusen AHW, Liu X, Van Dingenen R, Dentener F, Yao Q, Xu B, Ran X, Yu Z, Bouwman AF (2020) Spatially explicit inventory of sources of nitrogen inputs to the Yellow Sea, East China Sea, and South China Sea for the period 1970–2010. Earths Future 8: e2020EF001516. https://doi.org/10.1029/2020EF001516

Wang X, Baskaran M, Su K, Du J (2018) The important role of submarine groundwater discharge (SGD) to derive nutrient fluxes into river dominated ocean margins–the East China Sea. Mar Chem 204:121–132

Wear SL, Thurber RV (2015) Sewage pollution: mitigation is key for coral reef stewardship. Ann N Y Acad Sci 1355:15–30

West K, van Woesik R (2001) Spatial and temporal variance of river discharge on Okinawa (Japan): inferring the temporal impact on adjacent coral reefs. Mar Pollut Bull 42:864–872

Wu Z, Lin T, Hu L, Guo T, Guo Z (2020) Atmospheric legacy organochlorine pesticides and their recent exchange dynamics in the Northwest Pacific Ocean. Sci Total Environ 727:138408

Ya M, Wang X, Wu Y, Li Y, Yan J, Fang C, Zhao Y, Qian R, Lin X (2017) Seasonal variation of terrigenous polycyclic aromatic hydrocarbons along the marginal seas of China: input, phase partitioning, and ocean-current transport. Environ Sci Technol 51: 9072–9079

Ya M, Wu Y, Wu S, Li Y, Mu J, Fang C, Yan J, Zhao Y, Qian R, Lin X, Wang X (2019) Impacts of seasonal variation on organochlorine pesticides in the East China Sea and northern South China Sea. Environ Sci Technol 53:13088–13097

Yamamuro M, Komuro T, Kamiya H, Kato T, Hasegawa H, Kameda Y (2019) Neonicotinoids disrupt aquatic food webs and decrease fishery yields. Science 366:620–623

Yamano H, Sugihara K, Nomura K (2011) Rapid poleward range expansion of tropical reef corals in response to rising sea surface temperatures. Geophys Res Lett 38:L04601. https://doi.org/10.1029/2010GL046474

Yamano H, Satake K, Inoue T, Kadoya T, Hayashi S, Kinjo K, Nakajima D, Oguna H, Ishiguro S, Okagawa A, Suga S, Horie T, Nohara K, Fukayama N, Hibiki A (2015) An integrated approach to tropical and subtropical Island conservation. J Ecol Environ 38: 271–279

Yamashiro H, Wada N, Tang SL (2023) Succession and spread of coral diseases and coral-killing sponges with special reference to microbes in Southeast Asia and adjacent waters. In: Takeuchi I, Yamashiro H (eds) Coral reefs of eastern Asia under anthropogenic impacts. Coral reefs of the world, vol 17

Yang L, Mou S, Li H, Zhang Z, Jiao N, Zhang Y (2021) Terrestrial input of herbicides has significant impacts on phytoplankton and bacterioplankton communities in coastal waters. Limnol Oceanogr 66:4028–4045

Yasuda N (2018) Distribution expansion and historical population outbreak patterns of crown-of-thorns starfish, *Acanthaster planci* sensu lato, in Japan from 1912 to 2015. In: Iguchi A, Hongo C (eds) Coral reef studies of Japan, coral reefs of the world 13, Springer, Singapore, pp 125–148

Yoshida K, Tamura Y, Sato T, Hanyu T, Usui Y, Chang Q, Ono S (2022) Variety of the drift pumice clasts from the 2021 Fukutoku-Oka-no-Ba eruption, Japan. Island Arc 31:e12441. https://doi.org/10.1111/iar.12441

Yoshimoto S, Asai K, Tsuchihara T, Shirahata K, Ishida S (2016) Hydrogeochemical approach to estimate residence time of the groundwater in the freshwater lens of the Ryukyu limestone aquifer: application of sulfur hexafluoride and nitrate isotopes. Trans Japan Soc Irrigat Drainage Rural Eng 302:I_159–I_174. (In Japanese with English absract)

Yukioka S, Tanaka S, Suzuki Y, Echigo S, Kärrman A, Fujii S (2020) A profile analysis with suspect screening of per-and polyfluoroalkyl substances (PFASs) in firefighting foam impacted waters in Okinawa, Japan. Water Res 184:116207

Zhang R, Du J, Dong X, Huang Y, Xie H, Chen J, Li X, Kadokami K (2021) Occurrence and ecological risks of 156 pharmaceuticals and 296 pesticides in seawater from mariculture areas of Northeast China. Sci Total Environ 792:148375

Zhou F, Chai F, Huang D, Wells, M Ma, X, Meng Q, Xue H, Xuan J, Wang P, Ni X, Zhao Q, Liu C, Su J, Li, H (2020) Coupling and decoupling of high biomass phytoplankton production and hypoxia in a highly dynamic coastal system: the Changjiang (Yangtze River) estuary. Front Mar Sci 7: 259. doi:https://doi.org/10.3389/fmars.2020.00259

Development of a Compact Experimental System for Ecotoxicological Experiments on *Acropora* spp. 8

Ichiro Takeuchi and Kotaro Takayama

Abstract

The present chapter describes a compact experimental system for ecotoxicological experiments that use small clones of *Acropora* spp. Our system comprises widely accessible inexpensive equipment and devices. Tap water provided by the local government waterworks agency and coastal seawater in Japan is potentially contaminated with various herbicides and pesticides. Thus, we prepared artificial seawater using tap water filtered more than twice through water purifier. To avoid contamination by anionic surfactants, Petri dishes and other experimental devices were washed using acetone, hexane, and HCl. A rapid, straightforward method for attaching small clones of *Acropora* spp. to polycarbonate bolts using an ultraviolet light-curable oligomer-based adhesive is described. This method enables the prompt and precise measurements of the maximum effective quantum yield ($\Delta F/F_m'$) using a pulse amplitude modulated (PAM) fluorometer. The evaluation of coral colour based on RGB values (R, red; G, green; B, blue; each ranging from 0 to 255) of digital images acquired manually by a digital camera is described. This experimental system enables us to conduct various experiments, including recording of the bleaching process in corals exposed to high temperature and herbicides and gene expression analysis of *Acropora* exposed to herbicides, in a laboratory located far from coral reefs in a subtropical-tropical zone. Our system will advance the analysis of corals exposed to various anthropogenic factors.

I. Takeuchi (✉)
Graduate School of Agriculture, Ehime University, Matsuyama, Ehime, Japan

Center of Advanced Technology for the Environment, Graduate School of Agriculture, Ehime University, Matsuyama, Ehime, Japan
e-mail: takeuchi@agr.ehime-u.ac.jp

K. Takayama
Graduate School of Agriculture, Ehime University, Matsuyama, Ehime, Japan

Electronics-Inspired Interdisciplinary Research Institute (EIIRIS), Toyohashi University of Technology, Toyohashi, Aichi, Japan

Keywords

Acropora spp. · Artificial seawater · Colour evaluation · Coral bleaching · Experimental system · Filtered tap water

8.1 Introduction

Coral reefs are widely recognised as highly dynamic ecosystems with spectacular diversity, estimated at more than three million species, and covering only 0.07% of the ocean surface (Spalding and Brown 2015). Coral reefs provide us with various eco-services, including support to over a quarter of the world's small-scale fishers, job creation, and foreign exchange through tourism, and critical sea defence against storms, flooding, and land erosion (Spalding and Brown 2015). The coral reefs around the East China Sea are exposed to multiple anthropogenic factors and stresses, such as increasing seawater temperature, eutrophication, coastal development, various anthropogenic chemicals, and red soil run-off (Takeuchi 2023a, 2023b). Analysis of region-scale coral coverage and species abundance on the reefs around Okinawa Island indicated a drastic decrease in coral cover, from 24.4% to 7.5% (1.1% per year), from 1995 to 2009 (Hongo and Yamano 2013). This dramatic disappearance of corals around Okinawa Island has affected in particular several species and species complexes of *Acropora* (*A. digitifera*, *A. hyacinthus/cytherea*, and *A. intermedia/muricata*). This once dominant genus in the coral reefs before 2000, with approximately 20% abundance, experienced a significant change in abundance throughout this period (Hongo and Yamano 2013).

The genus *Acropora* is a highly diverse genus with unclear species boundaries; currently, the World Register of Marine Species (WoRMs) database records 135 valid species, 51 species as *taxon inquirendum*, and 6 species as

I. Takeuchi, H. Yamashiro (eds.), *Coral Reefs of Eastern Asia under Anthropogenic Impacts*, Coral Reefs of the World 17,
https://doi.org/10.1007/978-3-031-27560-9_8

nomen nudum (Hoeksema and Cairns 2021). The current classification of hermatypic corals, including the genus *Acropora*, is based on corallite morphology and colony forms (Fukami et al. 2010). These characteristics, however, show polymorphism, intraspecific variation, and phenotypic plasticity in response to environmental factors, leading to difficulty in quantifying skeletal morphologies of hermatypic corals and description and comparison of similar species based on skeletal morphology (Fukami et al. 2010). For example, monitoring of the genus *Acropora* in a field survey around Okinawa Island was conducted at a species/species complex level (Hongo and Yamano 2013). *Acropora* sp. 1, which superficially resembles *A. digitifera* in outer morphology, was reported from Sesoko Island, Okinawa (Ohki et al. 2015). Both species differ significantly in the minute morphology of the apical part of branches and the spawning season (Ohki et al. 2015). Recent maximum likelihood and Bayesian analyses of DNA sequences of Acroporidae and related families revealed that species-level relationships within the *Acropora* are incongruent with the macro-morphological characteristics traditionally used for taxonomic identification (Cowman et al. 2020). Over 50% of specimens in Cowman et al. (2020) could not be readily assigned to the previously described species of *Acropora*. Thus, clones from a single colony are preferred for various ecotoxicological and related experiments using *Acropora* spp.

Most experiments conducted before 2005 used 3–10 cm long colonies from a single branching colony (see Shafir et al. 2006). Then in 2006, Shafir et al. published a protocol for preparing small colonies, ca. 0.25 cm^2 in size, of *Stylophora pistillata* and *Pocillopora damicornis*.

Keeping hermatypic corals alive and healthy in closed, captive experimental systems was a challenging task until the early1980s. By the mid-1980s, technological innovations enabled the maintenance of hermatypic corals in the laboratory for various experiments (Bartlett 2013). Bartlett (2013) described a small-scale experimental system for various coral studies. He proposed the use of a common aquarium (20 cm high × 30 cm long × 15 cm wide to 40 cm high × 90 cm long × 30 cm wide), which can be purchased from local aquarium stores. In his small-scale system, seawater in the experimental tanks is supplied from a seawater reservoir and drained into waste containers (Bartlett 2013). He also proposed a chemical dosing experiment system, in which 1 L beakers are set in the holding tank with a water recirculation system that regulates the temperature in the beakers (Bartlett 2013). Rocha et al. (2015) presented a flexible culture system for ex situ experimental coral production based on the modular concept, which can be built using materials and equipment available from local suppliers almost everywhere in the world. The main module is composed of three experimental glass tanks (30 cm high × 60 cm long × 60 m wide), with a flow-through or recirculating regime, and a filtration sump (50 cm high × 70 cm long × 50 cm wide).

This section reports on the compact experimental system developed at our laboratory to conduct ecotoxicity experiments with small clones of *Acropora* spp. Our experimental system is much more compact than the previously reported system (see Bartlett 2013; Rocha et al. 2015). In addition, a prompt and straightforward evaluation method for coral colour based on the analysis of RGB values and the photo-efficiency of *Acropora* spp. are reviewed.

8.2 Setup of the Experimental Laboratory

The area surrounding the East China Sea is one of the areas with the heaviest usage of pesticides in the world, recording >10.0 kg/ ha (agricultural land area) in 2019 (Takeuchi 2023a). Of the 120 pesticides recorded in the 2013 Primary List of pesticides in the Japanese Drinking Water Quality Guidelines (JDWQG), 80 were detected in drinking water sources (raw water entering water treatment plants) in Japan (Narita et al. 2021). However, 5–43% of pesticides were detected in different regions of Japan, indicating a wide regional variation in pesticides detected in water (Narita et al. 2021). Monitoring of 162 pesticides at 12 main water supply utilities in Japan conducted from April to October between 2012 and 2017 frequently detected 31 substances (>100 times) in a total of 14,076 samples in 6 years (Kamata et al. 2020). Bromobutide, which is a herbicide not monitored in other Asian countries, and the widely applied herbicide bentazone, both used for paddy rice farming, were detected very frequently, in approximately 30% of samples in source water and approximately 10% of samples in treated (drinking) water, at concentrations >0.1 μg/L (Fig. 8.1) (Kamata et al. 2020). Sato et al. (2016) reported that the concentrations of imidacloprid and clothianidin, both neonicotinoid pesticides, in tap water were nearby consistent with their concentrations in the Sagami River, Kanagawa, Japan. This suggests that the chlorination of drinking water in the treatment plant does not remove these pesticides (Sato et al. 2016). Kadokami et al. (2021) analysed 484 chemicals, including 162 pharmaceuticals and personal care products (PPCPs), in 24-hour composite samples collected using a portable composite sampler before the first sedimentation tank (inflow) and after the final sedimentation tank (outflow) in eight activated sludge wastewater treatment plants across Japan. The number of detected chemicals and their mean total concentrations were 87 and 108.517 μg/L in the inflow to 92 and 31.537 μg/L in the outflow (Kadokami et al. 2021).

The use of tap water for control treatments in ecotoxicity experiments is not always possible (Nobukawa and Sanukida 1997; Mano et al. 2015). Drinking water from the Yodo River, western Japan, showed high mutagenicity as

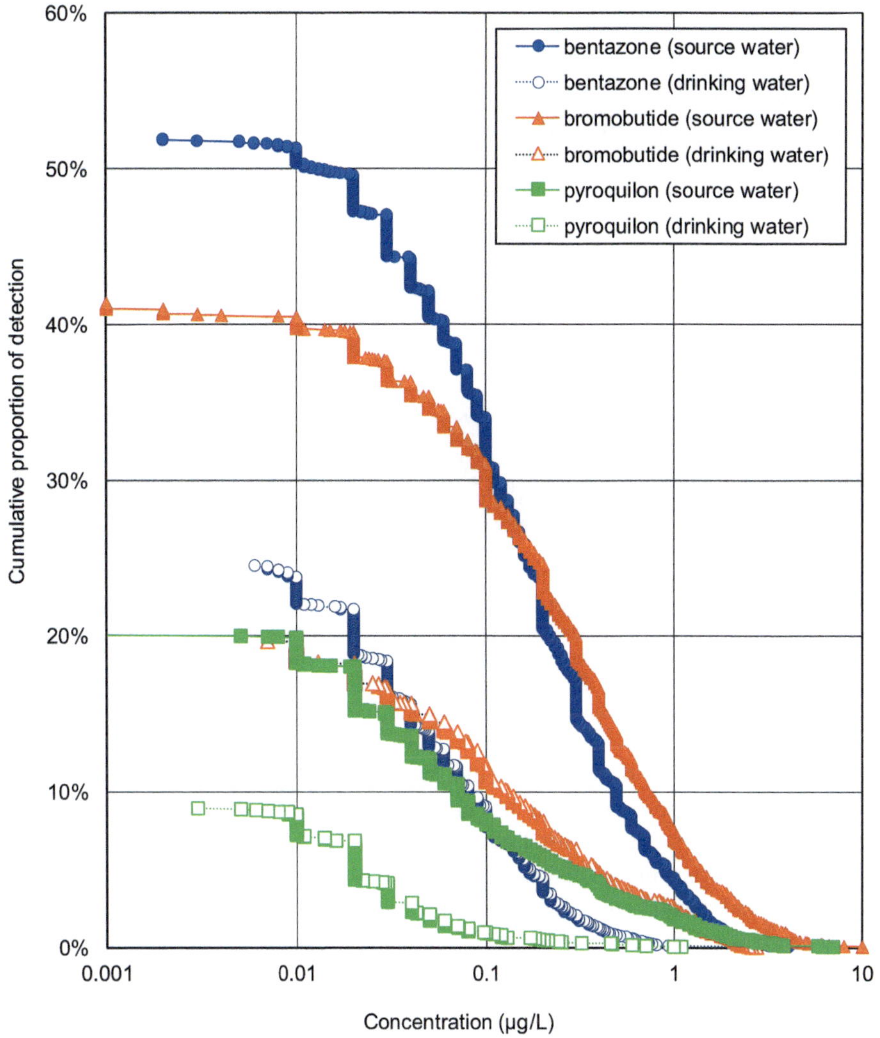

Fig. 8.1 Cumulative frequency distributions of three dominant herbicide concentrations in source water and treated water (drinking water) based on 14,076 samples from 12 water supply utilities in Japan during 2012–2017 (Kamata et al. 2020). Reprinted from Kamata et al. (2020, Sci Total Environ 744, 140930) with permission from Elsevier. Copyright (2020) Elsevier B.V.

chlorinated river water and sewage effluent based on Ames test using *Salmonella typhimurium* (TA100 and TA98) (Nobukawa and Sanukida 1997). The Ames test is a rapid bacterial reverse mutation assay for detecting a wide range of chemical mutagens (Mortelmans and Zeiger 2000). Mano et al. (2015) employed the algal growth inhibition test and short-term toxicity test using embryo and sac-fry stages of medaka fish *Oryzias latipes* to examine the toxicity of wastewater filtered by nanofiltration (NF) and reverse osmosis (RO) membranes in three different periods (June 2013, August 2013, and January 2014). The authors used dechlorinated tap water for these experiments (Mano et al. 2015), but for the experiments conducted in January 2014, they replaced it with pH-adjusted distilled water due to the high mortality rate of fertilised medaka fish eggs cultured in dechlorinated tap water. This suggested that the tap water used for the January 2014 experiment may have been contaminated with toxic substances other than pesticides used in rice cultivation. These studies (Nobukawa and Sanukida 1997; Mano et al. 2015; Kamata et al. 2020; Narita et al. 2021) suggest that various pesticides and PPCPs in tap water at Japanese universities and institutions enter the water system used for experiments.

To avoid contamination by pesticides and PPCPs, the tap water used in our system and supplied by the local government bureau of waterworks to Ehime University was filtered through an activated carbon filter (FAC-2; Organo Corporation, Koto-ku, Tokyo, Japan), polypropylene micro filter (5PF-1SA; Organo Corporation), and a water purifier (WG203; Yamato Scientific Co, Ltd., Chuo-ku, Tokyo, Japan) (Fig. 8.2). The filtration accuracy of the 5PF-1SA is 5 μm. The activated carbon filter and a 0.1 μm hollow fibre were installed before extracting ion-exchanged water by the ion exchange resin cartridge CPC-S in the water purifier WG203. Thus, the tap water, filtered through an activated carbon filter, a microfilter (5 μm), another activated carbon filter, and a microfilter (0.1 μm), was supplied to the ion exchange cartridge. The conductivity of the ion-exchanged water used for the experiments was $<0.5 \times 10^{-4}$ S/m according to the display on the water purifier. The usage of

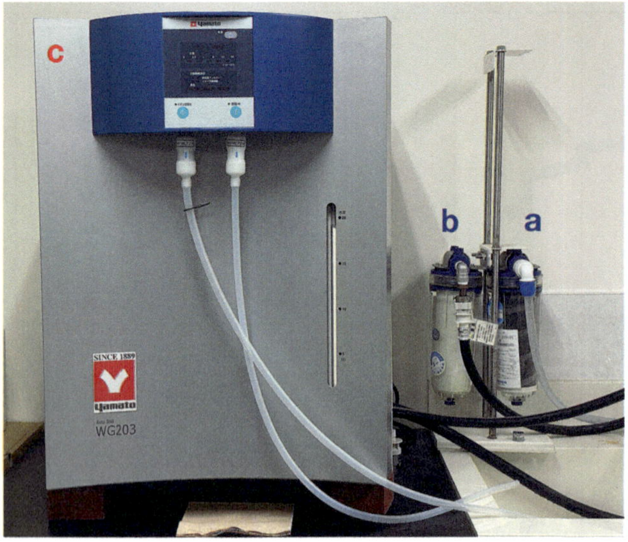

Fig. 8.2 Water treatment system used for coral ecotoxicological experiments. **a** Activated carbon filter (FAC-2; Organo Corporation, Koto-ku, Tokyo, Japan) set in the plastic housing (PF-VI; Organo Corporation); **b** Polypropylene micro filter (PF-1SA; Organo Corporation) set in the plastic housing (PF-VI; Organo Corporation); **c** Water purifier (WG203; Yamato Scientific Co, Ltd., Chuo-ku, Tokyo, Japan)

RO water purifier is recommended for rearing ornament corals including *Acropora* spp. in the aquarium fish industry sector in Japan (Kawasaki 2017; Tsuruta 2021).

Commercial cleaning agents were removed from the laboratory to avoid contamination with the experimental system. In general, alkaline, acid, or neutral laboratory cleaning agents are used to clean various pieces of research equipment (see Endo 2014). Most of these cleaners contain anionic surfactants and/or non-ionic surfactants for cleaning effects. Many popular commercial detergents also include linear alkylbenzene sulfonate (LAS), a representative anionic surfactant (see Katam et al. 2018), which has been reported to cause acute toxicity in the hermatypic coral genus *Pocillopora* (Kegler et al. 2015). The exposure of *Pocillopora verrucosa* fragments to 190 μg/L of LAS induced severe tissue loss after 84 h of exposure (Kegler et al. 2015). The median of the acute to chronic ratios ranges from 5.4 in algae to 7.0 in daphnia and 10.5 in fish, whereas those of individual chemicals vary substantially and may reach up to 4400 (Ahlers et al. 2006). Thus, chronic toxicity of LAS to *P. verrucosa* is estimated at concentrations from 43 ng/L to 35 μg/L. The 5th percentile hazardous concentrations of LAS have been reported for 129 to 259 μg/L (Belanger et al. 2016), which corresponds to the concentration range that induces tissue loss in *P. verrucosa* after <4 d exposure (Kegler et al. 2015). Therefore, it is safer to avoid using alkaline cleaning agents and popular commercial detergents until at least the thresholds for chronic toxicity of the major ingredients to hermatypic corals are reported.

Following the previously established experimental method for studying the chronic effect of tributyltin (TBT) on amphipod crustaceans (Aono and Takeuchi 2008), we cleaned Petri dishes and other laboratory glassware using acetone and hexane. The glassware was kept in a square container filled with hydrochloric acid (HCl) for more than 12 h and washed with distilled water through the above clearing system. Laboratory devices, including Petri dishes, made of glass were used whenever possible to prevent deterioration by hydrochloric acid.

8.3 Seawater Used for the Experiment

The coastal seawater around the Japanese Archipelago, including the Seto Island Sea, where the campus of the authors' university is located, is contaminated with various anthropogenic chemicals. For example, up to 39.9 ng/L of TBT was recorded from the Seto Inland Sea (see Takeuchi et al. 2004; Murai et al. 2005; Onduka et al. 2008), which is the amount that exceeds the concentration threshold for the chronic effects on the Pacific oyster *Crassostrea gigas* and skeleton shrimp *Caprella danievskii* (see Alzieu 2000; Aono and Takeuchi 2008).

Since the adoption of global ban on the use of TBT, several 'booster' agents, including the herbicide Irgarol 1051 (hereafter Irgarol), have been mixed with copper-based antifouling paints (Dafforn et al. 2011). Irgarol concentration of up to 35.1 ng/L has been reported off the coasts of Okinawa Island (Sheikh et al. 2009). Up to 1.85 μg/L and 2.18 μg/L of Irgarol and Diuron were reported from the Seto Inland Sea (see Okamura et al. 2003; Balakrishnan et al. 2012; Kaonga et al. 2015). Several micrograms per litre of Irgarol were also reported from international ports and resort islands in the South China Sea and western Japan (Basheer et al. 2002; Lam et al. 2005; Ali et al. 2013). Irgarol strongly affects the unicellular phytoplankton and hermatypic corals. Okamura et al. (2000) reported acute toxicity of Irgarol to *Raphidocelis subcapitata* (*Selenastrum capricornutum*), a unicellular alga in Chlorophyta ($EC_{50} = 1.6$ to 2.3 μg/L for 72 h of exposure). Jones and Kerswell (2003) reported a 50% reduction in the maximum effective quantum yield ($\Delta F/F_m'$) of hermatypic coral *Seriatopora hystrix* after a 10 h exposure to 0.7 μg/L. Recently, Kamei et al. (2020) recorded a significant reduction of 8% in the $\Delta F/F_m'$ of the coral *Acropora tenuis* after 7 d of exposure to 20 ng/L of Irgarol. The above concentrations overlapped with those reported in western Japan, including the Seto Inland Sea and Okinawa Islands (see Okamura et al. 2003; Sheikh et al. 2009; Balakrishnan et al. 2012; Kaonga et al. 2015). The coastal seawater of the Seto Inland Sea receives pesticides by runoff, especially during summer, from rice paddy fields (see Hano et al. 2019). A relatively higher concentration of the insecticide

Table 8.1 Concentration of principal components in 40.8 g of LiveSea salt per 1 L of water after Delphis Inc. (2021)

Principle component	Concentration (ppm = mg/L)
Cl^-	19,300
Na^+	10,710
SO_4^{2-}	2650
Mg^{2+}	1260
Ca^{2+}	455
K^+	396
Br^-	67.7
Sr^{2+}	13.5
$B(OH)_3$, $B(OH)_4^-$	5.20
F^-	1.73
I^-	0.07

fipronil, up to 0.322 μg/L, was detected in the estuaries of the Seto Inland Sea from June to July (Hano et al. 2019). Hano et al. (2019) estimated that the high concentration of fipronil mixed with simultaneously detected neonicotinoids will have significant effects on benthic crustaceans.

Thus, instead of the natural seawater, artificial seawater prepared from LIVESea salt (Delphis Inc., Itami, Hyogo, Japan) and distilled water has been used for various experiments in the author's laboratory. Table 8.1 shows the main components of the LIVESea salt according to the homepage of Delphis Inc. (Delphis Inc. 2021). LIVESea Salt is a powdered artificial salt made from highly purified raw materials without phosphate and nitrate (Delphis Inc. 2021). The LIVESea Salt was also used for the behavioural experiments on larvae of *Acropora tenuis* conducted at the National Institute for Basic Biology, Okazaki, Aichi, Japan (Sakai et al. 2020).

8.4 Transportation System

In most cases, set up for the laboratory experiments will not require a single large coral colony, but rather multiple small coral pieces (Bartlett 2013). Regarding the transportation of corals, two techniques have been developed, that is 'submerged method' and 'dry method' (see Petersen et al. 2004). In the submerged method, a colony is placed in a double plastic bag (one bag inside another bag) and sufficient amount of fresh saltwater is added to completely cover the coral. The bag is filled with oxygen to maintain the oxygen level high and to stabilise the bag shape during transport (Petersen et al. 2004). In the dry method, each coral (small 3–10 cm long fragments or relatively small colonies weighing <0.5 kg) is wrapped in wet plastic strips and placed in a plastic container to prevent damage from shipping (Petersen et al. 2004).

To minimise damage in various thermal and/or ecotoxicological experiments, small-sized colonies or fragmented colonies attached to polycarbonate bolts (see the next section) were transported to the laboratory as check-in baggage on air flights (Takeuchi et al. 2019) or by door-to-door parcel delivery by Yamato Transport Co., Ltd. (Chuo-ku, Tokyo, Japan). A colony or fragmented colonies were placed with approximately 14 L of seawater into a double or triple plastic bag inside a cooler. Crushed ice in summer or disposable body warmer in winter was included in the icebox to prevent temperature fluctuations of seawater in the icebox. The transport of corals from Okinawa to Matsuyama generally lasts <48 h. Takeuchi et al. (2019) used a 30 L portable cooler (Daiwa Light Trunk SU 3000RJ; Globeride Inc., Kurume, Tokyo, Japan) to transport corals from Sesoko Island to Matsuyama as check-in baggage on an air flight. A small plastic bag with approximately 1–2 kg of crushed ice included in the ice box kept the temperature <28 °C during most of the transportation period in summer (Takeuchi et al. 2019).

When the coral fragments were received, they were acclimated for 2–3 h by adjusting the salinity in the aquarium by 1 psu per hour (Takeuchi et al. 2019).

8.5 Fragmentation of *Acropora* spp.

Asexual propagation or 'fragmentation' is a simple method for obtaining large numbers of coral colonies (Osinga et al. 2012). New coral colonies are obtained by cutting a mother colony into smaller pieces and, if necessary, by attaching these fragments to a solid support (Osinga et al. 2012). A variety of techniques and tools are used to fragment corals, depending mainly on the size and shape of the parent colony (Bartlett 2013).

Recently, we developed a new, inexpensive, easy, and fast method for attaching fragmented colonies of *Acropora* spp. to small polycarbonate hexagon head bolts using an ultraviolet (UV)-curable oligomer-based adhesive agent (Takeuchi et al. 2019). Bondic® (Laser Bonding Tech, Inc., Aurora, ON, Canada), a dentist-invented UV-curable oligomer-based adhesive based on a dental composite, was also used for minute neurophysiological experiments to affix desert ants and fruit flies to a steel pen, tungsten wire, or filament (see Takeuchi et al. 2019). The method reported by Takeuchi et al. (2019) is summarised as follows:

1. Using a small diagonal cutting plier, the apical tip of each branch-like part (ca. 1.0 to 1.5 cm in length) of a donor coral is cut.
2. The moisture is absorbed from the cut section with sterilised filter paper.
3. A small drop of Bondic® is applied to the cut section of the coral fragment, and the fragment is immediately attached to a polycarbonate hexagon head bolt (Fig. 8.3).

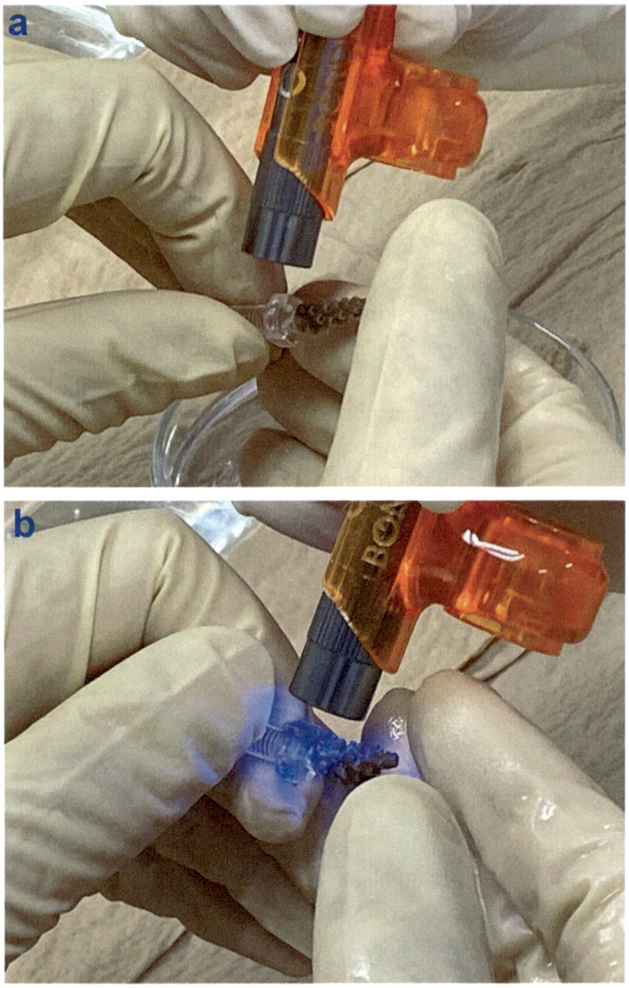

Fig. 8.3 Mounting of coral clone on a polycarbonate bolt using ultraviolet-curable oligomer-based adhesive according to Takeuchi et al. (2019). **a** Attaching coral clone to the polycarbonate bolt after applying Bondic® Evo (replaced a pistol grip and new LED from Bondic®) to the cut surface of the clone; **b** Curing of Bondic® Evo by ultraviolet LED irradiation

4. The UV radiation LED is used to irradiate the area around the attachment for approximately 5–10 s (Fig. 8.3).
5. The polycarbonate bolt with attached coral clone is transferred to a shallow aquarium.

8.6 Compact Aquarium System

A small colony or clone fragments of *Acropora* spp. mounted on polycarbonate bolts were placed in a compact commercial glass aquarium (60 cm wide × 30 cm deep × 22 cm high or 60 cm wide × 30 cm deep × 40 cm high) that could fit inside a normal low-temperature incubator. The polycarbonate bolts were set in a mesh substrate. In general, two filters, or one filter and one small protein skimmer, were used to circulate and filter seawater in the aquarium (see Fig. 8.4). White sand for coral rearing was spread over the bottom of the aquarium. A submersible pump for generating water flow was attached to the side of the aquarium using a magnetic holder. These devices attached to glass aquarium are inexpensive commercially available equipment sold to aquarists. Approximately 10 L of the artificial seawater was exchanged weekly. Nakamura and van Woesik (2001), in their experiment using a Perspex flume (10 cm high × 10 cm wide × 140 cm long), recorded that high water flow (50–70 cm/s) resulted in higher survivorship of *Acropora digitifera* than did low water flow (2–3 cm/s) under high seawater temperature (26.22–33.65 °C). Several to approximately 10 individuals of cowries (*Cypraea annuli* and/or *Monetaria moneta*) and/or *Holothuria* spp. were added to the aquarium to remove microorganisms attached to glass wall and microalgae flourishing on benthic sand. Figure 8.4 shows our most recent compact aquarium system for keeping clones of *Acropora* spp. in a low-temperature incubator (103.5 cm high × 90 cm wide × 55 cm long; IS-2000; Advantec Toyo Co., Ltd., Chiyoda-ku, Tokyo, Japan). Inside the low-temperature incubator, the seawater temperature of the aquarium was regulated by an air circulation system. Light was supplied with LED light (Power Shot; Kotobuki Co., Ltd., Matsubara, Osaka, Japan). The light intensity was set at approximately 50–100 μmol m^{-2} s^{-1} of photosynthetic photon flux density (PPFD) on the glass cover of the aquarium. The photoperiod of 12 h:12 h light:dark cycle was regulated using a 24-hour rewind timer.

8.7 Colour Evaluation Using RGB Values

Colour evaluation of corals is one of the most straightforward tools for identifying the health condition of corals. To assess the extent of coral bleaching, waterproof reference cards ('Coral Colour Reference Card') in six hues from darkest to brightest were used to code the colour of coral to record changes in bleaching state (Siebeck et al. 2006). This method is used as the primary tool for the citizen-based CoralWatch program, in which the chart colours are matched to the coral colour, the codes are recorded, and the data are entered on a chart via a website (Marshall et al. 2012). Recently, Hawaiian Ko'a Card in 10 hues was developed, mirroring the Australian Coral Colour Reference Card (Bahr et al. 2020).

We developed an evaluation method for coral bleaching by measuring the RGB values on coral digital images using Adobe Photoshop (Adobe Systems Incorporated, San Jose, CA, USA), one of the representative commercially available graphic software (see Hirayama et al. 2017). The same

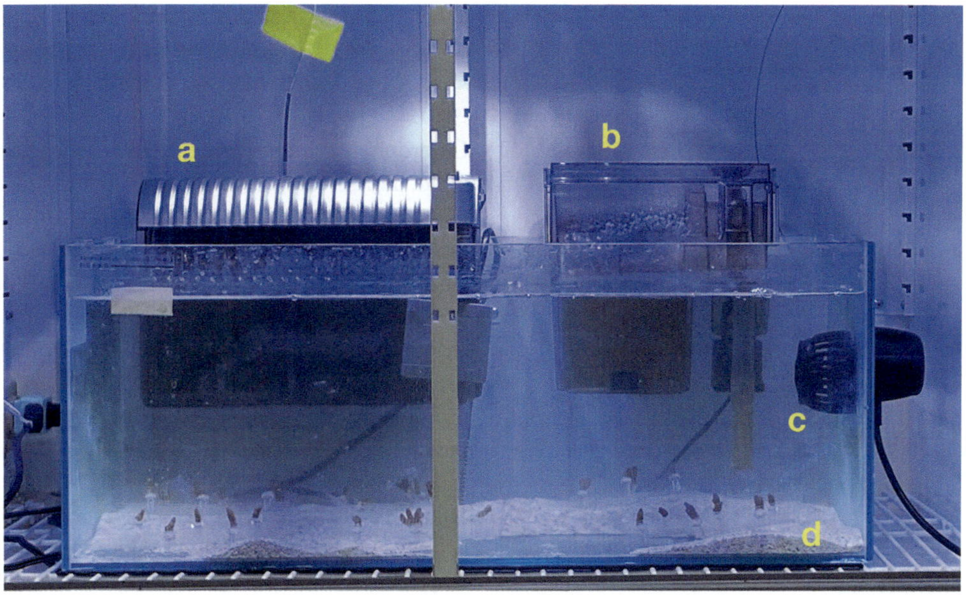

Fig. 8.4 Compact aquarium system installed inside a low-temperature incubator (IS-2000; Advantec Toyo Co., Ltd., Chiyoda-ku, Tokyo, Japan). Two filters, Kaido Kappa Filter Large (b; Kamihata Fish Industries Ltd., Himeji, Hyogo, Japan), Tetra Auto OneTouch Filter AT-75 W (a; Spectrum Brands Japan Co. Ltd., Yokohama, Kanagawa, Japan), and a circulator pump (c; MP10wQD; VorTEch, EcoTech, LCC, Bethlehem, PA, USA) were installed in the aquarium. The live aragonite reef sand (d; Fiji Pink, Arag-Alive!™; CalibSea, Inc., Fort Pierce, FL, USA) was spread on the bottom of the aquarium

technique has been used as one of the most effective techniques for monitoring plant growth in a greenhouse before it was applied to the evaluation of coral colour (see Omasa et al. 1987; Takayama and Nishina 2009). The RGB values specify every colour by combining the R, G, and B values, which range from 0 to 255 (Fig. 8.5). Black is represented by R = G = B = 0, and white by R = G = B = 255. Expressing colour based on RGB values is today one of the most commonly used approaches in digital graphic software and various computer displays.

In commercial digital cameras, brand- and model-specific processing of captured images changes the colour, contrast, and white balance, altering the intensity of each pixel in captured images (Akkaynak et al. 2014). Thus, the RGB values of JPEG images photographed by commercial cameras are transformed irreversibly from those in the original RAW images (Akkaynak et al. 2014). We developed an RGB evaluation system for corals in the laboratory based on NEF (RAW) files captured by a commercially available digital single-lens reflex camera equipped with a wireless speed light system in manual mode (Hirayama et al. 2017). Corals were photographed together with a colour chart in the field of view using a manual mode of the camera. The main settings of our method are summarised in Table 8.2 (see Hirayama et al. 2017).

The RGB values of the photographed corals were evaluated in Photoshop using NEF (RAW) files (Hirayama et al. 2017). The basic procedure was performed as described by Hirayama et al. (2017). First, the surface of the coral colony on the images was selected using the 'Quick Selection Tool' (Fig. 8.6). If the selection extended beyond the outline of the coral surface, the selected area was manually adjusted. Then, RGB values of the selected area were obtained using the Filter > Blur > Average function. Finally, the RGB

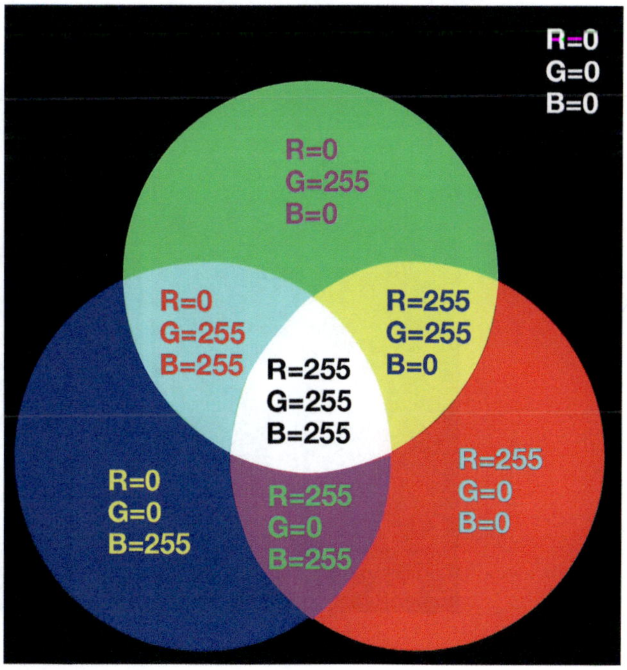

Fig. 8.5 Schematic diagram of RGB values

Table 8.2 Settings for recording RGB values of *Acropora* spp. using a commercial digital single-lens reflex camera, Nikon df and AF-S Nikkor 50 mm, and related wireless light system (Nikon Co. Ltd., Minato-ku, Tokyo, Japan)

Setting	
ISO sensitivity	ISO 100
Shutter speed	1/125 s
Aperture	f10
White balance	Flash (Colour temperature 5400 K)
Image quality	NEF (Nikon Electronic Format)[a] + JPEG fine
Image size	Large 4928 × 3280 pixels
Focus mode	Manual
Picture control	Standard
Manual flash output level	M1/1
Distance from lens to object	33 cm

Reprinted from Hirayama et al. (2017, Mar Pollut Bull 124, 678–686) with permission from Elsevier. Copyright (2017) Elsevier Ltd.
[a]NEF is one of the RAW formats

values of the coral surface were calibrated using the R, G, and B values of the colour chart.

Nishiguchi et al. (2018) applied the RGB analysis method for analysing in situ coral bleaching of *Acropora cytherea* and *Porites* sp. using a continuous recording of ca. 50 days obtained by an underwater digital camera equipped with a lens wiper. Although the images were captured only as JPEG files, the succession of RGB values showed a contrast in the response of the two corals to high seawater temperature; while the RGB values of *A. cytherea* were gradually elevated, moving to the 'white' end until the beginning of September, those of *Porites* sp. were stable throughout the monitoring season (Nishiguchi et al. 2018).

8.8 Compact Measurement of Photosynthetic Efficiency of Hermatypic Corals

Chlorophyll *a* fluorescence measurement has been used as a rapid, non-invasive, sensitive, and convenient indicator of photosynthetic performance in marine organisms, including the symbiotic dinoflagellates of hermatypic corals (see Bhagooli et al. 2021). Fluorescence signals can be measured at different scales from microscopic organelles such as chloroplasts to entire organisms (Bhagooli et al. 2021). The use of pulse amplitude modulated (PAM) fluorometers has prominently revolutionised the measurements of photosynthetic efficiency since the mid-1980s (Maxwell and Johnson 2000; Bhagooli et al. 2021). In PAM fluorometry, a 3 μs pulse of light is released to induce pulsed chlorophyll fluorescence emission. Pulsed chlorophyll fluorescence is amplified and measured with a synchronised recording system, which allows us to eliminate background noise (Maxwell and Johnson 2000; Bhagooli et al. 2021).

The attachment of coral fragments to polycarbonate bolts led to the establishment of more rapid and precise measurements of the $\Delta F/F_m'$ of *Acropora* spp. using a PAM fluorometer (MINI-PAM II; Heinz Walz GmbH, Effeltrich, Germany) and the accessories prepared by the manufacturer (Kamei et al. 2020; Gushi et al. 2021).

Of the various indices of photosynthetic efficiency, $\Delta F/F_m'$ or ϕPSII is suitable for precise measurements under light conditions in a standard laboratory. The photosynthetic photochemical reaction to light energy, $\Delta F = (F_m' - F)/F_m'$, was estimated based on two parameters: F, the steady-state level of fluorescence under light, and F_m', the maximal level of

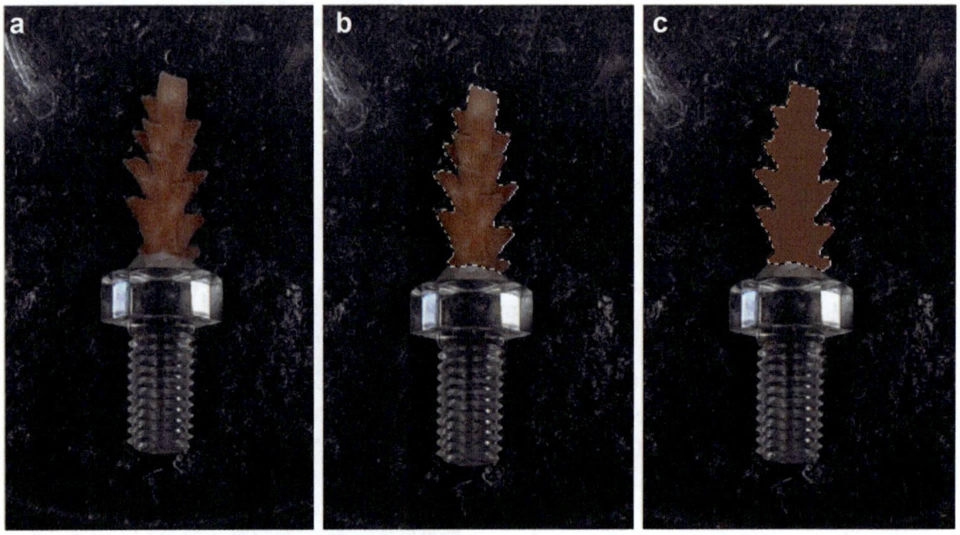

Fig. 8.6 Processing of screen-captured images using Photoshop CS6 (Adobe Systems Incorporated, San Jose, CA, USA) into RGB values according to Hirayama et al. (2017). **a** Clone of *Acropora tenuis* attached to polycarbonate bolt; **b** Coral area encircled by dotted line using "Quick Selection Tool"; **c** Coral area encircled by dotted line after selecting "Average" in "Blur" under the "Filter" tab in Photoshop CS6

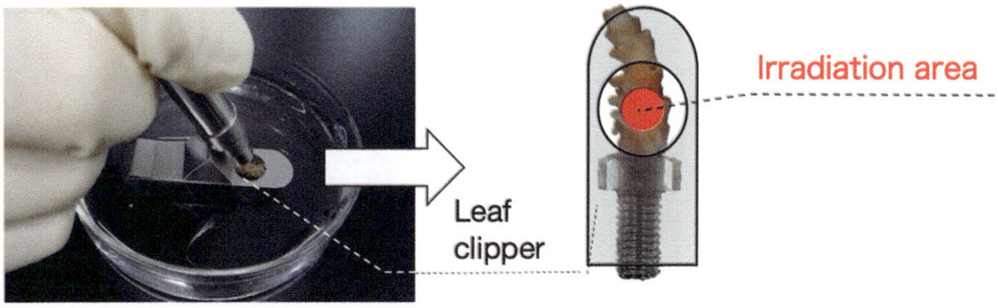

Fig. 8.7 Measurement method using a photosynthesis yield analyser MINI-PAM II with fibreoptics MINI-PAM/F1 and distance leaf-clip 2010-A (Heinz Walz GmbH). Reprinted from Gushi et al. (2021, Reg Stud Mar Sci 47, 101957) with permission from Elsevier. Copyright (2021) Elsevier B.V.

fluorescence in the light-adapted state under light conditions (see Murchie and Lawson 2013). This method does not require a dark environment to measure the maximum efficacy of PSII, F_v'/F_m' (Murchie and Lawson 2013), which is also used for chlorophyll fluorescence measurements of various symbiotic marine invertebrates (Bhagooli et al. 2021). The same part of the coral surface could be measured using MINI-PAM II set with fibreoptics MINI-PAM/F1 and a distance leaf-clip 60° 2010-A (Heinz Walz GmbH, Effeltrich, Germany) (Gushi et al. 2021). The end piece of the fibreoptics was set 5 mm from the coral surface at 60° angle (Fig. 8.7) (Gushi et al. 2021). This method enables repeated and precise measurements at the exact position of corals by visually confirming the irradiation area on the coral surface under the same light conditions in the laboratory. Furthermore, preparing the printed photos of each coral taken at the beginning of the experiment helps more prompt and accurate measurements of $\Delta F/F_m'$ using MINI-PAM II.

8.9 Coral Bleaching Process Under Anthropogenic Chemicals and High Seawater Temperature

The compact experimental system enables a detailed report on the bleaching process of *Acropora* spp. during 1- to 2-week experiments in a laboratory located far from the coral reefs (Takeuchi et al. 2020; Gushi et al. 2021; Shitaoka et al. 2021). Due to the intermediate stage of the system presented here, coral pieces in previous studies were either not attached to polycarbonate bolts (Hirayama et al. 2017; Ishibashi et al. 2018, 2021) or were affixed using alone-alpha (Toagosei Co. Ltd., Minato-ku, Tokyo, Japan) (Kamei et al. 2020).

Kamei (2020) conducted a 7-day experiment using apical branches separated from a cultured *A. tenuis* exposed to extremely low concentrations of Irgarol at 20 ng/L and 200 ng/L. Their experiment revealed that 20 ng/L of Irgarol significantly reduced $\Delta F/F_m'$ by 8%, while the RGB values of coral were stable throughout the experiment (Kamei et al. 2020). Hirayama et al. (2017) exposed apical branches of cultured *A. tenuis* to 1.0 and 10.0 μg/L Irgarol for 14 days at 27.5 °C. In 1.0 μg/L treatments, the RGB values were nearly constant until day 5, and then increased toward white until day 14, the end of the experiment (Hirayama et al. 2017) (Fig. 8.8). In the 10 μg/L treatment, RGB values gradually increased from the beginning of the experiment. All apical branches of *A. tenuis* were estimated to bleach around day 14 (1 μg/L) and day 10 (10 μg/L) (Hirayama et al. 2017). The gene expression of heat shock protein 90 (HSP90) was downregulated after 7 days of exposure to 1 μg/L Irgarol (Ishibashi et al. 2018). Heat shock proteins, including HSP 90, are molecular chaperones that regulate protein structure and function during and after stress treatment. Ishibashi et al. (2021) investigated the molecular mechanisms underlying herbicide-induced bleaching of *A. tenuis* and its symbiotic dinoflagellates exposed to 1.0 and 10.0 μg/L Irgarol for 7 days (see Ishibashi and Takeuchi 2023). Gushi et al. (2021) studied the combined effects of 30.0 °C and 1.0 μg/L Irgarol on *Acropora tenuis* by measuring the RGB values and $\Delta F/F_m'$ for 7 d. Although the RGB values and $\Delta F/F_m'$ in the control (no herbicide) under 27.5 °C (temperature control) and 30.0 °C were nearly stable throughout the experiment, the RGB values and $\Delta F/F_m'$ in treatments with Irgarol significantly differed from those in the control under both temperatures (Gushi et al. 2021).

The ecotoxicity experiment system reported here is composed of general research and aquarium equipment and devices. Our measurement method for coral health, including photosynthetic efficiency, is similarly based on general research equipment and software. Such a versatile and inexpensive system and the measuring methods could elucidate the details of the exposure of *Acropora* to Irgarol at environmentally detected concentrations.

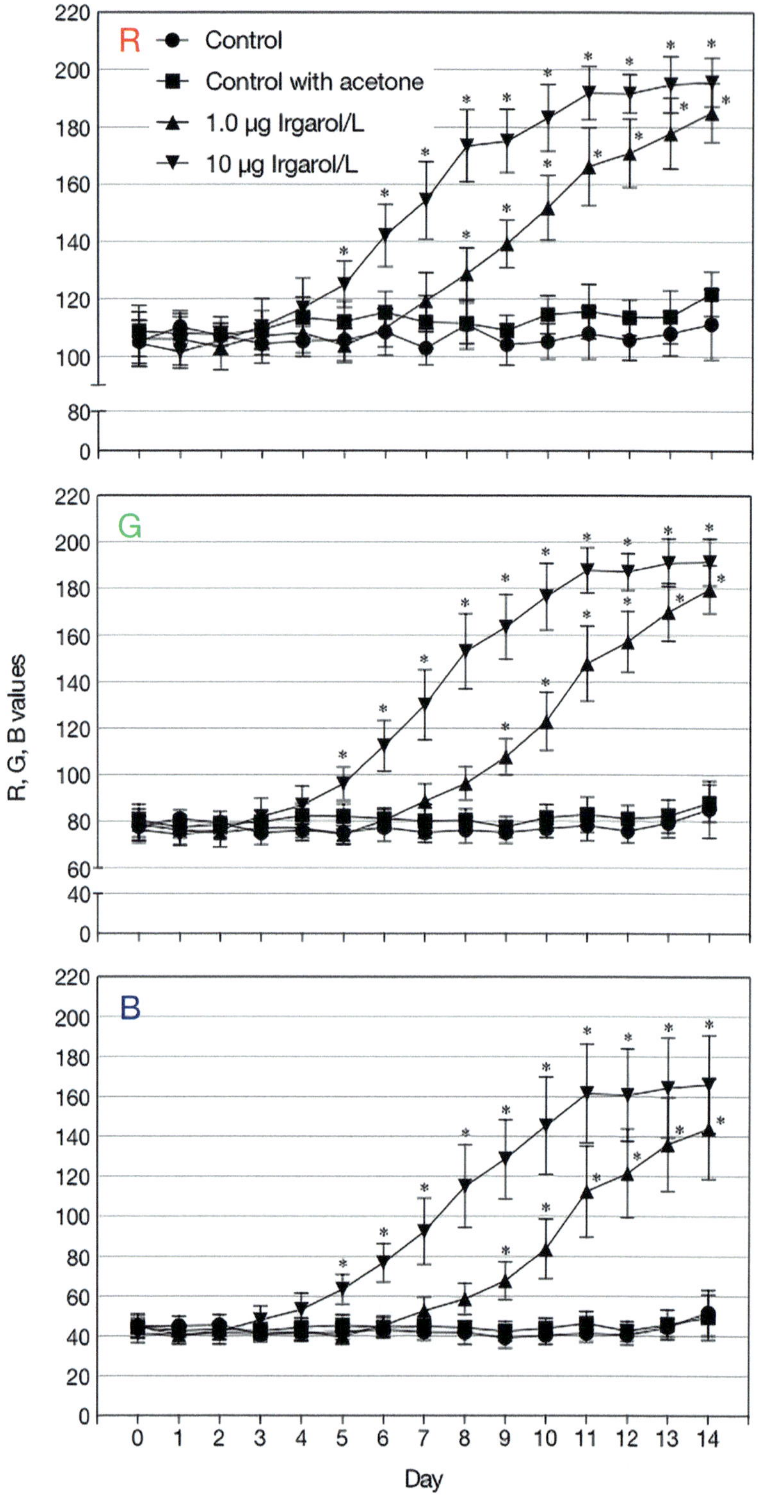

Fig. 8.8 Succession of RGB values of *Acropora tenuis* exposed to control, control with solvent, 1.0 µg/L Irgarol 1051, and 10 µg/L Irgarol 1051. Reprinted from Hirayama et al. (2017, Mar Pollut Bull 124, 678–686) with permission from Elsevier. Copyright (2017) Elsevier Ltd

8.10 Conclusion

Coral reefs in tropical and subtropical zones are recognised as highly dynamic ecosystems with spectacularly high diversity. Climate change and anthropogenic impacts are threatening coral reefs worldwide, especially around the East China Sea (Takeuchi 2023a, b). Recently, ecotoxicity experiments on the effect of anthropogenic chemicals on corals have been carried out in advanced large-scale aquariums close to coral reefs, such as the National Sea Simulator (SeaSim) at the Australian Institute of Marine Science in Townsville, QLD,

Australia (see Nordborg et al. 2018, 2021; Flores et al. 2020, 2021). The present compact method using widely available research and aquarium equipment and devices will allow scientists to conduct various ecotoxicity experiments, including transcriptome analysis, at any university or research institute laboratory located far from coral reefs in subtropical and tropical areas. More detailed experiments, such as those investigating the effects of triple anthropogenic factors and/or common garden experiments using corals from different localities to elucidate genetic differences in corals, could be conducted by assembling multiple sets of our system into a large-scale system in the near future.

Acknowledgements We express our sincere gratitude to Dr. H. Ishibashi and Prof. H. Yamashiro for their cooperation and comments on the establishment of the present system. The concepts of the present method were established mainly through several research grants from the Japan Society for the Promotion of Science (JSPS) Grants-in-Aid for Scientific Research (KAKENHI; Grant nos. 26281048, 18H03387, 21H03619). The following undergraduate and graduate students greatly contributed to the development of the methodologies and conducting the experiments: Hirayama, S.; Gushi, K.; Kamei, M.; and Iba, H. We express our thanks to Editage (www.editage.com) for English language editing.

References

Ahlers J, Riedhammer C, Vogliano M, Ebert R-U, Kühne R, Schüürmann G (2006) Acute to chronic ratios in aquatic toxicity—variation across trophic levels and relationship with chemical structure. Environ Toxicol Chem 5:2937–2945

Akkaynak D, Treibitz T, Xiao B, Gürkan UA, Allen JJ, Demirci U, Hanlon RT (2014) Use of commercial off-the-shelf digital cameras for scientific data acquisition and scene-specific color calibration. J Optic Soc Am A 31:312 321

Ali HR, Arifin MM, Sheikh MA, Shazili NAM, Bachok Z (2013) Occurrence and distribution of antifouling biocide Irgarol-1051 in coastal waters of Peninsular Malaysia. Mar Pollut Bull 70: 253–257

Alzieu C (2000) Environmental impact of TBT: the French experience. Sci Total Environ 258:99–102

Aono A, Takeuchi I (2008) Effects of tributyltin at concentrations below ambient levels in seawater on *Caprella danilevskii* (Crustacea: Amphipoda: Caprellidae). Mar Pollut Bull 57:515–523

Bahr KD, Severino SJL, Tsang AO, Han JJ, Dona AR, Stender YO, Weible RM, Graham A, McGowan AE, Rodgers KS (2020) The Hawaiian Ko'a card: coral health and bleaching assessment color reference card for Hawaiian corals. SN Appl Sci 2:1706. https://doi.org/10.1007/s42452-020-03487-3

Balakrishnan S, Takeda K, Sakugawa H (2012) Occurrence of Diuron and Irgarol in seawater, sediments and planktons of Seto Inland Sea, Japan. Geochem J 46:169–177

Bartlett TC (2013) Small Scale Experimental Systems for Coral Research: Considerations, Planning, and Recommendations, NOAA Technical Memorandum NOS NCCOS 165 and CRCP 18, v + 68 pp

Basheer C, Tan KS, Lee HK (2002) Organotin and Irgarol 1051 contamination in Singapore coastal waters. Mar Pollut Bull 44:697–703

Belanger SE, Brill JL, Rawlings JM, Price BB (2016) Development of acute toxicity quantitative structure activity relationships (QSAR) and their use in linear alkylbenzene sulfonate species sensitivity distributions. Chemosphere 155:18–27

Bhagooli R, Mattan-Moorgawa S, Kaullysing D, Louis YD, Gopeechund A, Ramah S, Soondur M, Pilly SS, Beesoo R, Wijayanti DP, Bachok ZB, Monrás VC, Casareto BE, Suzuki Y, Baker AC (2021) Chlorophyll fluorescence–a tool to assess photosynthetic performance and stress photophysiology in symbiotic marine invertebrates and sea plants. Mar Pollut Bull 165:112059

Cowman PF, Quattrini AM, Bridge TCL, Watkins-Colwell GJ, Fadli N, Grinblat M, Roberts TE, McFadden CS, Miller DJ, Baird AH (2020) An enhanced target-enrichment bait set for Hexacorallia provides phylogenomic resolution of the staghorn corals (Acroporidae) and close relatives. Mol Phylogenet Evol 153:106944

Dafforn KA, Lewis JA, Johnston EL (2011) Antifouling strategies: history and regulation, ecological impacts and mitigation. Mar Pollut Bull 62:453–465

Delphis Inc (2021) LIVE sea salt. https://delphis.co.jp/product/prod_lss.html. Accessed 28 Sept 2021. (In Japanese)

Endo T (2014) The teaching of washing method of chemical glassware. Bull Tsurumi College Pt 4, Stud Humanit Soc Nat Sci 51:1–3

Flores F, Kaserzon S, Elisei G, Ricardo G, Negri AP (2020) Toxicity thresholds of three insecticides and two fungicides to larvae of the coral *Acropora tenuis*. PeerJ 8:e9615. https://doi.org/10.7717/peerj.9615

Flores F, Marques JA, Uthicke S, Fisher R, Patel F, Kaserzon S, Negri AP (2021) Combined effects of climate change and the herbicide diuron on the coral *Acropora millepora*. Mar Pollut Bull 169:112582

Fukami H, Tachikawa H, Suzuki G, Nagata S, Sugihara K (2010) Current status and problems with the identification and taxonomy of zooxanthellate scleractinian corals in Japan. J Japan Coral Reef Soc 12:17–31. (In Japanese with English Abstract)

Gushi M, Ishibashi H, Takayama K, Yamashiro H, Takeuchi I (2021) Changes in the colour and photosynthetic efficiency of the hermatypic coral *Acropora tenuis* exposed to Irgarol 1051 at 30 °C seawater temperature. Reg Stud Mar Sci 47:101957

Hano T, Ito K, Ohkubo N, Sakaji H, Watanabe A, Takashima K, Sato T, Sugaya T, Matsuki K, Onduka T, Ito M, Somiya R, Mochida K (2019) Occurrence of neonicotinoids and fipronil in estuaries and their potential risks to aquatic invertebrates. Environ Pollut 252:205–215

Hirayama K, Takayama K, Haruta S, Ishibashi H, Takeuchi I (2017) Effect of low concentrations of Irgarol 1051 on RGB (R, red; G, green; B, blue) colour values of the hard-coral *Acropora tenuis*. Mar Pollut Bull 124:678–686

Hoeksema BW, Cairns S (2021) World List of Scleractinia. *Acropora* Oken, 1815. Accessed through: World Register of Marine Species at: http://marinespecies.org/aphia.php?p=taxdetails&id=205469 on 2021-09-27

Hongo C, Yamano H (2013) Species-specific responses of corals to bleaching events on anthropogenically turbid reefs on Okinawa Island, Japan, over a 15-year period (1995–2009). PLoS One 8: e60952. https://doi.org/10.1371/journal.pone.0060952

Ishibashi H, Takeuchi I (2023) Effects of anthropogenic chemicals to hermatypic corals with special reference to gene expression. In: Takeuchi I, Yamashiro H (eds) Coral reefs of eastern Asia under anthropogenic impacts. Coral reefs of the world, vol 17. Springer Nature Switzerland AG, Cham, pp 151–163

Ishibashi H, Minamide S, Takeuchi I (2018) Identification and characterization of heat shock protein 90 (HSP90) in the hard coral *Acropora tenuis* in response to Irgarol 1051. Mar Pollut Bull 133: 773–780

Ishibashi H, Takaichi D, Takeuchi I (2021) Effects of the herbicide Irgarol 1051 on the transcriptome of hermatypic coral *Acropora tenuis* and its symbiotic dinoflagellates. Sci Total Environ 780: 146542

Jones RJ, Kerswell AP (2003) Phytotoxicity of photosystem II (PSII) herbicides to coral. Mar Ecol Prog Ser 261:149–159

Kadokami K, Miyawaki T, Iwabuchi K, Takagi S, Adachi F, Iida H, Watanabe K, Kosugi Y, Suzuki T, Nagahora S, Tahara R, Orihara T, Eguchi A (2021) Inflow and outflow loads of 484 daily-use chemicals in wastewater treatment plants across Japan. Environ Monitor Contamin Res 1:1–16

Kamata M, Matsui Y, Asami M (2020) National trends in pesticides in drinking water and water sources in Japan. Sci Total Environ 744: 140930

Kamei M, Takayama K, Ishibashi H, Takeuchi I (2020) Effects of ecologically relevant concentrations of Irgarol 1051 in tropical to subtropical coastal seawater on hermatypic coral *Acropora tenuis* and its symbiotic dinoflagellates. Mar Pollut Bull 150:110734

Kaonga CC, Takeda K, Sakugawa H (2015) Antifouling agents and fenitrothion contamination in seawater, sediment, plankton, fish and selected marine animals from the Seto Inland Sea, Japan. Geochem J 49:23–37

Katam K, Maetani K, Shimizu T, Nakajima J, Bhattacharyya D (2018) Study of aerobic biodegradation of surfactants and fluorescent whitening agents in detergents of a few selected Asian countries (India, Indonesia, Japan, and Thailand). J Water Environ Technol 16:18–29

Kawasaki K (ed) (2017) Sakura Mook 02. That's what we want to know! Coral keeping manual 2017. Kasakura Publishing, Daido-ku, Tokyo, 170 pp. (In Japanese)

Kegler P, Baum G, Indriana LF, Wild C, Kunzmann A (2015) Physiological response of the hard coral *Pocillopora verrucosa* from Lombok, Indonesia, to two common pollutants in combination with high temperature. PLoS One 10:e0142744. https://doi.org/10.1371/journal.pone.0142744

Lam K-H, Cai Z, Wai H-Y, Tsang VW-H, Lam MH-W, Cheung RY-H, Yu H, Lam PK-S (2005) Identification of a new Irgarol-1051 related *s*-triazine species in coastal waters. Environ Pollut 136:221–230

Mano H, Takeda F, Kitamura T, Okamoto S, Kobayashi K, Takabatake H, Yamashita N, Tanaka H (2015) Evaluation of toxicity reduction of treated wastewater by UF and RO membrane technologies based on algal growth inhibition test and short term toxicity test on embryo and sac-fry stages of medaka. J Japan Soc Civil Eng Ser G Environ Res 74:III_179–III_187. (In Japanese with English Abstract)

Marshall NJ, Kleine DA, Dean AJ (2012) CoralWatch: education, monitoring, and sustainability through citizen science. Front Ecol Environ 10:332–334

Maxwell K, Johnson GN (2000) Chlorophyll fluorescence—a practical guide. J Exp Bot 51:659–668

Mortelmans K, Zeiger E (2000) The Ames *salmonella*/microsome mutagenicity assay. Mutat Res - Fundam Mol Mech Mutagen 455:29–60

Murai R, Takahashi S, Tanabe S, Takeuchi I (2005) Status of butyltin pollution along the coasts of Western Japan in 2001, 11 years after partial restrictions on the usage of tributyltin. Mar Pollut Bull 51: 940–949

Murchie EH, Lawson T (2013) Chlorophyll fluorescence analysis: a guide to good practice and understanding some new applications. J Exp Bot 64:3983–3998

Nakamura T, van Woesik R (2001) Water-flow rates and passive diffusion partially explain differential survival of corals during the 1998 bleaching event. Mar Ecol Prog Ser 212:301–304

Narita K, Matsui Y, Matsushita T, Shirasaki N (2021) Selection of priority pesticides in Japanese drinking water quality regulation: validity, limitations, and evolution of a risk prediction method. Sci Total Environ 751:141636

Nishiguchi S, Wada N, Yamashiro H, Ishibashi H, Takeuchi I (2018) Continuous recordings of the coral bleaching process on Sesoko Island, Okinawa, Japan, over about 50 days using an underwater camera equipped with a lens wiper. Mar Pollut Bull 131:422–427

Nobukawa T, Sanukida S (1997) Genotoxicity of drinking water from the Yodo River using chromosomal aberration tests in cultured mammalian cells. J Japan Soc Water Environ 11:705–709. (In Japanese with English Abstract)

Nordborg FM, Flores F, Brinkman DL, Agustí S, Negri AP (2018) Phototoxic effects of two common marine fuels on the settlement success of the coral *Acropora tenuis*. Sci Rep 8:8635. https://doi.org/10.1038/s41598-018-26972-7

Nordborg FM, Brinkman DL, Ricardo GF, Agustí S, Negri AP (2021) Comparative sensitivity of the early life stages of a coral to heavy fuel oil and UV radiation. Sci Total Environ 781:146676

Ohki S, Kowalski RK, Kitanobo S, Morita M (2015) Changes in spawning time led to the speciation of the broadcast spawning corals *Acropora digitifera* and the cryptic species *Acropora* sp. 1 with similar gamete recognition systems. Coral Reefs 34:1189–1198

Okamura H, Aoyama I, Liu D, Maguire RJ, Pacepavicius GJ, Lau YL (2000) Fate and ecotoxicity of the new antifouling compound Irgarol 1051 in the aquatic environment. Water Res 34:3523–3530

Okamura H, Aoyama I, Ono Y, Nishida T (2003) Antifouling herbicides in the coastal waters of western Japan. Mar Pollut Bull 47:59–67

Omasa K, Shimazaki K, Aiga I, Larcher W, Onoe M (1987) Image analysis of chlorophyll fluorescence transients for diagnosing the photosynthetic system of attached leaves. Plant Physiol 84: 748–752

Onduka T, Kono K, Ichihashi H, Tanaka H (2008) Distribution of organotin compounds in seawaters and sediments in Hiroshima Bay, Japan. J Environ Chem 18:9–17

Osinga R, Schutter M, Wijgerde T, Rinkevich B, Shafir S, Shpigel M, Luna GM, Danovaro R, Bongiorni L, Deutsch A, Kuecken M, Hiddinga B, Janse M, McLeod A, Gili C, Lavorano S, Henard S, Barthelemy D, Westhoff G, Baylina N, Santos E, Weissenbacher A, Kuba M, Jones R, Leewis R, Petersen D, Laterveer M (2012) The CORALZOO project: a synopsis of four years of public aquarium science. J Mar Biol Ass UK 92:753–768

Petersen D, Laterveer M, van Bergen D, Kuenen M (2004) Transportation techniques for massive scleractinian corals. Zoo Biol 23: 165–176

Rocha RJ, Bontas B, Cartaxana P, Leal MC, Ferreira JM, Rosa R, Serôdio J, Calado R (2015) Development of a standardized modular system for experimental coral culture. J World Aquacult Soc 46: 235–251

Sakai Y, Kato K, Koyama H, Kuba A, Takahashi H, Fujimori T, Hatta M, Negri AP, Baird AH, Ueno N (2020) A step-down photophobic response in coral larvae: implications for the light-dependent distribution of the common reef coral, *Acropora tenuis*. Sci Rep 10: 17680. https://doi.org/10.1038/s41598-020-74649-x

Sato M, Uemura H, Kosaka K, Asami M, Kamata M (2016) Survey of pesticide concentrations, including neonicotinoids in the Sagami River, its tributaries and tap water. J Jpn Soc Water Environ 39: 153–162. (In Japanese with English Abstract)

Shafir S, Van Rijn J, Rinkevich B (2006) Coral nubbins as source material for coral biological research: a prospectus. Aquaculture 259:444–448

Sheikh MA, Higuchi T, Fujimura H, Imo TS, Miyagi T, Oomori T (2009) Contamination and impacts of new antifouling biocide Irgarol-1051 on subtropical coral reef waters. Int J Environ Sci Technol 6:353–358

Shitaoka R, Ishibashi H, Takeuchi I (2021) Thermal tolerance of the hermatypic coral *Acropora tenuis* elucidated by RGB analysis and expression of heat shock proteins in coral and symbiotic dinoflagellates. Mar Pollut Bull 162:111812

Siebeck UE, Marshall NJ, Klüter A, Hoegh-Guldberg O (2006) Monitoring coral bleaching using a colour reference card. Coral Reefs 25: 453–460

Spalding MD, Brown BE (2015) Warm-water coral reefs and climate change. Science 350:769–771

Takayama K, Nishina H (2009) Chlorophyll fluorescence imaging of chlorophyll fluorescence induction phenomenon for plant health monitoring. Environ Control Biol 47:101–109

Takeuchi I (2023a) Introduction–The relevance of anthropogenic factors to coral reef conservation in the coastal areas of the East China Sea. In: Takeuchi I, Yamashiro H (eds) Coral reefs of eastern Asia under anthropogenic impacts. Coral reefs of the world, vol 17. Springer Nature Switzerland AG, Cham, pp 1–5

Takeuchi I (2023b) Anthropogenic stresses in coral reefs and adjacent ecosystem around East China Sea. In: Takeuchi I, Yamashiro H (eds) Coral reefs of eastern Asia under anthropogenic impacts. Coral reefs of the world, vol 17. Springer Nature Switzerland AG, Cham, pp 137–149

Takeuchi I, Takahashi S, Tanabe S, Miyazaki N (2004) Butyltin concentrations along the Japanese coast from 1997 to 1999 monitored by *Caprella* spp. (Crustacea: Amphipoda). Mar Env Res 57: 397–414

Takeuchi I, Yamashiro H, Gushi M (2019) Usage of UV-curable oligomer-based adhesive agent in hermatypic coral experimental research. MethodsX 6:1600–1607

Takeuchi I, Takaichi D, Katsumata M, Ishibashi H (2020) Succession of delayed fluorescence correlated with coral bleaching in the hermatypic coral *Acropora tenuis*. Mar Pollut Bull 154:111008

Tsuruta K (2021) Coral research and aquaculture in the classroom. Coral Fleaks 33:42–45. (In Japanese)

Effects of Anthropogenic Chemicals on Hermatypic Corals with Special Reference to Gene Expression

9

Hiroshi Ishibashi and Ichiro Takeuchi

Abstract

Many studies have reported the adverse effects of various anthropogenic activities, including climate change and ocean acidification, on coral reefs. While the molecular effects of these environmental stressors on corals and symbionts have been well studied, the effect of anthropogenic chemicals on corals remains poorly understood. In this review, we investigated the potential toxic effects including bleaching by heavy metals (copper and cadmium), organotins (TBT, tributyltin), polycyclic aromatic carbons (PAHs) (anthracene, phenanthrene, and benzo[a] pyrene), polychlorinated biphenyls (PCBs, PCB118, representative congener of PCBs, and Aroclor 1254, representative PCB products), photosystem II herbicides (Diuron and Irgarol 1051), and other contaminants (caffeine, 4-nonylphenol, and oxybenzone) on gene expression biomarkers. The potential biomarkers mainly included chaperones (heat shock proteins HSP70 and HSP90 and calreticulin), components of the cytochrome P450 monooxygenase system, antioxidant enzymes (glutathione S-transferase [GST], catalase [CAT], and superoxide dismutase [SOD]), and green and red fluorescent proteins (GFP and RFP). Another survey using DNA microarray and RNA sequencing (RNA-seq) identified hundreds to thousands of differentially expressed genes (DEGs) in corals and the symbiotic dinoflagellates. Bioinformatics analyses using DEGs also documented the potential effects of various gene ontology terms, pathways, and protein interaction networks. These findings using transcriptomic approach provide new insights into the potential molecular mechanisms underlying the toxic and bleaching effects of anthropogenic chemicals on corals and symbiotic dinoflagellates.

Keywords

Acropora spp. · Anthropogenic chemicals · Biomarker genes · Bleaching mechanism · Coral · Zooxanthellae

9.1 Introduction

Coral reefs have been adversely affected by various anthropogenic factors, such as climate change, ocean acidification, overfishing, disease, coastal development, and chemical pollutants. Many studies have described the potential molecular mechanisms of coral bleaching by thermal stress at the molecular level (Portune et al. 2010; Moya et al. 2012; Palumbi et al. 2014; Kaniewska et al. 2015). Louis et al. (2017) reviewed the gene expression biomarkers (GEBs) of thermal stress in scleractinian corals. Many candidate GEBs, including certain heat shock protein genes, metabolic genes, oxidative stress genes, immune response genes, ion transport genes, and structural genes have been investigated, and some genes, such as *hsp16*, *Cacna1*, *MnSOD*, *SLC26*, and *Nf-kB*, have been identified as reliable indicators of thermal stress in corals (Louis et al. 2017).

Contamination by anthropogenic chemicals, including pesticides and other environmental pollutants, has been reported in marine ecosystems around the East China Sea (Takeuchi 2023). These anthropogenic chemicals likely cause several ecotoxicological effects to hermatypic coral and its symbiotic flagellates. High Pb/Ca ratio was detected in the hermatypic coral *Porites* from a river mouth on Okinawa Island, Japan, suggesting the anthropogenic input of run-off from urban areas (Ramos et al. 2004). However, the molecular effects of anthropogenic chemicals on corals and symbionts are poorly understood. Although many studies have investigated the toxic effects of anthropogenic chemicals on corals and symbiotic dinoflagellates, few

H. Ishibashi · I. Takeuchi (✉)
Graduate School of Agriculture, Ehime University, Matsuyama, Ehime, Japan

Center of Advanced Technology for the Environment, Graduate School of Agriculture, Ehime University, Matsuyama, Ehime, Japan
e-mail: takeuchi@agr.ehime-u.ac.jp

I. Takeuchi, H. Yamashiro (eds.), *Coral Reefs of Eastern Asia under Anthropogenic Impacts*, Coral Reefs of the World 17,
https://doi.org/10.1007/978-3-031-27560-9_9

studies have examined their mode of action at the molecular level. In this section, we synthesize the current knowledge of coral GEBs and highlight the gaps in our understanding to identify directions for future work.

9.2 Effects of Anthropogenic Chemicals on Gene Expression

9.2.1 Heavy Metals

Exposure to heavy metals induces oxidative stress in many organisms through cellular biochemical reactions that produce reactive oxygen species (ROS) such as the superoxide anion radical (O_2^-), hydrogen peroxide (H_2O_2), hydroxide radical (OH), and singlet oxygen ($^1O_2$) (Lesser 2006; Weis 2008). Under stable conditions, cells detoxify ROS through a variety of enzymatic and scavenging responses that directly detoxify reactive oxygen. Cells also upregulate cellular processes that repair cellular damage, including the production of chaperone heat shock proteins (HSPs) and DNA damage repair proteins (De Nadal et al. 2011). However, the molecular effects of heavy metals on coral are not completely understood. Copper (Cu) is a common chemical contaminant in coastal environments including coral reefs; it has been detected in seawater (ranging from 0.13 to 9.5 μg/L) (Kozelka and Bruland 1998) and polluted areas (ranging from 0.12 to 18.4 μg/L) (Jones 2010). Venn et al. (2009) investigated the impact of Cu on the reef coral *Orbicella faveolata*, previously known as *Montastraea franksi*, by analyzing the expression of several stress-related genes (*P-glycoprotein*, *HSP70*, and *HSP90*) in the coral host using real-time quantitative reverse transcription-PCR (qRT-PCR) (Table 9.1). They detected a significant increase in *P-glycoprotein* expression in corals exposed to 100 μg/L Cu for 4 and 8 h and to 30 μg/L Cu for 8 h. Similarly, *HSP70* was significantly upregulated after 4 h of exposure to 100 μg/L Cu, and both *HSP70* and *HSP90* were significantly upregulated following 8 h exposure to 30 and 100 μg/L Cu. These data suggest that *P-glycoprotein*, *HSP70*, and *HSP90* are inducible forms in *O. faveolata* and are activated under stressful conditions such as Cu exposure. Moreover, the response of *P-glycoprotein* and *HSP70* was significant after 4 h and that of *HSP90* after 8 h of exposure to 100 μg/L Cu. These findings suggest that *P-glycoprotein* and *HSP* genes are important markers that should be added to the suite of molecular indicators developed to identify cellular damage by chemical stressors in *O. faveolata*.

Schwarz et al. (2013) investigated the potential stress and detrimental effects in both *O. faveolata* and its algal symbionts exposed to Cu using ecotoxicological and microarray studies (Table 9.1). Corals exposed to 30 μg/L Cu for 48 h experienced significant DNA damage and displayed changes in the expression patterns of genes known to play roles in cellular and oxidative stress responses. Of the 1311 cDNAs printed on the arrays, 34 cDNAs from Cu-exposed corals showed significant changes in abundance relative to the controls. More than half of these genes exhibited no similarity to known genes according to BLAST annotation. Among the ten most highly upregulated genes were *peroxidasin* (a peroxidase), *aldose reductase* (a metabolic enzyme), *melanocortin receptor 4* (a G-protein coupled receptor), and *ETS1* (a transcription factor). Among the ten most highly downregulated genes were *galaxin* (a skeletonization protein), *aquarius* (an RNA processing protein), *isopeptidase T* (ubiquitination/protein degradation), and a protein that has been identified in breast carcinoma in humans. Of the 34 genes that were differentially expressed, at least 24% (8 genes) were known to play roles in cellular and oxidative stress. Oxidative stress genes that showed significant changes in expression included the chaperone *HSP90*, the detoxification enzyme *glutathione-S-transferase* (*GST*), *thioredoxin glutathione reductase*, *catalase* (*CAT*), *peroxidasin*, *cysteine protease cathepsin B*, DNA damage repair enzyme *rad24*, and ETS-domain-containing transcription factor related to *ETS1*. Coral-specific *SCRIP2* expression was also significantly downregulated in response to Cu exposure.

Yuyama et al. (2012) investigated the potential antifoulant tributyltin chloride (TBT-Cl)-responsive genes in *Acropora tenuis* juvenile polyps using high-coverage expression profiling (HiCEP) and qRT-PCR (Table 9.1). Tributyltin is one of the dominant organotin compounds used in antifouling paints (Dafforn et al. 2011), whereas *A. tenuis* is among the representative species of hermatypic corals in the Indo-Pacific region including southern Japan. In their experiment, it was unclear whether the genes that showed decreased expression were actually downregulated in response to stress or merely reflected a reduced number of algal cells. Therefore, Yuyama et al. (2012) focused on the upregulated genes in stressed polyps. HiCEP analysis revealed 25 upregulated genes in juvenile corals after exposure to 1 μg/L TBT-Cl for 11 days. Of those, five genes that matched animal genes using the BLASTx homology search (GenBank/EMBL/DDBJ), namely *oxidative stress-responsive protein* (*Ox stress*), *HSP90*, *calreticulin* (*CRT*), *aquaglyceroporin*, and *methionine adenosyltransferase*, and two genes matching the algal genes *light-harvesting complex* (*LHC*) and *DnaJ-like protein 1* were quantified by qRT-PCR. The results showed that the expression of *CRT* and *Ox stress* was induced by TBT-Cl, suggesting that TBT-Cl induces oxidation and consequently suppresses the growth of corals. In endosymbiotic algae, the expression patterns of *DnaJ-like protein 1*, but not those of *LHC*, were altered by TBT-Cl, suggesting that DnaJ-like proteins may be sensitive to

Table 9.1 List of chemical names, dose, and exposure period, species, life stage, toxic effect, and gene expression response in hermatypic corals exposed to anthropogenic chemicals

Chemical (cas. no.)	Dose	Exposure period	Species	Life stage	Toxic effect	Gene, protein, activity, and others	Technique	Reference
Copper (7440-50-8)	3, 10, 30, 100 μg/L	4, 8 h	*Orbicella franksi*	–	Algal density→	HSP90↑ HSP70↑ P-glycoprotein↑	qRT-PCR	Venn et al. (2009)
	<1, 8, 30 μg/L	2 d	*Orbicella franksi*	Parent colony	DNA damage↑	GST↑ PXDN↑ CAT↑ TGR↓ Rad24↑ HSP90↑ ETS1↑ cathepsin B↑ PTP↓ PTR↑ SCRIP2↓ aldose reductase↑ melanocortin receptor 4↑ galaxin↓ aquarius↓ isopeptidase T↓ breast carcinoma seq. 2↓	Microarray, qRT-PCR	Schwarz et al. (2013)
Tributyltin chloride (1461-22-9)	1 μg/L	11 d	*A. tenuis* *Symbiodinium*	Juvenile	Pigmentation↓	Ox stress↑ HSP90→ CRT↑ aquaglyceroporin→ methionine adenosyltransferase↑ LHC→ DnaJ-like protein 1↑	qRT-PCR	Yuyama et al. (2012)
Cadmium (7440-43-9)	20 μg/L	6, 12, 24 h	*Pocillopora damicornis*	Nubbin	Algal density↓	SOD activity↓ CAT activity↓ GST activity→ caspase 3 activation↑ GSH concentration↓ 3538 genes↑ 8048 genes↓	RNA sequencing	Zhou et al. (2018)
Standard mixture of 13 PAHs	100 μg/L	1 d	*Scleronephthya gracillimum*	Polyps	–	20 genes↑ 17 genes↓	Differential display PCR, qRT-PCR	Woo et al. (2014)
Anthracene (120-12-7)	4, 17 μg/L	2, 3 d	*Acropora tenuis*	Larvae	Survival↓ Metamorphosis↓	GFP↑ RFP↑ Ox stress↑ HSP90↑ HSP70↑ MnSOD→ CAT↑ p53→	qRT-PCR	Overmans et al. (2018)

(continued)

Table 9.1 (continued)

Chemical (cas. no.)	Dose	Exposure period	Species	Life stage	Toxic effect	Gene, protein, activity, and others	Technique	Reference
Phenanthrene (85-01-8)	39, 309 μg/L	2, 3 d	*Acropora tenuis*	Larvae	Survival↓ Metamorphosis↓	GFP↑ RFP↓ Ox stress→ HSP90→ HSP70→ MnSOD→ CAT→ p53→	qRT-PCR	Overmans et al. (2018)
Benzo[a]pyrene (50-32-8)	0.01, 0.1 mg/L	1, 3 d	*Orbicella franksi*	Polyps	Pigmentation→	CAT activity↑ SOD activity ↑ GST activity→ MFO content↑		Ramos and García (2007)
	10, 40 μg/L	1, 3 d	*Acropora formosa* *Acropora nasuta*		Chlorophyll a↓	SOD activity↓ HSP70↓ P-glycoprotein↑ SOD activity↑ HSP70→ P-glycoprotein↑	qRT-PCR	Xiang et al. (2019)
PCB118 (31508-00-6)	Injection with 1 ml/10 g tissue (0.1 mg/ml in corn oil)	2 d	*Dendronephthya klunzingeri*	–	–	(HSP90↑)	Western blotting	Wiens et al. (2000)
Aroclor 1254 (11097-69-1)	300 ng/L (=total PCBs)	4 h, 4 d	*Stylophora pistillata*	Adult	Photosynthesis→ survival→ growth→	HSP70-like↑ dipeptidase↑ laminin G domain-like↓ ARP↓ ARL6↓ Arf1↓ Rab7↓ GRX↓	Suppression subtractive hybridization	Chen et al. (2012)
Diuron (330-54-1)	10 μg/L	13 d	*Acropora tenuis* *Symbiodinium*	Juvenile	Pigmentation↓	Ox stress↑ HSP90→ CRT↑ aquaglyceroporin↑ methionine adenosyltransferase→ LHC↓ DnaJ-like protein 1↑	qRT-PCR	Yuyama et al. (2012)
Irgarol 1051 (28159-98-0)	1, 10 μg/L	7 d	*Acropora tenuis*	Adult	RGB values↑	HSP90↓ CRT↓	qRT-PCR	Ishibashi et al. (2018)
	0.02, 0.2 μg/L	7 d 7 d	*Acropora tenuis* *Symbiodinium*	Adult	RGB values→ photosynthesis↓	HSP90→ HSP90↑ HSP70↑	qRT-PCR	Kamei et al. (2020)

	1, 10 μg/L	7 d	*Acropora tenuis Symbiodinium*	Adult	RGB values↑	100 genes↑, 42 genes↓ in 1 μg/L 40 genes↑, 69 genes↓ in 10 μg/L 97 genes↑, 14 genes↓ in 1 μg/L 15 genes↑, 101 genes↓ in 10 μg/L	RNA sequencing	Ishibashi et al. (2021)
Caffeine (58-08-2)	50 μg/L–100 mg/L	22–35 d	Clade B *Symbiodinium* sp. from *Pseudoterogorgia bipinnata* (B7) Clade B *Symbiodinium* sp. from *Aiptasia pallida* (B6) Clade C *Symbiodinium goreaui* (C)	–	Growth↓	(HSP70↑) (HSP90↓) (HSP70↓) (HSP60↓) (HSP90↓) (HSP70↓)	Peptide mass spectrometry	Pollack et al. (2009)
4-nonylphenol (104-40-5)	1 μg/L	42 d	*Pocillopora damicornis*	Adult	Number of planulae↓	Total cholesterol→ free cholesterol→ 17β-estradiol→ estrone→ progesterone→ testosterone→ CYP17 activity→ CYP19 activity→ 3βHSD activity→ 3-β-gydroxysteroid dehydrogenase activity→ GST activity→ UGT activity↑ β-glucuronidase activity→ STS activity→ steroid sulfatase activity→	ELISA Enzyme assay	Rougée et al. (2021)

HSP, heat shock protein; CRT, calreticulin; GFP, green fluorescent protein; RFP, red fluorescent protein; Ox stress, oxidative stress-responsive protein; MnSOD, superoxide dismutase; CAT, catalase; p53, tumor protein p53; LHC, light-harvesting complex; GST, glutathione S-transferase; PXDN, peroxidasin-homolog-like; CAT, catalase; TGR, thioredoxin glutathione reductase; Rad24, DNA damage repair; ETS1, ETS1 transcription factor; PTP, protein tyrosine phosphatase; PTR, protein tyrosine receptor; SCRIP2, scleractinian cysteine-rich protein 2; ARP, actin-related protein 2/3; ARL6, ADP ribosylation factor 6-like; Arf1, ADP ribosylation factor 1; GRX, glutaredoxin

TBT-Cl and other stressors. These data confirm the negative effects of TBT-Cl on endosymbiotic algae.

Negri and Heyward (2001) demonstrated the effects of TBT and Cu on the inhibition of fertilization (4-hour treatment) and larval metamorphosis (24-hour treatment) in the coral *Acropora millepora* using laboratory-based bioassays. The respective 50% inhibitory nominal concentrations (IC_{50}) of TBT and Cu were 17.4 and 200 μg/L during fertilization and 110 and 2.0 μg/L during larval metamorphosis. These results suggest that the toxic effects of TBT and Cu vary at different stages of the life cycle, such as fertilization and larval metamorphosis, of *A. millepora*.

Zhou et al. (2018) investigated the density of symbiotic zooxanthellae, the levels of crucial physiological activities, and the transcriptome of the hermatypic coral *Pocillopora damicornis* after acute exposure to elevated cadmium (Cd) concentrations (Table 9.1). The density of symbiotic zooxanthellae decreased significantly during the 12–24 h period and reached its lowest level at 24 h after acute Cd stress. No significant changes were observed in GST activity during the stress exposure, whereas the activities of caspase3 significantly increased and that of superoxide dismutase (SOD) and CAT, as well as the concentration of glutathione, significantly decreased after Cd exposure. Moreover, transcriptome sequencing and bioinformatics analysis revealed 3538 significantly upregulated genes and 8048 significantly downregulated genes 12 h after the treatment. There were 12 overrepresented gene ontology (GO) terms for significantly upregulated genes, mostly related to unfolded protein response, endoplasmic reticulum stress, and apoptosis. In addition, 32 GO terms mainly correlated with macromolecular metabolic processes were overrepresented for significantly downregulated genes. These results suggest that acute Cd stress can induce apoptosis by repressing the production of antioxidants, elevating oxidative stress, and activating the unfolded protein response.

9.2.2 Polycyclic Aromatic Carbons

Polycyclic aromatic hydrocarbons (PAHs) consist of two or more benzene rings fused in various arrangements and are thought to be derived from three sources: fossil fuels, burning of organic matter, and the transformation of natural organic precursors by relatively rapid chemical/biological processes (see Neff et al. 2005; Takeuchi et al. 2009). The wide range of biological effects of PAHs so far reported include acute toxicity, carcinogenicity, mutagenicity, teratogenicity, and endocrine disrupting activity (see Takeuchi et al. 2009). Unlike polychlorinated biphenyls (PCBs), the concentrations of PAHs were reported to decrease with increasing trophic levels in the food web of shallow-water ecosystems (Takeuchi et al. 2009). PAHs pose a growing risk to coral reefs due to increasing shipping and petroleum extraction in tropical waters. The damaging effects of specific PAHs can be further enhanced by the presence of ultraviolet (UV) radiation, a condition known as phototoxicity (Yu et al. 2006). Woo et al. (2014) identified candidate genes whose transcript levels changed in response to PAH exposure using differential display polymerase chain reaction (DD-PCR) in a soft coral (*Scleronephthya gracillimum*). DD-PCR of coral colonies exposed to 100 μg/L of a standard mixture of PAHs for 24 h identified 37 types of candidate genes (20 upregulated and 17 downregulated genes), the functions of which included oxidative stress response, ribosomal structure maintenance, molecular chaperone activity, protein kinase activation and tumorigenesis, defense mechanisms, transcription, and other biological responses. These isolated gene candidates are thus potential molecular biomarkers for understanding coral responses to environmental stressors.

Overmans et al. (2018) conducted laboratory experiments to examine the effects of anthracene and phenanthrene, both consisting of three fused benzene rings in PAHs, on SOD enzyme activity, gene expression patterns, settlement success, and survival of *Acropora tenuis* planulae (Table 9.1). Another aim of their study was to establish whether co-exposure to environmentally relevant intensities of UVA (320–400 nm) increases the harmful effects of anthracene and phenanthrene through phototoxicity. The activity of SOD was reduced by anthracene, whereas phenanthrene and UVA exposure had no effect on the enzyme. The expression of *MnSOD* remained constant across all treatments. *CAT*, *HSP70*, and *HSP90* showed increased expression levels in larvae exposed to anthracene but not in those exposed to phenanthrene. The expression patterns of *red fluorescent protein* (*RFP*) and *green fluorescent protein* (*GFP*) showed little evidence of PAH-induced stress, apart from the significant downregulation of *GFP* after phenanthrene treatment. Gene expression of *p53* was upregulated in the presence of UVA but downregulated when exposed to PAHs. The influence of anthracene on stress-related biochemical pathways and gene expression in *A. tenuis* larvae was considerably greater than that of phenanthrene, and UVA-induced phototoxicity was evident only for anthracene. The combined effects of UVA and PAH exposure on larval survival and metamorphosis paralleled the sublethal stress responses, clearly highlighting the interaction of UVA with anthracene toxicity and, ultimately, affecting coral development.

Benzo[a]pyrene (BaP), a five-ring polycyclic aromatic hydrocarbon, is known to possesses many adverse biological effects, including tumor formation, immunosuppression, teratogenicity, hormonal effects, and disruption of protein–protein signaling pathways (Verma et al. 2012). BaP exerts pressure on corals owing to water discharge, oil spills, and coastal tourism. In earlier studies, Ramos and García (2007) investigated the components of the cytochrome P450 monooxygenase system (MFO) and antioxidant enzymes, such as CAT, SOD, and GST, in the coral *Orbicella faveolata* exposed to 0.01 and 0.1 mg/L BaP for 24 and 72 h (Table 9.1). They reported that the content of CAT and SOD in polyps and zooxanthellae and that of GST in polyps increased significantly at the highest BaP concentration and maximum exposure time. Cytochrome P420 was found in all colonies, and its content was the highest in colonies with the highest concentrations of the contaminant, whereas NADPH cytochrome c reductase activity and pigment concentrations did not vary between treatments (Ramos and García 2007). Their study is the first report on the induction of both detoxifying mechanisms, the MFO system and antioxidant enzymes, upon acute exposure to an organic contaminant in the reef-constructing coral species *O. faveolata*.

Xiang et al. (2019) investigated the physiological response, oxidative stress, and stress-related genetic expression in two species of *Acropora* (*A. formosa* and *A. nasuta*) exposed to 10 and 40 μg/L BaP for 24 h and 72 h. Their results showed that BaP affects the health of zooxanthellae in coral symbiosis after 72 h exposure due to a significant decline in chlorophyll *a* concentrations in *Acropora* spp. during this period. Exposure to 10 μg/L BaP for 24 h induced a serious oxidative damage in *Acropora* spp., causing a significant decline in SOD activity in *A. formosa* and its increase in *A. nasuta*. The *P-glycoprotein* was more sensitive in *A. formosa*, whereas *HSP70* was more sensitive in *A. nasuta*. Sensitivity to BaP exposure differed between these two coral species based on antioxidant defense patterns, with *A. formosa* having a lower ability to resist organic pollutants in coral reefs. Further ecotoxicological studies are needed to investigate the impact of chemical pollutants on corals and compare the different response mechanisms among coral species.

9.2.3 Polychlorinated Biphenyls

PCBs are a group of widespread contaminants listed as persistent organic pollutants (POPs) by the Stockholm Convention. PCBs are well known to biomagnify at successively higher trophic levels in food webs of shallow-water ecosystems (see Matsuo et al. 2009). Wiens et al. (2000) investigated the effects of thermal stress, Cd, and PCB118, on the expression of HSP90 in octocoral *Dendronephthya klunzingeri* (Table 9.1). PCB118, which is one of the high toxicity congeners of PCBs known as dioxin-like PCBs, were also one of the main congeners detected in fish in the Far East Asia (Matsumoto et al. 2014, 2016). Coral specimens were injected with 1 mL of PCB118 (0.1 mg/mL in corn oil) per 10 g tissue and samples were collected 2 days later. Western blot analysis revealed that the amount of HSP90 in *D. klunzingeri* was strongly increased after treatment with PCB118.

Chen et al. (2012) investigated the acute toxicity of Aroclor 1254 on the scleractinian coral *Stylophora pistillata*. Aroclor 1254 is one of the commercial PCB mixtures produced in the USA, and similar to Kaneclor 400 made in Japan and Copen A-40 in Germany (Takasuga et al. 2006). Coral nubbins were incubated in either control seawater or seawater dosed with PCBs (~300 ng/L) for 96 h. The effect of PCB exposure on coral gene expression at 4 h post-exposure was tested using suppression subtractive hybridization (SSH) and qRT-PCR. Photosystem II (PSII) activity of the zooxanthellae was measured at 96 h. After exposure, the nubbins were moved into clean seawater, and their survival and growth were observed for another 50 days. The results revealed that all nubbins survived during the exposure in the following 50-day recovery period. The PSII activity and coral growth were not affected by PCB exposure. SSH analysis identified 26 upregulated and 28 downregulated clones, which were sequenced for gene expression analysis; 15% of these sequences were identified, including genes involved in general stress response, peptide metabolism, cellular receptor, cytoskeleton organization, membrane trafficking, and oxidative stress response (Table 9.1). These results suggest that acute exposure of *S. pistillata* to Aroclor 1254 at 300 ng/L did not affect coral survival, photosynthesis, or growth but may have altered the expression of certain genes involved in various important cellular functions.

9.2.4 Herbicides

PSII herbicides such as Diuron have been shown to affect photosynthesis in corals at low environmentally relevant concentrations (Mercurio et al. 2018). PSII herbicides act by inhibiting photosynthetic electron transport in chloroplasts and are used to control redundant algal growth on submerged structures (Oettmeier 1999; Jones and Kerswell 2003).

Diuron (IUPAC name: 3-[3,4-dichlorophenyl]-1,1-dimethylurea) is commonly used in catchments that drain into the Great Barrier Reef lagoon and Okinawa Prefecture (Kitada et al. 2009; Kennedy et al. 2012; Takeuchi 2023). The copper-based antifouling paints mixed with several "booster" agents including Diuron have replaced butyltin-containing antifouling paints (Manzo et al. 2006; Dafforn et al. 2011). Thus, Diuron potentially reduces photosynthesis in *Symbiodinium*, a dinoflagellate symbiont associated with reef corals. Many studies on the effects of Diuron exposure on photosynthetic activity of zooxanthellae symbiotic with corals have been published previously (Oettmeier 1999; Jones and Kerswell 2003; Cantin et al. 2007; Katsumata and Takeuchi 2017). However, the molecular mechanisms of Diuron in corals are not completely understood. Yuyama et al. (2012) investigated the potential Diuron-responsive genes in *Acropora tenuis* juvenile polyps using the HiCEP and qRT-PCR methods (Table 9.1). HiCEP analysis revealed 49 upregulated genes in juvenile corals after exposure to 10 μg/L Diuron for 13 days. Of the 49 upregulated genes, five in corals (*Ox stress*, *HSP90*, *CRT*, *aquaglyceroporin*, and *methionine adenosyltransferase*) and two in algae (*LHC* and *DnaJ-like protein 1*) were measured by qRT-PCR (Table 9.1). Diuron caused an increase in the expression levels of *aquaglyceroporin* and *CRT* in the coral and *DnaJ-like protein* in *Symbiodinium*. The expression of neither *Ox stress* nor *HSP90* in coral was markedly changed by Diuron, suggesting a direct effect of Diuron on the coral. In contrast, the *LHC* of *Symbiodinium* was downregulated under Diuron-induced stress. Previous studies have reported that photobleaching of *Symbiodinium* cultures is associated with the loss of chlorophyll *a*–chlorophyll c_2–peridinin-protein complexes (acpPC), which constitute the majority of light-harvesting complex proteins (Takahashi et al. 2008). These results suggest that photoinhibition by Diuron causes damage to LHC in *Symbiodinium*. However, the only gene showing increased expression in response to all three stresses (thermal, TBT-Cl, and Diuron) was the coral gene coding for the oxidative stress-responsive protein (Ox stress), suggesting that oxidative stress generally occurs in each stress-exposed polyp.

Irgarol 1051 (IUPAC name: 2-*N*-*tert*-butyl-4-*N*-cyclopropyl-6-methylsulfanyl-1,3,5-triazine-2,4-diamine), as well as Diuron, is a widely used booster agent that is mixed with copper-dominated antifouling biocides (Manzo et al. 2006; Dafforn et al. 2011). To quantitatively evaluate the bleaching effects of Irgarol 1051 on corals, our research group developed a digital color imaging system based on NEF (RAW) files for studying corals in the laboratory setting using a commercially available digital single-lens reflex camera and a wireless speed light system (Hirayama et al. 2017). Using this assay system, we evaluated the RGB color of *Acropora tenuis* in response to Irgarol. After the exposure of *A. tenuis* branches to 1 and 10 μg/L Irgarol 1051 for 14 days, RGB values increased, moving toward the "white" end of the spectrum, with increasing Irgarol 1051 concentration. In contrast, a 7-day exposure of the coral to ecologically relevant concentrations (20 ng/L and 200 ng/L) of Irgarol 1051 decreased the maximum effective quantum yield ($\Delta F/F_m'$) in symbiotic dinoflagellates (Kamei et al. 2020). These results suggest that the threshold of chlorophyll fluorescence in symbiotic dinoflagellates is less than 20 ng/L Irgarol 1051, which is close to ecologically relevant concentrations in tropical to subtropical waters.

In contrast to the bleaching effects of Irgarol 1051, there are few reports on the molecular mechanism of Irgarol 1051 in bleached corals. In an earlier study, HSP70 protein synthesis was induced in *Madracis mirabilis* exposed to Irgarol 1051 (10 μg/L for 8 h and 24 h) for 24 h (Downs and Downs 2007). However, little information is available on the nucleotide sequence and characterization of HSPs in corals (Tom et al. 1999; Hashimoto et al. 2004). To elucidate the effects of Irgarol 1051 on gene expression in hard corals, we initially isolated full-length cDNA encoding HSP90 from *A. tenuis*, which has a deduced open reading frame of 732-amino acid residues and a predicted molecular mass of 84.5 kDa (Ishibashi et al. 2018) (Fig. 9.1). The complete sequence of *A. tenuis* HSP90 is available under GenBank accession no. LC314154. Phylogenetic analysis using the HSP90 amino acid sequences from vertebrates, insects, mollusks, crustaceans, cnidarians, flatworms, and bacteria revealed that *A. tenuis* HSP90 was positioned with the HSP90s of cnidarian species and showed high similarities

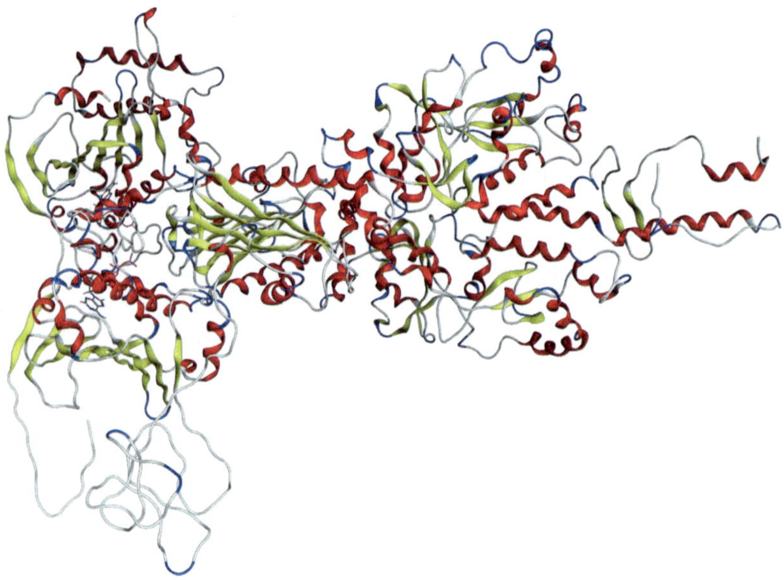

Fig. 9.1 Heat shock protein 90 (HSP90) in *Acropora tenuis*. In silico homology modeling of *A. tenuis* HSP90 based on the X-ray crystal structure of yeast Hsp90-Sba1 closed chaperone complex (Protein Data Bank ID: 2CG9; Ali et al. 2006) was constructed using the Molecular Operating Environment (MOE) program (Chemical Computing Group, Inc., Montreal, Canada) as previously reported (Ishibashi et al. 2019)

with *O. faveolata* HSP90β-like (97.3%), *A. digitifera* HSP90-α-like (96.5%), *Exaiptasia pallida* HSP90α-like (93.4%), and *Nematostella vectensis* HSP90α (93.1%). In addition, five conserved HSP90 family signatures (signatures 1–5), the MEEVD consensus sequence, ATP-binding GxxGxG motif, lysine-rich nuclear localization signals, and protein-binding leucine-zipper Lx6Lx5Lx6Lx6L motif, were identified and conserved in the deduced amino acid sequence of *A. tenuis* HSP90.

We further analyzed the gene expression of *HSP90* in corals after exposure to 1 or 10 μg/L Irgarol 1051 for 7 days. The body color of corals in the 10 μg/L treatment group was significantly whiter (bleached), whereas no such effects were observed in the 1 μg/L treatment group (Fig. 9.2A). However, the expression level of coral *HSP90* was significantly downregulated after exposure to both 1 and 10 μg/L Irgarol 1051 (Fig. 9.2A). We also measured *HSP90* and *HSP70* in symbiotic algae with corals (Kamei et al. 2020). The expression of *HSP90* and *HSP70* in symbiotic dinoflagellates was upregulated after 7 days of exposure to both Irgarol 1051 concentrations (20 ng/L and 200 ng/L), whereas *HSP90* in the coral was not upregulated (Fig. 9.2B). These results suggest that *HSP90* and *HSP70* in corals and symbiotic dinoflagellates may be useful molecular biomarkers for predicting bleaching caused by environmental stressors such as the herbicide Irgarol 1051.

We also investigated the transcriptomes of *A. tenuis* and its symbiotic dinoflagellates to elucidate the molecular mechanisms underlying Irgarol-induced bleaching (Ishibashi et al. 2021). RNA sequencing (RNA-Seq) revealed differentially expressed genes (DEGs) in corals and symbiotic dinoflagellates following a 7-day Irgarol exposure (1 and 10 μg/L) (Fig. 9.3). We identified 142 (100 up- and 42 downregulated) and 109 (40 up- and 69 downregulated) DEGs in the coral after exposure to 1 and 10 μg/L Irgarol 1051, respectively. A comparison of these DEGs revealed that 58 genes, including *GFP*, *blue-light-sensing photoreceptor* (*cryptochrome*: *cry1*), *chromoprotein*, *caspase 8*, and *nuclear receptor* (*AmNR7*), were shared by the 1 and 10 μg/L Irgarol treatment groups. In symbiotic dinoflagellates, we identified 111 (97 up- and 14 downregulated) and 116 (15 up- and 101 downregulated) DEGs after exposure to 1 and 10 μg/L Irgarol 1051, respectively. A comparison of these DEGs showed that 55 genes, including light-harvesting protein-

A

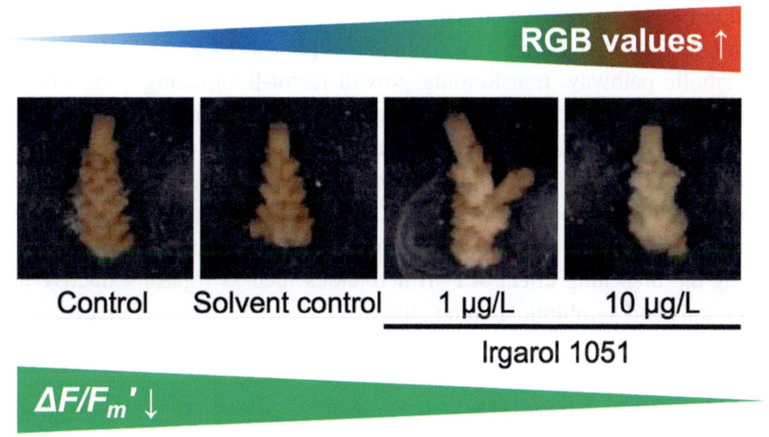

B

		Irgarol 1051 (μg/L)			
		0.02	0.2	1	10
Coral	RGB	→	→	→	↑
(*A. tenuis*)	*HSP90*	→	→	↓	↓
	CRT	-	-	↓	↓
Symbiodinium	*ΔF/Fm'*	↓	↓	↓	↓
	HSP90	↑	↑	↑	↓
	HSP70	↑	↑	↑	↓

Fig. 9.2 Photographs, RGB values, and gene expression of heat shock protein 90 (*HSP90*) and calreticulin (*CRT*) in the coral *Acropora tenuis*, and maximum effective quantum yield ($\Delta F/F_m'$) and gene expression of *HSP70* and *HSP90* in symbiotic dinoflagellates exposed to Irgarol 1051 for 7 days. Based on Ishibashi et al. (2018) and Kamei et al. (2020)

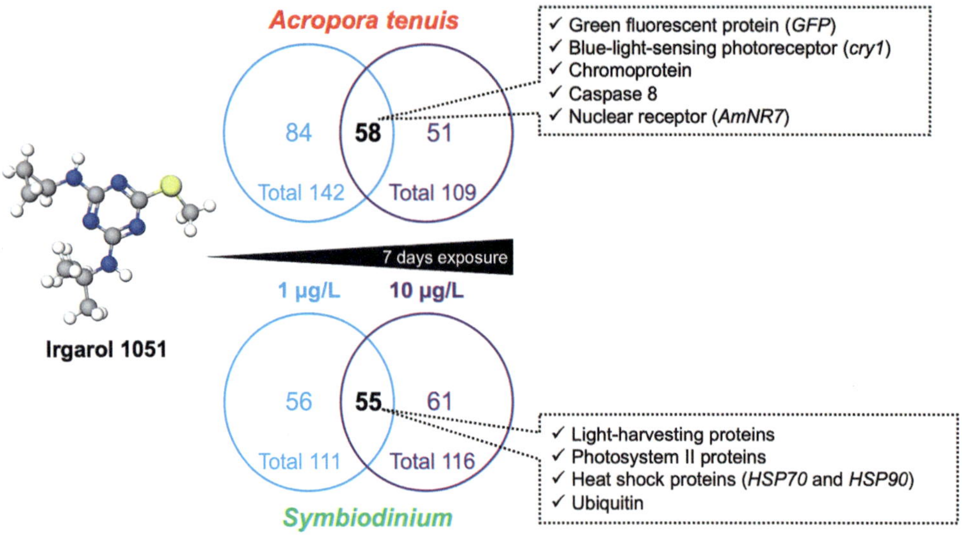

Fig. 9.3 Venn diagram showing differentially expressed genes (DEGs) in the coral *Acropora tenuis* and symbiotic dinoflagellates after 7-day exposures to 1 μg/L (light blue) and 10 μg/L (purple) Irgarol 1051. Based on Ishibashi et al. (2021)

coding, *PSII*, *heat shock proteins* (*HSP70* and *HSP90*), and ubiquitin-related genes, were present in both 1 and 10 μg/L Irgarol 1051 treatment groups. Bioinformatic analyses revealed that both Irgarol 1051 treatments disrupted various GO terms, pathways, and protein interaction networks, which were different in corals (e.g., oxidative phosphorylation, metabolic pathway, transforming growth factor-β signaling pathway, adherens junction, and apoptosis) and symbiotic dinoflagellates (e.g., protein processing in endoplasmic reticulum, carbon fixation in photosynthetic organisms, metabolic pathway, and photosynthesis) (Fig. 9.4). Our study provides new insights into the potential molecular mechanisms underlying the bleaching effect of PSII herbicides such as Irgarol on corals and symbiotic dinoflagellates.

9.2.5 Other Chemicals

Nonylphenol is found in various pharmaceuticals and *personal* care products (PPCPs). It is one of the major components of alkylphenol production, which is used as raw material in industrial detergents, paint ingredients, wetting agents, cosmetics, and domestic cleaning compounds (Takeuchi et al. 2009). Nonylphenol concentration decreases with increasing trophic levels in the food web of shallow-water ecosystems (Takeuchi et al. 2009). Ubiquitous nonylphenol ethoxylate contaminants in marine environments are of great concern. Rougée et al. (2021) investigated the effects of 4-nonylphenol (4-NP), a xenoestrogen, on endocrine and metabolic homeostasis in the hermatypic coral *Pocillopora damicornis* (Table 9.1). 4-NP is the most produced nonylphenol isomer. Coral colonies were chronically exposed (42 days) to a sublethal concentration (1 μg/L) of 4-NP and sampled during the coral's lunar reproductive cycle. Although the activity of steroidogenic enzymes (cytochrome P450 [CYP] 17, CYP19, and 3-β-hydroxysteroid dehydrogenase) and the conjugation enzyme GST were not altered, significant increases in the activity of the steroid-clearing enzyme UDP-glycosyltransferase (UGT) were observed. The natural fluctuation of UGT activity with the lunar cycle was replaced by consistently high UGT activity throughout the reproductive cycle during 4-NP exposure. These results suggest that environmentally relevant levels of 4-NP may affect the reproductive health of corals and threaten the persistence of coral reefs.

Caffeine (IUPAC name: 1,3,7-trimethylpurine-2,6-dione) is a widely consumed psychoactive drug ubiquitous in urban wastewater. Pollack et al. (2009) employed two-dimensional polyacrylamide gel electrophoresis and peptide mass spectrometry to identify proteins sensitive to caffeine exposure in endosymbionts of four species of coral algae belonging to three widely distributed clades: clade A *Symbiodinium microadriaticum* (A), clade B *Symbiodinium* sp. from *Aiptasia pallida* (B6), clade B *Symbiodinium* sp. from *Pseudoterogorgia bipinnata* (B7), and clade C *Symbiodinium goreaui* (C). LaJeunesse et al. (2018) proposed that evolutionarily divergent *Symbiodinium* "clades" are equivalent to genera in the family Symbiodiniaceae; thus clade B *Symbiodinium* sp. and clade C *Symbiodinium goreaui* correspond to *Breviolum* sp. and *Cladocopium goreaui*, respectively. Exposure of corals to 60 mg/L caffeine upregulated some and downregulated other polypeptides in algal species, and commonly affected HSPs such as HSP90, HSP70, and

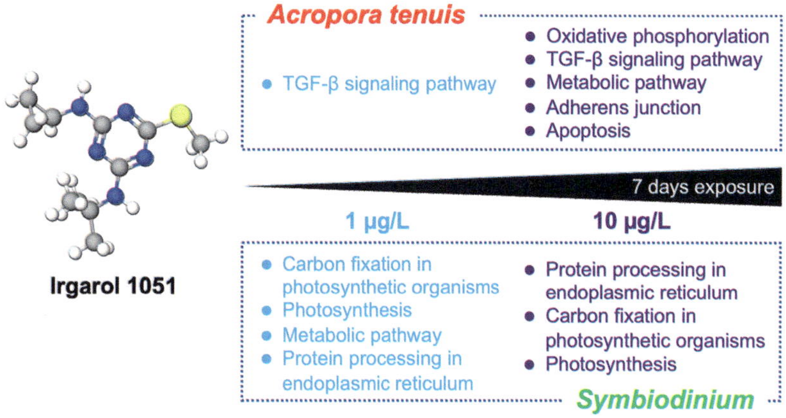

Fig. 9.4 Kyoto Encyclopedia of Genes and Genomes (KEGG) pathways changed in the coral *Acropora tenuis* and symbiotic dinoflagellates after 7-day exposures to 1 μg/L (light blue) and 10 μg/L (purple) Irgarol 1051. Based on Ishibashi et al. (2021)

HSP60. Although the results have important implications for wastewater processing and management and coral reef management globally, Pollack et al. (2009) concluded that more work is needed to better understand the role of HSPs in corals and their algal symbionts. Aguirre-Martínez et al. (2015) reported that 0.1 μg/L caffeine upregulated the activity of dibenzyl fluorescein dealkylase and glutathione S-transferase in bivalve *Corbicula fluminea*.

Benzophenone-3 (IUPAC name: [2-hydroxy-4-methoxyphenyl]-phenylmethanone; BP-3; oxybenzone) and related BPs, which are categorized as organic UV filters and used in various PPCPs, have been detected in seawater at nanogram and microgram per liter levels and have been suggested as an emerging class of chemical contaminants. Downs et al. (2016) reported the effects of oxybenzone on the larval form (planula) of the coral *Stylophora pistillata*. The 50% and 20% lethal concentration response (LC_{50} and LC_{20}) values (4 h in the light) for cells of seven different coral species ranged from 8 to 340 μg/L and from 0.062 to 8 μg/L, respectively. In contrast, contamination of oxybenzone ranged from 75 μg/L to 1.4 mg/L in the U.S. Virgin Islands coral reef and from 0.8 to 19.2 μg/L in Hawaiian sites. Furthermore, He et al. (2019) demonstrated the toxicity and bioaccumulation of 2,4-dihydroxybenzophenone (BP-1), BP-3, sulisobenzone (BP-4), and dioxybenzone (BP-8) in larvae and adults of two coral species, *Pocillopora damicornis* and *Seriatopora caliendrum* (concentrations ranging from 0.1 to 1000 μg/L). They revealed that BP-1 and BP-8 were up to 10 times more toxic to corals than BP-3 and were generated as metabolites of BP-3 in coral tissues. In addition, toxic sensitivities to BP exposure differed between coral species and at different life stages. BP-3, BP-1, and BP-8 pose high to medium risks to the health of corals in popular recreational areas in Taiwan and Hong Kong. Another survey demonstrated the acute toxic effects of BP-3 on the colonial hermatypic coral *Galaxea fascicularis* (Conway et al. 2021). The LC_{50} values for the three acute toxicity tests were similar and averaged at 6.53 ± 0.47 mg/L nominal concentration of BP-3 (4.45 mg/L measured dissolved BP-3). Using the calculated LC_{50} with measured concentrations from a high-quality UV filter monitoring study in Hawaii, the preliminary conservative risk quotient for BP-3 was calculated to be 0.032 (Conway et al. 2021). These results suggest that BP-3 likely does not pose an acute mortality risk to corals; however, additional testing is required to determine the sublethal effects of BP-3 under environmentally relevant concentrations and long-term exposures and to elucidate the molecular mechanism of the adverse effects of BP-3 in corals.

9.3 Future Directions

This review presented the potential toxic effects and endpoints including bleaching by heavy metals (copper and cadmium), organotins (tributyltin), polycyclic aromatic carbons (anthracene, phenanthrene, and benzo[a]pyrene), polychlorinated biphenyls (PCB118 and Aroclor 1254), PSII herbicides (Diuron and Irgarol 1051), and other contaminants (4-nonylphenol, caffeine, and oxybenzone) on corals and symbiotic dinoflagellates, and summarized the potential biomarkers such as HSPs, GFP and RFP, cry1, GST, CAT, SOD, light-harvesting proteins, and PSII proteins for both species (Fig. 9.5). In contrast, some of the genes affected by anthropogenic chemicals also respond to heat stress (Louis et al. 2017). However, limited information is available on the difference in the potential bleaching mechanisms between chemical exposure and thermal stress in corals and symbiotic dinoflagellates. In addition to bleaching, further studies need to investigate the relationship between other toxic effects/endpoints (e.g., oxidative stress, DNA damage, growth, and survival) and the responses of potential biomarker genes to anthropogenic chemicals in corals and symbiotic dinoflagellate (Fig. 9.5). The findings presented here provide new insights into the potential molecular mechanisms underlying the toxic and bleaching effects of anthropogenic chemicals on corals and symbiotic dinoflagellates.

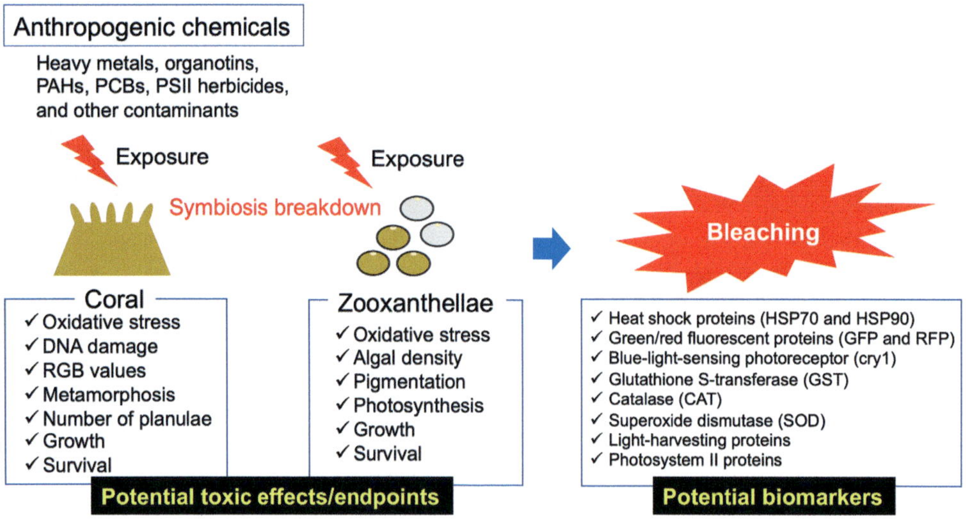

Fig. 9.5 Schematic diagram of the relationship between potential toxic effects/endpoints and biomarker gene expression responses caused by exposure to anthropogenic chemicals in corals and symbiotic dinoflagellates. Based on the following literatures (Wiens et al. 2000; Ramos and García 2007; Pollack et al. 2009; Venn et al. 2009; Chen et al. 2012; Yuyama et al. 2012; Schwarz et al. 2013; Woo et al. 2014; Ishibashi et al. 2018; Overmans et al. 2018; Zhou et al. 2018; Xiang et al. 2019; Kamei et al. 2020; Ishibashi et al. 2021; Rougée et al. 2021)

Shinzato et al. (2011) were the first to report the draft genome sequence of the coral *Acropora digitifera* (eggs, gastrulae, planulae, polyps, and adults) using next-generation genome sequencing. In contrast, Shoguchi et al. (2013) reported a draft genome sequence of *Symbiodinium minutum* (currently treated as *Breviolum minutum*) (LaJeunesse et al. 2018). Molecular resources are critical for understanding resilience to the changing environment and chemical stress responses in coral species, and yet such resources are unavailable for most scleractinian corals. In a recent study, Shinzato et al. (2021) reported the complete genome sequences of 15 *Acropora* species and three other acroporid taxa belonging to the genera *Montipora* and *Astreopora* to examine genomic novelties that explain their evolutionary success. Zhang et al. (2019) reported comprehensive transcriptome resources for 14 common species of scleractinian corals in the South China Sea, including a few structure-forming species, and constructed a publicly accessible database (www.comp.hkbu.edu.hk/~db/CoralTBase). A transcriptomic approach also revealed sex- and phase-specific gene expression profiles in ovaries and testes at different developmental phases from a gonochoric coral, *Euphyllia ancora* (Chiu et al. 2020). These resources will facilitate not only functional studies using these corals to understand the molecular basis of stress responses and adaptation to anthropogenic chemicals, but will also assist comparative transcriptomic studies on other species of corals and more distantly related cnidarians.

Cell culture of corals may be useful for elucidating the molecular mechanisms of bleaching by chemical stress. Recently, Roger et al. (2021) conducted a thorough review and tested available methods (i.e., cell dissociation, isolation, sorting, attachment, and proliferation) of coral cell culture. These authors combined the most effective and reproducible techniques to consolidate the culture methods and generate uncontaminated coral cell cultures for ~7 days (10 days maximum). Moreover, they suggested that the development of methods for cell separation and identification in conjunction with further investigations into coral cell type-specific metabolic requirements will allow us to tailor growth media for optimized monocultures as a tool for studying essential reef-building coral traits such as symbiosis, wound healing, and calcification at multiple scales.

A recently created new genome-editing technology, CRISPR/Cas9, can be applied to many model and non-model organisms not only to generate loss-of-function mutations, but also for introduction of more subtly modified genes, tagging of proteins, and large-scale genomic restructuring. A CRISPR/Cas9 system has been developed for cnidarians, such as the sea anemone *Nematostella vectensis* and the hydrozoan *Hydractinia echinata*; however, this system has not been available for coral species. Cleves et al. (2018) demonstrated efficient genome editing using the CRISPR/Cas9 system that targets genes encoding fibroblast growth factor 1a (*FGF1a*), *GFP*, and *RFP* in the coral *Acropora millepora*. CRISPR/Cas9-based gene editing, as well as genomic and transcriptomic analyses, are useful technologies that allow rigorous testing of gene function in coral species to understand the bleaching mechanism of corals and their symbiotic dinoflagellates by anthropogenic chemicals.

Acknowledgments The concept of the present review was formed mainly through several research grants from the Japan Society for the Promotion of Science (JSPS) Grants-in-Aid for Scientific Research (KAKENHI; grant nos. 26281048, 18H03387, 21H03619).

References

Ali MM, Roe SM, Vaughan CK, Meyer P, Panaretou B, Piper PW, Prodromou C, Pearl LH (2006) Crystal structure of an Hsp90-nucleotide-p23/Sba1 closed chaperone complex. Nature 440:1013–1017

Cantin NE, Negri AP, Willis BL (2007) Photoinhibition from chronic herbicide exposure reduces reproductive output of reef-building corals. Mar Ecol Prog Ser 344:81–93

Chen TH, Cheng YM, Cheng JO, Ko FC (2012) Assessing the effects of polychlorinated biphenyls (Aroclor 1254) on a scleractinian coral (*Stylophora pistillata*) at organism, physiological, and molecular levels. Ecotoxicol Environ Saf 75:207–212

Chiu YL, Shikina S, Yoshioka Y, Shinzato C, Chang CF (2020) De novo transcriptome assembly from the gonads of a scleractinian coral, *Euphyllia ancora*: molecular mechanisms underlying scleractinian gametogenesis. BMC Genomics 21:732. https://doi.org/10.1186/s12864-020-07113-9

Cleves PA, Strader ME, Bay LK, Pringle JR, Matz MV (2018) CRISPR/Cas9-mediated genome editing in a reef-building coral. Proc Natl Acad Sci U S A 115:5235–5240

Conway AJ, Gonsior M, Clark C, Heyes A, Mitchelmore CL (2021) Acute toxicity of the UV filter oxybenzone to the coral *Galaxea fascicularis*. Sci Total Environ 796:148666

Dafforn KA, Lewis JA, Johnston EL (2011) Antifouling strategies: history and regulation, ecological impacts and mitigation. Mar Pollut Bull 62:453–465

De Nadal E, Ammerer G, Posas F (2011) Controlling gene expression in response to stress. Nat Rev Genet 12:833–845

Downs C, Downs A (2007) Preliminary examination of short-term cellular toxicological responses of the coral *Madracis mirabilis* to acute Irgarol 1051 exposure. Arch Environ Contam Toxicol 52:47–57

Downs CA, Kramarsky-Winter E, Segal R, Fauth J, Knutson S, Bronstein O, Ciner FR, Jeger R, Lichtenfeld Y, Woodley CM, Pennington P, Cadenas K, Kushmaro A, Loya Y (2016) Toxicopathological effects of the sunscreen UV filter, oxybenzone (benzophenone-3), on coral planulae and cultured primary cells and its environmental contamination in Hawaii and the U.S. Virgin Islands. Arch Environ Contam Toxicol 70:265–288

Hashimoto K, Shibuno T, Murayama-Kayano E, Tanaka H, Kayano T (2004) Isolation and characterization of stress-responsive genes from the scleractinian coral *Pocillopora damicornis*. Coral Reefs 23:485–491

He T, Tsui MMP, Tan CJ, Ng KY, Guo FW, Wang LH, Chen TH, Fan TY, Lam PKS, Murphy MB (2019) Comparative toxicities of four benzophenone ultraviolet filters to two life stages of two coral species. Sci Total Environ 651:2391–2399

Hirayama K, Takayama K, Haruta S, Ishibashi H, Takeuchi I (2017) Effect of low concentrations of Irgarol 1051 on RGB (R, red; G, green; B, blue) colour values of the hard-coral *Acropora tenuis*. Mar Pollut Bull 124:678–686

Ishibashi H, Minamide S, Takeuchi I (2018) Identification and characterization of heat shock protein 90 (HSP90) in the hard coral *Acropora tenuis* in response to Irgarol 1051. Mar Pollut Bull 133: 773–780

Ishibashi H, Hirano M, Kim EY, Iwata H (2019) *In vitro* and *in silico* evaluations of binding affinities of perfluoroalkyl substances to Baikal seal and human peroxisome proliferator-activated receptor α. Environ Sci Technol 53:2181–2188

Ishibashi H, Takaichi D, Takeuchi I (2021) Effects of the herbicide Irgarol 1051 on the transcriptome of hermatypic coral *Acropora tenuis* and its symbiotic dinoflagellates. Sci Total Environ 780: 146542

Jones R (2010) Environmental contamination associated with a marine landfill ('seafill') beside a coral reef. Mar Pollut Bull 60:1993–2006

Jones RJ, Kerswell AP (2003) Phytotoxicity of photosystem II (PSII) herbicides to coral. Mar Ecol Prog Ser 261:149–159

Kamei M, Takayama K, Ishibashi H, Takeuchi I (2020) Effects of ecologically relevant concentrations of Irgarol 1051 in tropical to subtropical coastal seawater on hermatypic coral *Acropora tenuis* and its symbiotic dinoflagellates. Mar Pollut Bull 150:110734

Kaniewska P, Chan CK, Kline D, Ling EY, Rosic N, Edwards D, Hoegh-Guldberg O, Dove S (2015) Transcriptomic changes in coral holobionts provide insights into physiological challenges of future climate and ocean change. PLoS One 10:e0139223. https://doi.org/10.1371/journal.pone.0139223

Katsumata M, Takeuchi I (2017) Delayed fluorescence as an indicator of the influence of the herbicides Irgarol 1051 and diuron on hard coral *Acropora digitifera*. Mar Pollut Bull 124:687–693

Kennedy K, Schroeder T, Shaw M, Haynes D, Lewis S, Bentley C, Paxman C, Carter S, Brando VE, Bartkow M, Hearn L, Mueller JF (2012) Long term monitoring of photosystem II herbicides–correlation with remotely sensed freshwater extent to monitor changes in the quality of water entering the great barrier reef, Australia. Mar Pollut Bull 65:292–305

Kitada Y, Kawahata H, Suzuki A, Oomori T (2009) Environmental pollution by toxic chemicals - contamination of pesticides within coral reefs and adjacent river mouths in the Ryukyu islands. Chishitsu News 656:39–44. (In Japanese)

Kozelka PB, Bruland KW (1998) Chemical speciation of dissolved Cu, Zn, Cd, Pb in Narragansett Bay, Rhode Island. Mar Chem 60:267–282

LaJeunesse TC, Parkinson JE, Gabrielson PW, Jeong HJ, Reimer JD, Voolstra CR, Santos SR (2018) Systematic revision of Symbiodiniaceae highlights the antiquity and diversity of coral endosymbionts. Curr Biol 28:2570–2580

Lesser MP (2006) Oxidative stress in marine environments: biochemistry and physiological ecology. Annu Rev Physiol 68:253–278

Louis YD, Bhagooli R, Kenkel CD, Baker AC, Dyall SD (2017) Gene expression biomarkers of heat stress in scleractinian corals: promises and limitations. Comp Biochem Physiol C Toxicol Pharmacol 191: 63–77

Manzo S, Buono S, Cremisini C (2006) Toxic effects of irgarol and diuron on sea urchin *Paracentrotus lividus* early development, fertilization, and offspring quality. Arch Environ Contam Toxicol 51: 61–68

Matsumoto R, Tu NPC, Haruta S, Kawano M, Takeuchi I (2014) Polychlorinated biphenyl (PCB) concentrations and congener composition in masu salmon from Japan: a study of all 209 PCB congeners by high-resolution gas chromatography/high-resolution mass spectrometry (HRGC/HRMS). Mar Pollut Bull 85:549–557

Matsumoto R, Tu NPC, Haruta S, Kawano M, Takeuchi I (2016) Analysis of all 209 polychlorinated biphenyl (PCB) congeners with special reference to dioxin-like PCB congeners in Japanese seabass and related species by high-resolution gas chromatography/high-resolution mass spectrometry (HRGC/HRMS). Reg Stud Mar Sci 3:119–130

Matsuo H, Kawano M, Omori K, Nakajima K, Takeuchi I (2009) Stable isotope-guided analysis of congener-specific PCB concentrations in a Japanese coastal food web. Mar Pollut Bull 58:1615–1623

Mercurio P, Eaglesham G, Parks S, Kenway M, Beltran V, Flores F, Mueller JF, Negri AP (2018) Contribution of transformation products towards the total herbicide toxicity to tropical marine organisms. Sci Rep 8:4808. https://doi.org/10.1038/s41598-018-23153-4

Moya A, Huisman L, Ball EE, Hayward DC, Grasso LC, Chua CM, Woo HN, Gattuso JP, Forêt S, Miller DJ (2012) Whole transcriptome analysis of the coral *Acropora millepora* reveals complex responses to CO_2-driven acidification during the initiation of calcification. Mol Ecol 21:2440–2454

Neff JM, Stout SA, Gunster DG (2005) Ecological risk assessment of polycyclic aromatic hydrocarbons in sediments: identifying sources and ecological hazard. Integr Environ Assess Manag 1:22–33

Negri AP, Heyward AJ (2001) Inhibition of coral fertilisation and larval metamorphosis by tributyltin and copper. Mar Environ Res 51:17–27

Oettmeier W (1999) Herbicide resistance and supersensitivity in photosystem II. Cell Mol Life Sci 55:1255–1277

Overmans S, Nordborg M, Díaz-Rúa R, Brinkman DL, Negri AP, Agustí S (2018) Phototoxic effects of PAH and UVA exposure on molecular responses and developmental success in coral larvae. Aquat Toxicol 198:165–174

Palumbi SR, Barshis DJ, Traylor-Knowles N, Bay RA (2014) Mechanisms of reef coral resistance to future climate change. Science 344:895–898

Pollack K, Balazs K, Ogunseitan O (2009) Proteomic assessment of caffeine effects on coral symbionts. Environ Sci Technol 43:2085–2091

Portune KJ, Voolstra CR, Medina M, Szmant AM (2010) Development and heat stress-induced transcriptomic changes during embryogenesis of the scleractinian coral *Acropora palmata*. Mar Genomics 3: 51–62

Ramos AA, Inoue Y, Ohde S (2004) Metal contents in *Porites* corals: anthropogenic input of river run-off into a coral reef from an urbanized area, Okinawa. Mar Pollut Bull 48:281–294

Ramos R, García E (2007) Induction of mixed-function oxygenase system and antioxidant enzymes in the coral *Montastraea faveolata* on acute exposure to benzo(a)pyrene. Comp Biochem Physiol C Toxicol Pharmacol 144:348–355

Roger LM, Reich HG, Lawrence E, Li S, Vizgaudis W, Brenner N, Kumar L, Klein-Seetharaman J, Yang J, Putnam HM, Lewinski NA (2021) Applying model approaches in non-model systems: a review and case study on coral cell culture. PLoS One 16:e0248953. https://doi.org/10.1371/journal.pone.0248953

Rougée LRA, Collier AC, Richmond RH (2021) Chronic exposure to 4-nonylphenol alters UDP-glycosyltransferase and sulfotransferase clearance of steroids in the hard coral, *Pocillopora damicornis*. Front Physiol 12:608056. https://doi.org/10.3389/fphys.2021.608056

Schwarz JA, Mitchelmore CL, Jones R, O'Dea A, Seymour S (2013) Exposure to copper induces oxidative and stress responses and DNA damage in the coral *Montastraea franksi*. Comp Biochem Physiol C Toxicol Pharmacol 157:272–279

Shinzato C, Shoguchi E, Kawashima T, Hamada M, Hisata K, Tanaka M, Fujie M, Fujiwara M, Koyanagi R, Ikuta T, Fujiyama A, Miller DJ, Satoh N (2011) Using the *Acropora digitifera* genome to understand coral responses to environmental change. Nature 476:320–323

Shinzato C, Khalturin K, Inoue J, Zayasu Y, Kanda M, Kawamitsu M, Yoshioka Y, Yamashita H, Suzuki G, Satoh N (2021) Eighteen coral genomes reveal the evolutionary origin of *Acropora* strategies to accommodate environmental changes. Mol Biol Evol 38:16–30

Shoguchi E, Shinzato C, Kawashima T, Gyoja F, Mungpakdee S, Koyanagi R, Takeuchi T, Hisata K, Tanaka M, Fujiwara M, Hamada M, Seidi A, Fujie M, Usami T, Goto H, Yamasaki S, Arakaki N, Suzuki Y, Sugano S, Toyoda A, Kuroki Y, Fujiyama A, Medina M, Coffroth MA, Bhattacharya D, Satoh N (2013) Draft assembly of the *Symbiodinium minutum* nuclear genome reveals dinoflagellate gene structure. Curr Biol 23:1399–1408

Takahashi S, Whitney S, Itoh S, Maruyama T, Badger M (2008) Heat stress causes inhibition of the de novo synthesis of antenna proteins and photobleaching in cultured *Symbiodinium*. Proc Natl Acad Sci U S A 105:4203–4208

Takasuga T, Senthilkumar K, Matsumura T, Shiozaki K, Sakai S (2006) Isotope dilution analysis of polychlorinated biphenyls (PCBs) in transformer oil and global commercial PCB formulations by high resolution gas chromatography–high resolution mass spectrometry. Chemosphere 62:469–484

Takeuchi I (2023) Anthropogenic stresses in coral reefs and adjacent ecosystem around the East China Sea. In: Takeuchi I, Yamashiro H (eds) Coral reefs of the world. Coral reefs of eastern Asia under anthropogenic impacts, vol 17. Springer Nature Switzerland AG, Cham, pp 111–135

Takeuchi I, Miyoshi N, Mizukawa K, Tadata H, Ikemoto T, Omori K, Tsuchiya K (2009) Biomagnification profiles of polycyclic aromatic hydrocarbons, alkylphenols and polychlorinated biphenyls in Tokyo Bay elucidated by $\delta^{13}C$ and $\delta^{15}N$ isotope ratios as guides to trophic web structure. Mar Pollut Bull 58:663–671

Tom M, Douek J, Yankelevich I, Bosch TC, Rinkevich B (1999) Molecular characterization of the first heat shock protein 70 from a reef coral. Biochem Biophys Res Commun 262:103–108

Venn AA, Quinn J, Jones R, Bodnar A (2009) P-glycoprotein (multi-xenobiotic resistance) and heat shock protein gene expression in the reef coral *Montastraea franksi* in response to environmental toxicants. Aquat Toxicol 93:188–195

Verma N, Pink M, Rettenmeier AW, Schmitz-Spanke S (2012) Review on proteomic analyses of benzo[a]pyrene toxicity. Proteomics 12: 1731–1755

Weis VM (2008) Cellular mechanisms of Cnidarian bleaching: stress causes the collapse of symbiosis. J Exp Biol 211:3059–3066

Wiens M, Ammar MS, Nawar AH, Koziol C, Hassanein HM, Eisinger M, Müller IM, Müller WE (2000) Induction of heat-shock (stress) protein gene expression by selected natural and anthropogenic disturbances in the octocoral *Dendronephthya klunzingeri*. J Exp Mar Biol Ecol 245:265–276

Woo S, Lee A, Denis V, Chen CA, Yum S (2014) Transcript response of soft coral (*Scleronephthya gracillimum*) on exposure to polycyclic aromatic hydrocarbons. Environ Sci Pollut Res 21:901–910

Xiang N, Jiang C, Huang W, Nordhaus I, Zhou H, Drews M, Diao X (2019) The impact of acute benzo(a)pyrene on antioxidant enzyme and stress-related genes in tropical stony corals (*Acropora* spp.). Sci Total Environ 694:133474

Yu H, Xia Q, Yan J, Herreno-Saenz D, Wu YS, Tang IW, Fu PP (2006) Photoirradiation of polycyclic aromatic hydrocarbons with UVA light - a pathway leading to the generation of reactive oxygen species, lipid peroxidation, and DNA damage. Int J Environ Res Public Health 3:348–354. https://doi.org/10.3390/ijerph2006030045

Yuyama I, Ito Y, Watanabe T, Hidaka M, Suzuki Y, Nishida M (2012) Differential gene expression in juvenile polyps of the coral *Acropora tenuis* exposed to thermal and chemical stresses. J Exp Mar Biol Ecol 430–431:17–24

Zhang Y, Chen Q, Xie JY, Yeung YH, Xiao B, Liao B, Xu J, Qiu JW (2019) Development of a transcriptomic database for 14 species of scleractinian corals. BMC Genomics 20:387. https://doi.org/10.1186/s12864-019-5744-8

Zhou Z, Yu X, Tang J, Wu Y, Wang L, Huang B (2018) Systemic response of the stony coral *Pocillopora damicornis* against acute cadmium stress. Aquat Toxicol 194:132–139

Perspective for the Conservation of Coral Reefs in the East China Sea

10

Ichiro Takeuchi

Abstract

Climate change, including increasing seawater temperature combined with various local stresses, leads to coral bleaching and changes in the native genera of coral reefs of the Ryukyu Archipelago in the East China Sea. The high density of gross domestic projects, excessive use of pesticides, and other agricultural activities have led to a decline in terrestrial water quality on the Ryukyu Archipelago Islands. In addition to red-soil problems, the establishment of systematic monitoring across major sources of pollution, including river water, treated wastewater, and military bases, as well as the development of a cost-efficient water treatment system, is necessary. The upwelling of deeper waters near the Ryukyu Archipelago might mitigate the increase in seawater temperature. The occurrence of local internal tides and protection from continental anthropogenic pressure would promote coral reef conservation. Prompt monitoring of coral reefs using advanced technologies, as well as the implementation of recently developed cutting-edge technologies in ecotoxicology, including "coral-on-a-chip," will aid coral reef management. Various technologies for the establishment of artificial coral reefs are required in response to the rising sea level.

Keywords

Advanced technologies · Anthropogenic chemicals · Climate changes · Coastal oceanography · Systematic monitoring · Water treatment

10.1 Introduction

Recent worldwide bleaching events driven by unprecedented global heat waves have changed the approach to coral reef management and policy (Anthony et al. 2020). To sustain coral reefs in the long term, reducing greenhouse gas emissions will be necessary (Anthony et al. 2020). However, IPCC (2018) estimated that coral reefs of the world would decline by 70–90% with global warming of 1.5 °C above the pre-industrial level, and even larger losses (>99%) are projected at 2.0 °C. Moreover, anthropogenic pressures on coral reefs in the near future are a combination of global and local pressures (Burke et al. 2011). Thus, the number of years left for planning and implementing the conservation measures of coral reef ecosystems is limited.

Ultimately, solutions to the climate challenge for coral reefs should consider what our society wants, what we can achieve technically and economically, and what we can implement rapidly (Anthony et al. 2020). However, defining multiple objectives is often conflicting for complex social-ecological systems, including coral reefs (Anthony et al. 2020). The future conditions of coral reefs will depend on the trajectory of global emissions and their decreasing capacity to build resilience to more frequent bleaching events through local stresses (Hughes et al. 2018).

The coral reefs of the Ryukyu Archipelago provide various ecoservices: they serve as natural breakwater and fishing grounds, are used for building materials, decorations, souvenirs, as tourist attraction, and training grounds for various educational activities, and are important elements of regional cultures (The Ministry of the Environment, Japan, 2017 [2016]). Climate change, including increasing seawater temperature combined with various local stresses, has led to coral bleaching and changes in the native genera of coral reefs in the East China Sea (Hongo and Yamano 2013; Harii et al. 2014; Takeuchi 2023b). Takeuchi (2023b) summarized the exposure of coral reefs in the Ryukyu Archipelago to various terrestrial anthropogenic impacts released into the sea

I. Takeuchi (✉)
Graduate School of Agriculture, Ehime University, Matsuyama, Ehime, Japan

Center of Advanced Technology for the Environment, Graduate School of Agriculture, Ehime University, Matsuyama, Ehime, Japan
e-mail: takeuchi@agr.ehime-u.ac.jp

I. Takeuchi, H. Yamashiro (eds.), *Coral Reefs of Eastern Asia under Anthropogenic Impacts*, Coral Reefs of the World 17,
https://doi.org/10.1007/978-3-031-27560-9_10

Fig. 10.1 Runoff of terrestrial waters into coastal area after rainfall along the west coast of Ogimi-son located in the northern part of Okinawa Island. The image was captured on 21 November 2021 during a regular flight from Matsuyama Airport, Ehime, Japan to Naha Airport, Okinawa, Japan

as a result of high precipitation and short river lengths (Fig. 10.1). The prevalence and geographic distribution of coral diseases have increased along the islands of the Ryukyu Archipelago, at least since 2000 (Yamashiro et al. 2023). Ocean acidification caused by elevated CO_2 concentrations indicates that coral calcification may have already decreased from a pre-industrial level, and further progress of acidification is expected to increase the risk for coral reef degradation and accompanied species compositional changes under climate change scenarios (Suzuki et al. 2023).

The selection of priority localities for coral reef conservation in the Ryukyu Archipelago has been mainly based on the evaluation by Large-scale Reef Monitoring Team of the Japanese Coral Reef Society's Conservation Committee, the status of coral coverage surveys conducted from 1989 to 1992 and in 2006 and the assessment of "potential" coral growth based on physical environmental data analysis (see WWF Japan 2010). The Action Plan to Conserve Coral Reef Ecosystems in Japan 2016–2020 addressed the following three priority issues: (1) promotion of measures against red soil sediments and nutrient salts from the land, (2) promotion of sustainable tourism in coral reef ecosystems, and (3) establishment of a relationship between community life and coral reef ecosystems (The Ministry of the Environment, Japan 2017[2016]). However, to consider the conservation of coral reefs under various anthropogenic pressures, selecting prioritized methodologies and localities for coral reef conservation is necessary, because resources that could be used for conservation widely differ from resources necessary for conservation (Anthony et al. 2020). Moreover, the coral reefs in the East China Sea are located in the most economically developed areas in the world (Takeuchi 2023a, b). Maintaining adequate economic growth with developed industries and technologies is also one of the key factors for considering the quality of life of people inhabiting nearby coral reefs of the East China Sea.

In this chapter, I present the perspective for the future conservation of coral reefs in the Ryukyu Archipelago under the present critical pressure by climate change and the accumulation of various anthropogenic impacts.

10.2 Monitoring and Reducing Terrestrial and Coastal Impacts

The "local" (viz. water quality improvement) as well as the "global" (viz. CO_2 reduction) should be considered as a function in the model to reduce the future likelihood of mass coral bleaching (Wooldridge 2009). Wooldridge (2009) estimated that the reduction in end-of-river dissolved inorganic nitrogen (DIN) of >50–80% is equivalent to 2.0–2.5 °C of the upper thermal bleaching limit in the high-risk inshore areas of the Great Barrier Reef (GBR), Australia. Numerous rivers in the catchment areas of the GBR discharge pollutants from agricultural, urban and mining activities (Brodie et al. 2012). The major pollutants are suspended sediments from cattle grazing areas, nitrate from fertilizer applications on crop lands, and herbicides from various land uses (Brodie et al. 2012). Shitaoka et al. (2021) suggested that seawater temperature of 30 °C, which was previously considered a moderate heat stress temperature (Fujise et al. 2014), is a normally inhabitable environmental temperature for *Acropora tenuis* and *A. digitifera* along the coast of southern Japan. The seawater temperature of 30 °C is 1 °C above the monthly mean of the hottest month at Okinawa Island, southern Japan (Kayanne 2017). Hongo and Yamano (2013) mentioned that high turbidity caused by terrestrial runoff in the post-World War II era could have reduced the resilience of tabular and branching *Acropora* spp., including *A. digitifera*, to severe thermal stress events around Okinawa Island. Thus, reducing the terrestrial and other impacts would increase the survival of *Acropora* spp. at ~30 °C in the Ryukyu Archipelago.

To protect the GBR, the State of Queensland and Australian governments set a reef water quality protection plan to minimize pollutants from diffuse sources and reduce their entry into the GBR (The State of Queensland and Commonwealth of Australia 2003). The two major objectives were: (1) reducing the load of pollutants in the waters entering the GBR and (2) rehabilitating and conserving the Reef catchment areas involved in removing waterborne pollutants. The actions undertaken by the Reef Plan were not effective in solving the declining water quality in the reef (The State of Queensland 2009). Contaminants derived from adjacent lands, including suspended sediments, nutrients, and pesticides, were in the range of concentrations that were likely to cause environmental harm (The State of Queensland 2009). Therefore, the Reef Water Quality Protection Plan was updated to reduce the runoff of photosystem II-inhibiting herbicides, including Diuron, which is one of the dominant herbicides used in Okinawa (Takeuchi 2023b), decrease nitrogen, phosphorus, and total suspended solids from agricultural land, and promote improved land management practices (The State of Queensland 2009).

Takeuchi (2023a, b) reported that gross domestic product (GDP) density values and pesticide use are high in countries around the East China Sea. These intensive economic and agricultural activities lead to a decline in terrestrial water quality on the islands of the Ryukyu Archipelago (Yamano et al. 2015; Takeuchi 2023b) (Fig. 10.2). The biological oxygen demand (BOD) and suspended particulate matter (SPM) at river mouths is indicative of strong positive relationships with human population density within river catchment areas along the coast of Okinawa Island (West and Van Woesik 2001).

Sewage is discharged in waters surrounding more than 90% of coral reefs in the world (Wear and Thurber 2015). Wear and Thurber (2015) reviewed the occurrence of multiple stressors in sewage, such as freshwater, inorganic nutrients, pathogens, endocrine disrupters, suspended solids, sediments, and heavy metals, on coral growth and/or reproduction. In Florida Keys, Florida, USA, the release of sewage induces white pox in hermatypic coral *Acropora palmata* via the fecal enterobacterium *Serratia marcescens* (Sutherland et al. 2010). Most public sewage systems in Japan are operated by a conventional activated sludge system, which is composed of an aeration tank (first tank) for biological degradation and a sedimentation tank (second clarifier) for separating solid sludge from wastewater. Takeda et al. (2018) estimated that the treatment waters of Okinawa Prefecture contain several anthropogenic chemicals including linear alkylbenzene sulfonate (LAS) at high levels (see Takeuchi 2023b). Thus, the present public sewage system seems to be therefore insufficient for avoiding ecological risks for the coral reefs on the islands of Okinawa Prefecture. The sewerage population rate of Okinawa Prefecture is 72.0% and lower than the average of 79.7% estimated for Japan (Okinawa Prefecture 2021). High concentrations of pharmaceutical and personal care products (PPCPs) have been detected in the rivers of southern Okinawa Island (Mano et al. 2017). This indicates inefficient onsite treatment facilities for areas that are not connected to public systematic sewage systems. Sewerage development in Ishigaki City, Okinawa Prefecture, started not in the downtown area of southern Ishigaki Island but in the Kabira District (Ishigaki City 2020). This district is located in the northern part of the island facing Kabira Bay, which is known for black pearl aquaculture and is a major tourist destination for its coral reefs with >200 species of corals (Ishigaki City 2020). The local governments, at least that of Ishigaki City, are aware of the challenges coral reef conservation faces in relation to onsite treatment facilities. Thus, as a first step, promoting a network of systematic public sewage systems. is urgently required, while the current public sewage system is insufficient (see Takeuchi 2023b).

Mitchell (2020) recently mentioned serious environmental threats to Okinawa associated with per- and polyfluoroalkyl substances (PFASs) and other chemicals originating from military bases. The tourism sector has grown considerably over the last two decades in the Pulau Redang and Pulau Tioman coral resort islands located off the east coast of Peninsular Malaysia, and minimal sewage treatment could threaten the resilience of coral reefs (Lachs et al. 2019). Similar situation is expected for the islands of the Ryukyu Archipelago, where tourism is one of the dominant economic sectors (Okinawa Prefecture 2018) and the sewerage rate of Okinawa Prefecture is lower than the Japan's average (Okinawa Prefecture 2021). Livestock wastes are also

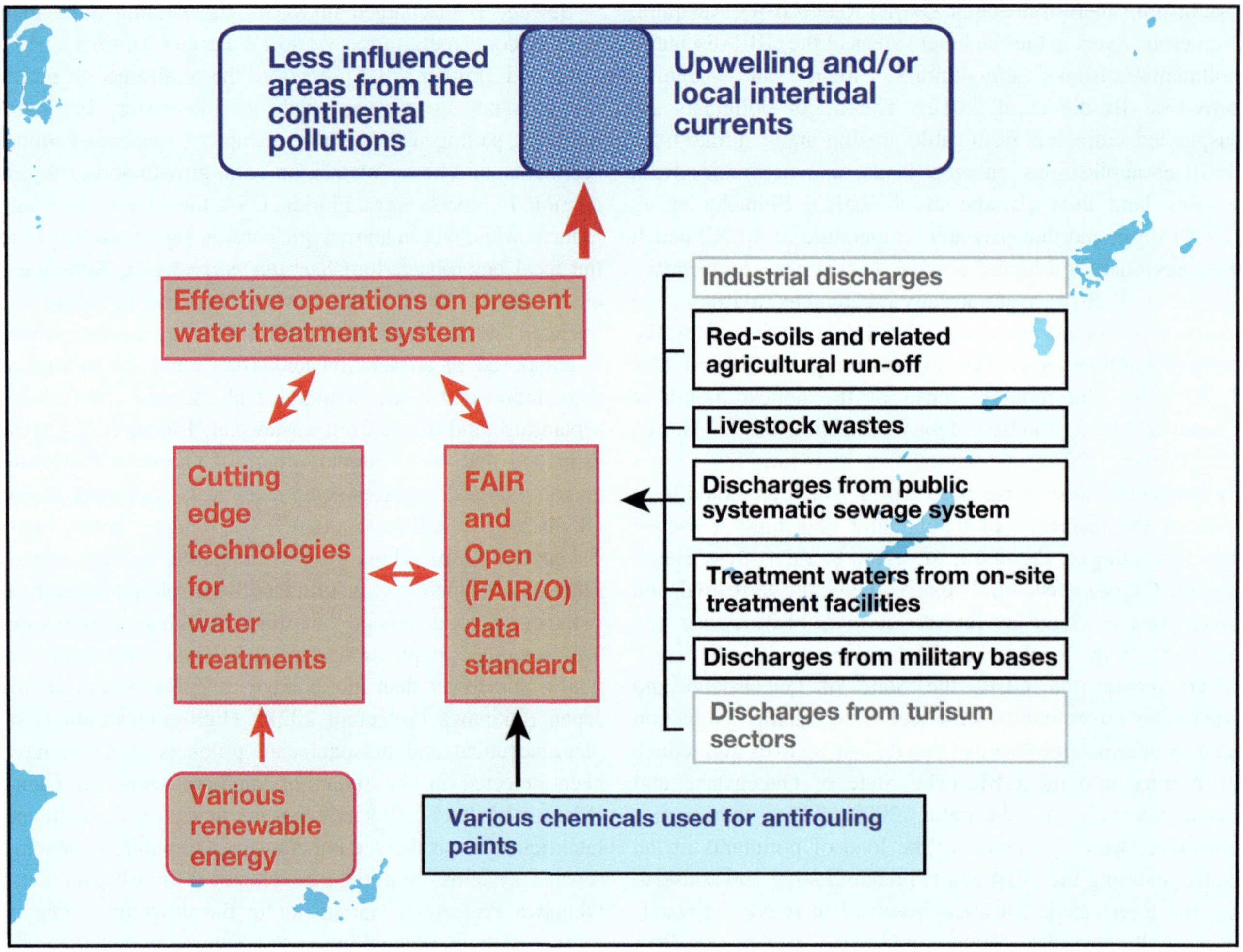

Fig. 10.2 Concept for the coral reef conservation along the coasts of the Ryukyu Archipelago in the East China Sea. FAIR, findable, accessible, interoperable, and reusable data, was advocated by Quay et al. (2022)

considered the source of pollutants in rivers and tidal flats of Okinawa Island (Tashiro et al. 2003; Mano et al. 2017). In addition to these terrestrial anthropogenic impacts, Takeuchi (2023b) confirmed anthropogenic contamination originating from antifouling paints in the coastal regions of the Ryukyu Archipelago.

There are two cross-disciplinary movements revolutionizing the storage and management of scientific data: "Open Science" initiative for reducing restrictions to accessing scientific data, and making data findable, accessible, interoperable, and reusable (FAIR) (Quay et al. 2022). Quay et al. (2022) advocates for FAIR and open (FAIR/O) data standards in water treatment research. Data integrity in the water treatment research sector is often limited by a small number of observational datasets and inadequate data documentation (Quay et al. 2022). Strong integration research requires the use of underlying research data, easy recombination with other interoperable research datasets, and holistic evaluation of integrated data (Quay et al. 2022). This situation, as defined by Quay et al. (2022) along with the analysis of ecotoxicity of anthropogenic chemicals, would be applicable to water quality monitoring of coral reefs under various anthropogenic stresses.

Recently, Zhang et al. (2021) studied the occurrence and distribution of 484 chemicals, including 296 pesticides and 156 pharmaceuticals, in mariculture areas and rivers of Dailan flowing into the Yellow Sea, north of the East China Sea. The authors utilized liquid chromatography-quadrupole time-of-flight mass spectrometry (LC-QTOF-MS) with sequential window acquisition of all theoretical fragment-ion spectra aquation (SWATH) to identify targeted anthropogenic chemicals. LC-QTOF-MS is one of the best analytical methods for simultaneous analysis of various chemicals, and SWATH is one of the three data acquisition methods for simultaneous measurements in LC-QTOF-MS (Kadokami and Ueno 2019).

Since the first report on membrane bioreactor (MBR) technology in 1969, MBRs have been increasingly used in

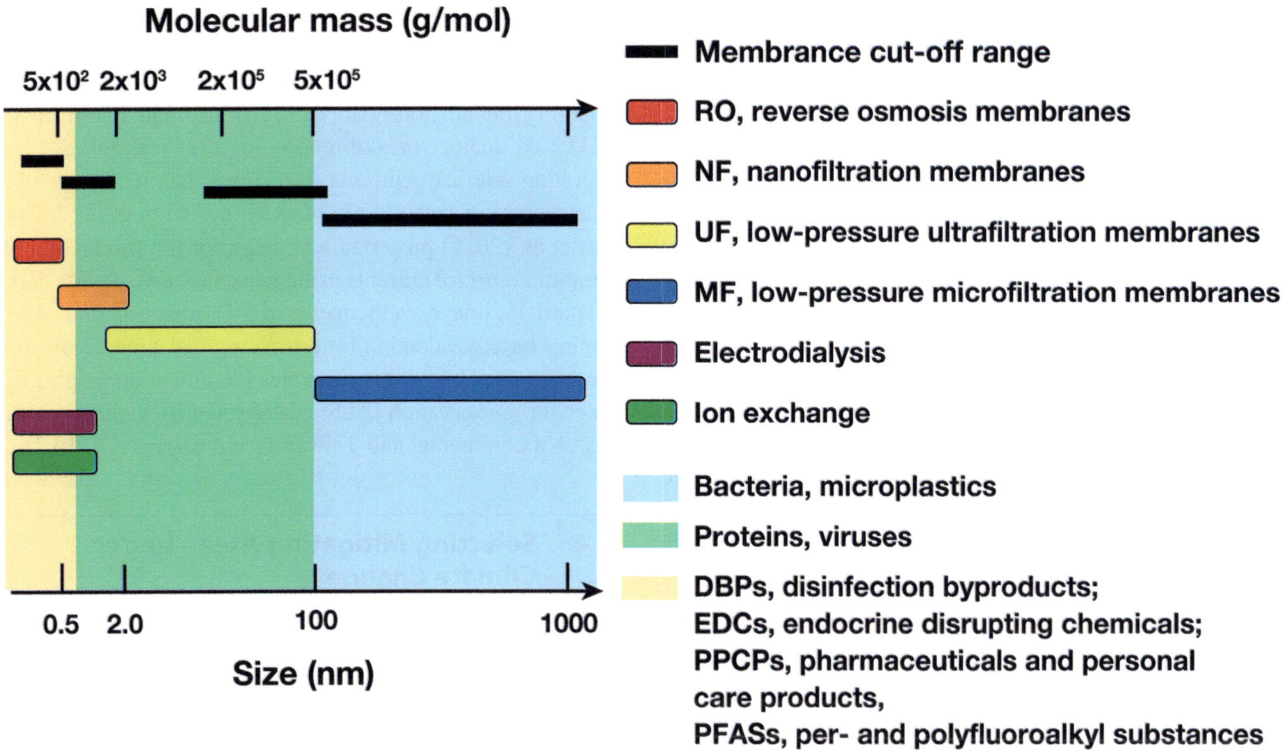

Fig. 10.3 Size and molecular mass of pollutants and pathogens that can be removed by different water filtering techniques. Adapted with permission from Keller et al. (2022, ACS EST Eng 2, 273–291). Copyright (2022) American Chemical Society

treatment of wastewater with a capacity of up to several 100,000 population equivalents (p.e.) owing to their capacity for controlled biomass retention, improved effluent quality, and decreased footprint (Kraume and Drews 2010; Al-Asheh et al. 2021). Population equivalent of one is equal to 60 g of BOD/24 h, which corresponds to the mean oxygen demand of the wastewater produced by one person per day. MBR combines biological processes with membrane filtration, such that treated wastewater is separated from microorganisms in a membrane module, and biomass is degraded inside a bioreactor tank (Al-Asheh et al. 2021). In China, MBR applications have been growing exponentially from the first pilot test of 10 m^3/d in 1999 to the first application with capacity of 110,000 m^3/d for municipal water 10 years later (Li et al. 2019). Beijing Huaifang Water Reclamation Plant complicated 2017 possesses the capacity of 600,000 m^3/d for municipal water (Li et al. 2019). To further improve the elimination capacity, hybrid MBRs, which are combined with wastewater technologies other than the conventional activated sludge process, have also been proposed (Kraume and Drews 2010).

The development of cutting-edge technologies for water treatment is required not only for conserving coral reefs but also for providing direct potable reuse of treatment water (Keller et al. 2022). Figure **10.3** shows the different removal performance of each water treatment technology per particle size based on Keller et al. (2022). Low-pressure microfiltration (MF) and ultrafiltration (UF) membranes are suitable for removing large (>200 Da) organic molecules after primary and secondary wastewater treatments (Keller et al. 2022). Reverse osmosis (RO) membranes remove most of the smaller organic molecules (down to 150–200 Da), divalent ions, and a fraction of the monovalent ions (Keller et al. 2022). Wang et al. (2018) reported that integrated membrane systems of MBR treatment with RO or nanofiltration (NF) membrane treatment can achieve even higher removal rates of >95% for most of PPCPs in municipal wastewater. At least several compounds of PPCPs induce damages to corals and other marine invertebrates at environmentally equivalent concentrations (Takeuchi 2023b). Mano et al. (2017) reported that the recharge of rivers with reclaimed water subjected to advanced treatment (RO or NF) could improve the water quality of the rivers in southern Okinawa Island by diluting the concentrations of PPCPs. The use of an ion-exchange resin cartridge or RO is recommended when making artificial seawater for eco-toxicological experiments on *Acropora* spp. and for rearing ornamental corals (Takeuchi and Takayama 2023).

As the energy required per unit of treated water increases significantly, it is critical to consider the energy consumption for operating water treatment systems (Keller et al. 2022). A treatment system with RO (or NF) has 80–90% higher energy requirements than alternatives such as microfiltration–ozone–biological activated carbon (MF–O_3–BAC) (Keller et al. 2022). While RO or equivalent filtering equipment is needed to control the concentration of PPCPs, the energy efficiency of these high-quality systems becomes an important factor in the selection of wastewater treatments. It is desirable to develop RO or an equivalent technology that can remove PPCPs with best economical running costs and energy consumption.

Recently, various renewable energy sources, such as solar thermal technologies, wind, hydroelectric, biomass and geothermal energy, as well as hybrid systems based on these technologies, have been used for desalination of seawater for industrial and household purposes, including drinking water, industry, and crop irrigation (Bundschuh et al. 2021). For islands, such as the Ryukyu Islands, the introduction of renewable energy will contribute to the development of effective and economical water treatment technologies and positively affect the conservation of coral reef environments.

Thus, the innovation of more effective technologies in water treatment systems and on-site treatment facilities should concentrate on two points, improvement of the removal efficiency and energy costs.

10.3 Selecting Area less Affected by Continental Anthropogenic Pressures

The coastal region of the East China Sea is the area with the highest frequency of pesticide use worldwide (see Takeuchi 2023a). Pesticide production and usage in China have been rapidly increasing since 1990, positioning China among the top producers and consumers of pesticides in the 2010s (Grung et al. 2015). The seasonal fluctuation in the concentration of hexachlorocyclohexanes (HCH) and dichlorodiphenyltrichloroethane and its metabolites (DDT) in the East China Sea and northern South China Sea suggests long-range oceanic transport into the Western Pacific from these marginal seas (Ya et al. 2019). Takeuchi (2023b) estimated that the prevailing northwest wind in winter drives anthropogenic contamination and plastic wastes originating from China over the Ryukyu Archipelago via the surface layer of the East China Sea. A survey conducted nearly 20 years ago confirmed that marine debris originating from southern China and Taiwan Island washes up on the coasts of the Ryukyu Archipelago (Fujieda et al. 2006). Thus, the analysis of stranded marine debris can help in the selection of coral reefs less affected by continental anthropogenic pressures. Takeuchi (2023b) recorded the stranding of pumice from a volcano near the Ogasawara Islands on the northern coast of Okinawa Island. Coral reefs along the coast where large amounts of pumice wash up are likely exposed to contaminants from the continent and Taiwan Island.

Surface currents on the ocean, however, are the most complex flows dominated by non-homogeneous, non-isotropic, and non-stationary processes with temporal variability on an hourly to yearly time scale (Choi et al. 2021). A major breakthrough in the technology for generating surface currents has been facilitated by the advancement of satellite remote sensing (Choi et al. 2021). Choi et al. (2021) proposed a strategy for the production of operational surface currents using geostationary satellite data, the particle image velocimetry (PIV) method, and deep learning-based evaluation for estimating real-time ocean surface currents. This strategy enables identification of priority places for conservation likely not affected by anthropogenic effects of continental and Taiwan Island origin (Fig. 10.2).

10.4 Selecting Mitigating Areas Under Climate Change

The coral reefs situated along the Ryukyu Archipelago are located in the areas where the progress of climate change is relatively slow due to the complicated topography of the islands of the Archipelago (Takeuchi 2023b). The rate of increase in seawater temperature in the Sakishima Islands locating western part of the Ryukyu Archipelago is lower than that in the seas around Japan (Japan Meteorological Agency 2021; Takeuchi 2023b). The slower acidification of seawater near the Ryukyu Archipelago has also been reported (Lui et al. 2015; Takeuchi 2023b).

A relatively large upwelling of deeper water is present in the Pacific near the Ryukyu Archipelago (Gopalakrishnan et al. 2013). The simulations of Gopalakrishnan et al. (2013) indicated the presence of upwelling and associated cyclonic circulation, including the Taiwan Strait outflow induced by the northward flow of the Kuroshio Current northeast of Taiwan. They estimated that a cold-water patch (3 °C colder than the ambient water at 50 m depth at the approximately 100 km spatial scale) extended to the Mien-Hua Canyon, nearly 100 km northeast of Taiwan on the continental shelf of the East China Sea. Their model also predicted that upwelling events occur more frequently in spring and summer, when the core of the Kuroshio tends to flow away from the east coast of Taiwan, than in autumn and winter, when the core of the Kuroshio is located closer to the east coast of Taiwan (Gopalakrishnan et al. 2013).

The Okinawa Trough, with a depth greater than 2000 m, is separated from the Pacific by a submarine ridge associated with the Ryukyu Archipelago (Nishina et al. 2016). The Kerama Gap, located south of Okinawa Island, is narrow (50 km wide), but an 1100 m deep channel connecting the Pacific to the Okinawa Trough (Nishina et al. 2016). The strong turbulent mixing in the Kerama Gap caused by the breaking of internal gravity waves and/or hydraulic jumps also supplies buoyancy for upwelling deep water over the southern Okinawa Trough (Nishina et al. 2016).

The Kuroshio Current across the Tokara Strait, north of the Amami Islands, flows from the East China Sea to the Pacific Ocean. The Kuroshio Current strongly interacts with the topography around the Tokara Strait (Liu et al. 2019). The lees of Tokara Islands drive in the upwelling current that is 0.3–0.5 °C colder than the surrounding waters (Liu et al. 2019).

In addition to these relatively large-scale upwellings, local internal tides have been reported from several places along the west coast of Okinawa Island (Nakaza et al. 2006; Rouf and Nakaza 2014). Rouf and Nakaza (2014) recorded the occurrence of a cross-shore intertidal current, together with a current flowing nearly parallel to the coastline at Genka Bay, northwest of Okinawa Island. The intertidal current changes the vertical temperature profile, even near the shore regions (Rouf and Nakaza 2014). Thus, intertidal currents are considered to form a unique cooling system that is stronger in the summer (Rouf and Nakaza 2014). The same cold intertidal current toward the coastlines was also observed near the Zampa Cave close to the middle of the west coast of Okinawa Island (Nakaza et al. 2006).

Thus, the complicated topography along the Ryukyu Islands would interfere with various sizes of upwelling and local intertidal currents, both of which bring cold waters, especially in summer (Gopalakrishnan et al. 2013; Rouf and Nakaza 2014). These deeper seawaters might be less contaminated with various anthropogenic chemicals than surface waters. Thus, the coral reefs there might avoid a drastic decline at seawater temperature of 1.5 °C above the pre-industrial level (IPCC 2018).

Detailed oceanographic observations for elucidating localities and scales of upwelling and intertidal currents lead to the identification of prioritized areas for coral reef conservation, where the increase in seawater temperature and decrease in pH are mitigated under climate change (Fig. 10.2).

The southern coast of Sesoko Island is one of the places that might be less influenced by continental pollutions and be exposed to local currents. When the sea level rises from low tide to high tide, seawater flows along the southern coast of Sesoko Island from the East China Sea side to Okinawa Island. Although *Acropora* is the most declined hermatypic coral along the coast of Okinawa Islands (Hongo and Yamano 2013; Harii et al. 2014), the dense community composed of large table-like and bush-like *Acropora* spp. continues to survive there (Fig. 10.4), unlike the coral reefs off the southeastern coast of Sesoko Island (Sakai et al. 2019; Takeuchi and Yamashiro 2022). An extremely large microatoll type of *Porites* colony >10 m in diameter was recently found near the *Acropora* community (Takeuchi and Yamashiro 2017).

10.5 Prompt Coral Reef Monitoring Technology

Recent developments in advanced technologies have enabled a previously unimaginable prompt and cost-effective monitoring of coral reefs.

Monitoring of coral reef ecosystems requires prompt feedback, especially during coral bleaching. Drones (unmanned aerial vehicles, UAV) can capture detailed imagery of coral reefs with pixel size on the order of centimeters and with high flexibility in timing and frequency of image capture (Joyce et al. 2019). A series of downward images captured from one or more drone flights can be stitched together to produce image mosaics or orthomosaics (see Joyce et al. 2019, Takeuchi and Yamashiro 2022). David et al. (2021) reconstructed shallow-water reefs (<1 m) using consumer-grade UAVs and structure from motion–multi view stereo (SfM-MVS) techniques and obtained the best results by placing ground control points in water with no refraction correction.

Underwater structure from motion (SfM) photogrammetry is another increasingly relevant technique for mapping the substrate complexity of shallow-water environments, including coral reefs (Bayley and Mogg 2020; Ventura et al. 2022). Bayley and Mogg (2020) reported a protocol for the collection, analysis, and display of 3D coral reef data using primarily open-source software. Ventura et al. (2022) provided an SfM-based protocol for monitoring seagrass communities based on the images obtained by a SCUBA diver aided by a diver propulsion vehicle (DPV) equipped with a de facto action camera, GoPro Hero 5 Black (Ventura et al. 2022). Mizuno et al. (2020) developed an efficient image collection system using the "Speedy Sea Scanner (SSS)" in which six digital cameras were set up on a towed body for coral reef survey. The SSS was towed like a glider by a boat at a speed of 2–3 knots during the survey (Mizuno et al. 2020). The recent rapid development of electric surfboards (Chakravarthi 2020) with action cameras aids the prompt monitoring of coral reefs.

Enhancing the comparability of nutrient concentrations using different methods in seawater is required to augment the quality and utility of measured concentrations in global databases (Daniel et al. 2020). Daniel et al. (2020) discussed the standardization of technical and analytical controls carried out on three dominant types of nutrient sensors currently

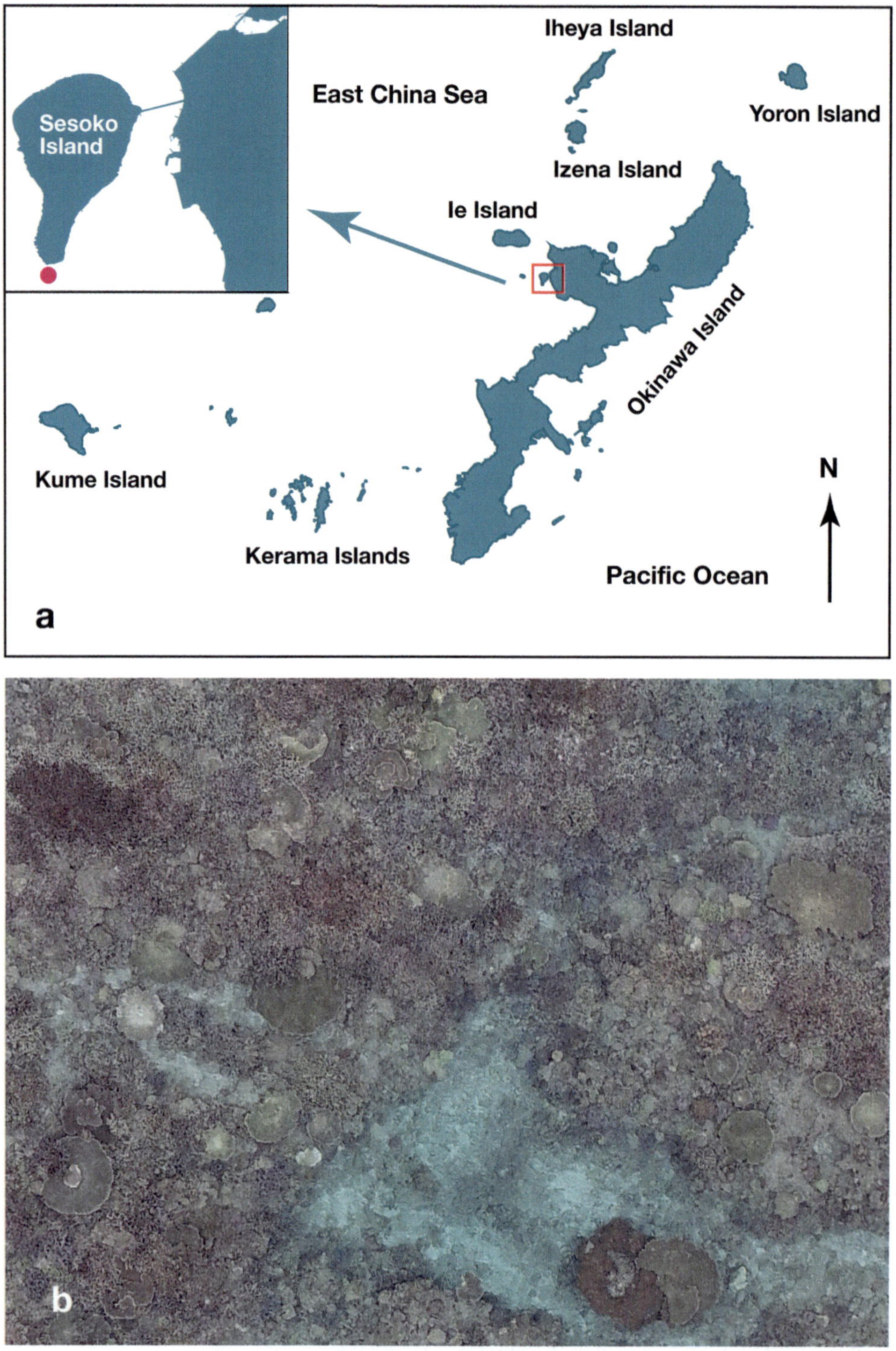

Fig. 10.4 Healthy *Acropora* community off the southern coast of Sesoko Island located northwest of Okinawa Island. a, Location on the map; b, Aerial image taken on 13 July 2018 using a DJI Phantom 4 Professional (DJI Co. Ltd., Guangdong, China)

available in situ: wet chemical analyzers, ultraviolet optical sensors, and electrochemical sensors. Wei et al. (2021) reported that electrochemical technologies are the most suitable candidates for marine nutrient monitoring systems because of their miniaturization, low reagent consumption, and easy combination with automated devices. Mowlem et al. (2021) reviewed the development of in situ lab-on-chip (LOC) technology using microfluidics. As of 2021, nitrate, nitrite, phosphate, silicate, iron, and pH sensors are currently commercially available (Mowlem et al. 2021). LOC sensors that can be used for the turbulence environment in shallow coral reefs around small islands have been developed.

10.6 Cutting-Edge Technologies of Coral Ecotoxicology Assays

Organisms are exposed to a mixture of thousands of natural and anthropogenic chemicals throughout their lifetime (Scholz et al. 2022). Monitoring anthropogenic chemicals in water based on chemical analysis alone cannot assess the toxicity of these chemicals to aquatic species and ecosystems (Schuijt et al. 2021). The anthropogenic chemicals and their mixtures induce various stresses at the sub-organism level, including xenobiotic metabolism, oxidative stress, genotoxicity, photosynthesis inhibition, neurotoxicity, endocrine disruption, cellular energy metabolism, immunotoxicity, hematological markers, and histopathology (Schuijt et al. 2021).

Recently, the behavioral response of bivalves to anthropogenic chemicals have been used as a biomarker to monitor various anthropogenic chemical substances and hypoxic water (Hartmann et al. 2016; Chmist et al. 2019; Oh et al. 2021; Mbah et al. 2021). Modern software and "Hall sensor" technology have enabled real-time monitoring of the mussel filtration behavior at a very high resolution (Hartmann et al. 2016). The Hall sensor is a magnetoelectric device for measuring unusual valve movements originally developed for early detection of noxious blooms of the dinoflagellate *Heterocapsa circularisquama* (Nagai et al. 2006). Oh et al. (2021) developed a monitoring system using shell valve movements (SVMs) of the Pacific oyster *Crassostrea gigas* for hypoxic water. The SVMs were continuously recorded in a data logger as the changes in the output voltages (mV) between the Hall element sensor attached to the left valve and the magnet on the right valve (Oh et al. 2021). The Okinawa Prefectural Fisheries Experiment Station has been conducting the seeding production of three species of giant clams, *Tridacna crocea*, *T. derasa*, and *T. squamosa*, and the distribution of their juveniles to local fishery associations for further aquaculture (Iwai et al. 2006). Thus, aquacultured giant clams would be suitable for biomonitoring of various stresses in coral reefs around the East China Sea.

Various cutting-edge biomonitoring technologies have been developed in recent times. A microbial fuel cell (MFC), a small device that converts chemical energy into electricity through the catalytic activities of microorganisms, is considered a prospective alternative energy source, wastewater treatment process, and biosensor for oxygen and pollutants (Kim et al. 2007). ElMekawy et al. (2018) reported that MFC-based biosensors have outstanding potential for the rapid and real-time monitoring of water source quality.

As Campana and Wlodkowic (2018) pointed out, microfluidic lab-on-a-chip technologies are one of the most remarkable outcomes of bioengineering, enabling the development of an entirely new generation of bioanalytical instrumentation. The "organ-on-a-chip" is a system containing engineered or natural miniature tissues grown inside microfluidic systems (Leung et al. 2022). Shapiro et al. (2016) designed a coral-on-a-chip using a silicone elastomer that could keep a single polyp of *Pocillopora damicornis* on a simple straight channel (4 cm × 3 mm × 0.1 mm) on the upper surface of a rectangular plate made of silicone elastomer (Fig. 10.5). The experimental system reported by Shapiro et al. (2016) is extremely compact compared with those reported in this book (Takeuchi and Takayama 2023). Pang et al. (2020) maintained bailed-out single polyps of *P. damicornis* for more than 15 days on a microfluidic chip they developed. A single polyp of *P. damicornis* bailed out due to an increase in salinity (Shapiro et al. 2016) or calcium-free artificial seawater (Pang et al. 2020). Once a general bailing or dividing method that can be implemented for other coral genera including *Acropora* is established, coral-on-a chip combined with comparative transcriptomic gene analysis (Ishibashi and Takeuchi 2023) will enable various types of experiments for prompt and high accuracy evaluation of suitable environments for corals.

10.7 Conservation of Coral Reefs Under Increasing Sea Level

Due to the climate change, the seawater level is estimated to rise from +0.30 m to +1.30 m over the twenty-first century according to the Representative Concentration Pathway (RCP) 2.6 and RCP 8.5, respectively (OECD 2019). The reinforcement of typhoons triggered by climate change will also increase the risk of coastal damage by increasing wave height over coral reefs at the coastline due to wave setup influenced by rising wave height in the outer ocean (Hongo and Kiguchi 2021). Hongo and Kiguchi (2021) estimated that reef production at a rate of 3–10 kg $CaCO_3/m^2/y$ by 2100 is required to maintain a healthy reef. This rate of reef production entails vigorous growth of corals and other calcareous organisms, including the green algae *Halimeda* spp. (Hongo and Kiguchi 2021). Moreover, the construction of artificial structures along shorelines has recently increased because of societal responses to reduce flood and erosion risks from rising sea levels and more extreme storms resulting from climate change (Morris et al. 2019).

Thus, the development of artificially enhanced healthy coral reefs for reducing the risks to seashore line and reinforcement typhoons caused by climate change is one of the matters that needs urgent action. The coral reefs along the coasts of the islands in the East China Sea are in the areas

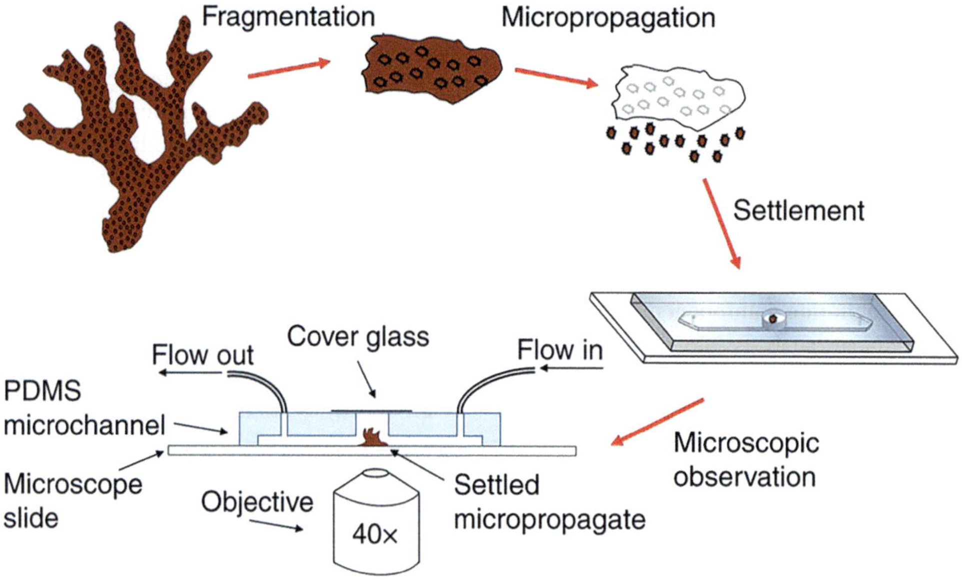

Fig. 10.5 Schematic diagram of the workflow for setting up the coral on-a-chip platform using bailing out polyp of *Pocillopora damicornis*. Figure is reprinted from Shapiro et al. (2016, Nat Commun 7, 10860) under Creative Commons CCBY

with high GDP (Takeuchi 2023a). The ecologically engineered shorelines should be designed to be multifunctional with potentially valuable benefits to multiple stakeholders, as each of them has their own objectives (see Morris et al. 2019).

Omori (2019) reviewed the research and technologies for coral restoration. A large number of small fragments separated from donor colonies should be attached to artificial substrata and kept in a nursery for some time until they grow to an adequate size for outplanting (Omori 2019). Two underwater farms for donor colonies were established off Onna Village on the west coast of Okinawa Island (Zayasu and Shinzato 2016; Higa et al. 2017). One of the major techniques employed in the farms is the coral pole culture method (Hibitate-shiki in Japanese), in which donor coral colonies are attached on forested iron poles 50 cm above the seafloor at a depth of 2–3 m (Higa et al. 2017; Omori 2019). Underwater farms become spawning hubs that produce millions of larvae near coastal areas (Zayasu and Shinzato 2016; Higa et al. 2017; Omori 2019).

Recently, 3D printers have been used to construct a wide range of substrates, from tiny bionic corals to artificial coral reefs, for coral reef conservation (Lange et al. 2020; Wangpraseurt et al. 2020; Albalawi et al. 2021). Lange et al. (2020) developed a method for creating artificial reef creatures on hexagonal, solid, 152–228 cm in diameter, clay tiles using a robotic 3D printing method. Albalawi et al. (2021) presented a new eco-friendly and sustainable approach, "3D CoraPrint," using 3D scanning based on natural coral skeletons (Fig. 10.6). They developed a coral-like printable calcium carbonate photoinitiated ink (CCP) by mixing calcium carbonate with a commercially available photocurable resin (Albalawi et al. 2021). A live coral was scanned using the Creaform scanning system and converted to the coral's physical structure using a computer-aided design (CAD) model (Albalawi et al. 2021). Then, the CCP coral model was made directly or by polylactic acid (PLA) molding (Albalawi et al. 2021). Wangpraseurt et al. (2020) developed 3D-printed bionic corals produced with a 3D bioprinting platform. The 3D printed bionic corals mimic the morphological features of living coral tissue and the underlying skeleton with micron resolution (Wangpraseurt et al. 2020). These bionic corals could maintain microalgae with high densities of up to 10^9 cells/mL (Wangpraseurt et al. 2020). Thus, manipulations using artificial substrates made by 3D printers will contribute to the faster development of coral reef ecosystems under rising sea surface conditions in the near future.

Studies on the production of coral reef ecosystems under the rising sea level and various cutting-edge technologies are a research project of imminent importance. These studies could also help to improve various aspects of the quality of life of residences on coral reef islands along the East China Sea.

Recent rapid advances in cutting-edge science and technology allow us to do things that were considered extremely challenging several decades ago. For example, the COVID-19 pandemic has led to a tech-celebration in healthcare, with the adoption of telemedicine taking place almost overnight (Cowie and Lam 2021). Remote consultation and monitoring, including mobile health and wearable tools, are essential to replace (or at least support) the traditional face-to-face

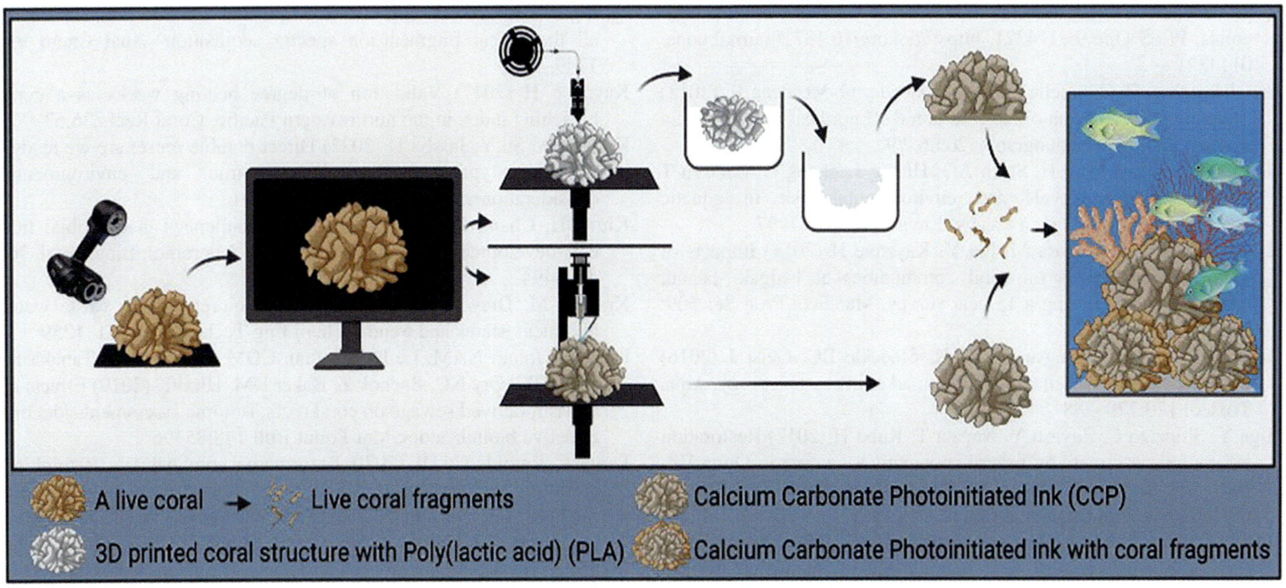

Fig. 10.6 Diagram of the 3D CoraPrint's fabrication methods. Reprinted from graphic abstract with permission from Albalawi et al. (2021, ACS Sustain Chem. Eng 9, 12634–12645). Copyright (2022) American Chemical Society

interaction between clinicians and their patients. Further innovations in digital technologies are required to ensure rapid feedback in various methods for coral reef monitoring in order to conserve reefs under different anthropogenic pressures and also to enable a broad range of stakeholders to prepare for increasing sea level.

Acknowledgments The present perspective was mainly formed by the studies carried by research grants from the Japan Society for the Promotion of Science (JSPS) Grants-in-Aid for Scientific Research (KAKENHI; grant nos. 26281048, 15K12253, 18H03387, and 21H03619). I thank the collaborators in these research grants, especially Prof. H. Yamashiro, Prof. S. Haruta, and Dr. H. Ishibashi, for various discussions, which led to the present perspective.

References

Al-Asheh S, Bagheri M, Aidan A (2021) Membrane bioreactor for wastewater treatment: a review. Case Stud Chem Environ Eng 4: 100109

Albalawi HI, Khan ZN, Valle-Pérez AU, Kahin KM, Hountondji M, Alwazani H, Schmidt-Roach S, Bilalis P, Aranda M, Duarte CM, Hauser CAE (2021) Sustainable and eco-friendly coral restoration through 3D printing and fabrication. ACS Sustain Chem Eng 9: 12634–12645

Anthony KRN, Helmstedt KJ, Bay LK, Fidelman P, Hussey KE, Lundgren P, Mead D, McLeod IM, Mumby PJ, Newlands M, Schaffelke B, Wilson KA, Hardisty PE (2020) Interventions to help coral reefs under global change—a complex decision challenge. PLoS One 15(8):e0236399. https://doi.org/10.1371/journal.pone.0236399

Bayley DT, Mogg AO (2020) A protocol for the large-scale analysis of reefs using structure from motion photogrammetry. Methods Ecol Evol 11(11):1410–1420

Brodie JE, Kroon FJ, Schaffelke B, Wolanski EC, Lewis SE, Devlin MJ, Bohnet IC, Bainbridge ZT, Waterhouse J, Davis AM (2012) Terrestrial pollutant runoff to the great barrier reef: an update of issues, priorities and management responses. Mar Pollut Bull 65:81–100

Bundschuh J, Kaczmarczyk M, Ghaffour N, Tomaszewska B (2021) State-of-the-art of renewable energy sources used in water desalination: present and future prospects. Desalination 508:115035

Burke L, Reytar K, Spalding M, Perry A (2011) Reefs at risk revisited. World Resources Institute, Washington, DC, x +114 pp

Campana O, Wlodkowic D (2018) Ecotoxicology goes on a chip: embracing miniaturized bioanalysis in aquatic risk assessment. Environ Sci Technol 52:932–946

Chakravarthi S (2020) Ranger. Surf your way. Master Thesis, School of Industrial Design, Lund University, Sweden, 129 pp

Chmist J, Szoszkiewicz K, Drożdżyński D (2019) Behavioural responses of *Unio tumidus* freshwater mussels to pesticide contamination. Arch Environ Contam Ttoxicol 77:432–442

Choi JM, Kim W, Hong TTM, Park Y-G (2021) Derivation and evaluation of satellite-based surface current. Front Mar Sci 8:695780. https://doi.org/10.3389/fmars.2021.695780

Cowie MR, Lam CS (2021) Remote monitoring and digital health tools in CVD management. Nat Rev Cardiol 18:457–458

Daniel A, Laës-Huon A, Barus C, Beaton AD, Blandfort D, Guigues N, Knockaert M, Munaron D, Salter I, Woodward MS, Greenwood N, Achterberg EP (2020) Toward a harmonization for using *in situ* nutrient sensors in the marine environment. Front Mar Sci 6:773. https://doi.org/10.3389/fmars.2019.00773

David CG, Kohl N, Casella E, Rovere A, Ballesteros P, Schlurmann T (2021) Structure-from-motion on shallow reefs and beaches: potential and limitations of consumer-grade drones to reconstruct topography and bathymetry. Coral Reefs 40:835–851

ElMekawy A, Hegab HM, Pant D, Saint CP (2018) Bio-analytical applications of microbial fuel cell–based biosensors for onsite water quality monitoring. J Appl Microbiol 124:302–313

Fujieda S, Kojima A, Kanehiro H (2006) Monitoring marine debris using disposable lighters as indicator. J Jpn Soc Waste Manage Experts 17:117–124. (In Japanese with English abstract)

Fujise L, Yamashita H, Suzuki G, Sasaki K, Liao LM, Koike K (2014) Moderate thermal stress causes active and immediate expulsion of photosynthetically damaged zooxanthellae (*Symbiodinium*) from corals. PLoS One 9:e114321. https://doi.org/10.1371/journal.pone.0114321

Gopalakrishnan G, Cornuelle BD, Gawarkiewicz G, McClean JL (2013) Structure and evolution of the cold dome off northeastern Taiwan: a numerical study. Oceanography 26:66–79

Grung M, Lin Y, Zhang H, Steen AO, Huang J, Zhang G, Larssen T (2015) Pesticide levels and environmental risk in aquatic environments in China – a review. Environ Int 81:87–97

Harii S, Hongo C, Ishihara M, Ide Y, Kayanne H (2014) Impacts of multiple disturbances on coral communities at Ishigaki Island, Okinawa, Japan, during a 15 year survey. Mar Ecol Prog Ser 509: 171–180

Hartmann JT, Beggel S, Auerswald K, Stoeckle BC, Geist J (2016) Establishing mussel behavior as a biomarker in ecotoxicology. Aqua Toxicol 170:279–288

Higa Y, Shinzato C, Zayasu Y, Nagata T, Kubo H (2017) Restoration efforts for coral reefs by fishery cooperatives~a case in Onna Village, Okinawa~. Galaxea J Coral Reef Stud 19:119–128. (In Japanese with English abstract)

Hongo C, Kiguchi M (2021) Assessment to 2100 of the effects of reef formation on increased wave heights due to intensified tropical cyclones and sea level rise at Ishigaki Island, Okinawa, Japan. Coast Eng J 63:216–226

Hongo C, Yamano H (2013) Species-specific responses of corals to bleaching events on anthropogenically turbid reefs on Okinawa Island, Japan, over a 15-year period (1995–2009). PLoS One 8(4): e60952. https://doi.org/10.1371/journal.pone.0060952

Hughes TP, Anderson KD, Connolly SR, Heron SF, Kerry JT, Lough JM, Baird AH, Baum JK, Berumen ML, Bridge TC, Claar DC, Eakin CM, Gilmour JP, Graham NAJ, Harrison H, Hobbs J-PA, Hoey AS, Hoogenboom M, Lowe RJ, McCulloch MT, Pandolfi JM, Pratchett M, Schoepf V, Torda G, Wilson S (2018) Spatial and temporal patterns of mass bleaching of corals in the Anthropocene. Science 359:80–83

IPCC (2018) Global Warming of 1.5 °C. An IPCC Special Report on the impacts of global warming of 1.5 °C above preindustrial levels and related global greenhouse gas emission pathways, in the context of strengthening the global response to the threat of climate change, sustainable development, and efforts to eradicate poverty. In: Masson-Delmotte V, Zhai P, Pörtner H-O, Roberts D, Skea J, Shukla PR, Pirani A, Moufouma-Okia W, Péan C, Pidcock R, Connors S, Matthews JBR, Chen Y, Zhou X, Gomis MI, Lonnoy E, Maycock T, Tignor M, Waterfield T (eds). Cambridge University Press, Cambridge, xi + 616 pp

Ishibashi H, Takeuchi I (2023) Effects of anthropogenic chemicals on hermatypic corals with special reference to gene expression. In: Takeuchi I, Yamashiro H (eds) Coral reefs of eastern Asia under anthropogenic impact. Coral reefs of the world, vol 17. Springer Nature Switzerland AG, Cham, pp 151–163

Ishigaki City (2020) Outline of Ishigaki City Sewerage System. https://www.city.ishigaki.okinawa.jp/soshiki/gesuido/1_1/349.html. Accessed 2 May 2022. (In Japanese)

Iwai K, Kubo H, Goya H, Takeuchi S, Takahashi N (2006) Projects on clam production. Okinawa Prefectural Fisheries Experimental Station Report for Heisei 14th. Heisei 14th, pp 185–195. (In Japanese)

Japan Meteorological Agency (2021) Long-term trend of sea surface temperature change in the sea around Japan published on 3 March 2021. https://www.data.jma.go.jp/kaiyou/data/shindan/a_1/japan_warm/japan_warm.html. Accessed 21 Aug 2021. (In Japanese)

Joyce KE, Duce S, Leahy SM, Leon J, Maier SW (2019) Principles and practice of acquiring drone-based image data in marine environments. Mar Freshw Res 70:952–963

Kadokami K, Ueno D (2019) Comprehensive target analysis for 484 organic micropollutants in environmental waters by the combination of tandem solid-phase extraction and quadrupole time-of-flight mass spectrometry with sequential window acquisition of all theoretical fragment-ion spectra acquisition. Anal Chem 91: 7749–7755

Kayanne H (2017) Validation of degree heating weeks as a coral bleaching index in the northwestern Pacific. Coral Reefs 36:63–70

Keller AA, Su Y, Jassby D (2022) Direct potable reuse: are we ready? A review of technological, economic, and environmental considerations. ACS EST Eng 2:273–291

Kim BH, Chang IS, Gadd GM (2007) Challenges in microbial fuel cell development and operation. Appl Microbiol Biotechnol 76: 485–494

Kraume M, Drews A (2010) Membrane bioreactors in waste water treatment–status and trends. Chem Eng Technol 33:1251–1259

Lachs L, Johari NAM, Le DQ, Safuan CDM, Duprey NN, Tanaka K, Hong TC, Ory NC, Bachok Z, Baker DM, Hirai K (2019) Effects of tourism-derived sewage on coral reefs: isotopic assessments identify effective bioindicators. Mar Pollut Bull 148:85–96

Lange C, Ratoi L, Co DL (2020) Reformative coral habitats. Rethinking artificial reef structures through a robotic 3D clay printing method. In: Holzer D, Nakapan W, Globa A, Koh I (eds) Anthropocene, Design in the Age of Humans - Proceedings of the 25th CAADRIA Conference - Volume 2, Chulalongkorn University, Bangkok, Thailand, 5–6 August 2020, pp 463–472

Leung CM, de Haan P, Ronaldson-Bouchard K, Kim G-A, Ko J, Rho HS, Chen Z, Habibovic P, Jeon NL, Takayama S, Shuler ML, Vunjak-Novakovic G, Frey O, Verpoorte E, Toh Y-C (2022) A guide to the organ-on-a-chip. Nat Rev Methods Primers 2:33. https://doi.org/10.1038/s43586-022-00118-6

Li P, Liu L, Wu J, Cheng R, Shi L, Zheng X, Zhang Z (2019) Identify driving forces of MBR applications in China. Sci Total Environ 647: 627–638

Liu Z-J, Nakamura H, Zhu X-H, Nishina A, Guo X, Dong M (2019) Tempo-spatial variations of the Kuroshio current in the Tokara Strait based on long-term ferryboat ADCP data. J Geophys Res Oceans 124:6030–6049

Lui H-K, Chen C-TA, Lee J, Wang S-L, Gong G-C, Bai Y, He X (2015) Acidifying intermediate water accelerates the acidification of seawater on shelves: an example of the East China Sea. Cont Shelf Res 111:223–233

Mano H, Takeda F, Kitamura T, Okamoto S, Suzuki Y, Park C-B, Yasui N, Kobayashi K, Tanaka Y, Yamashita N, Minamiyama M (2017) Water quality comparison of secondary effluent and reclaimed water to ambient river water of southern Okinawa Island via biological evaluation. Environ Monit Assess 189:1–14

Mbah JT, Chmist-Sikorska J, Szoszkiewicz K, Czekała W (2021) The effects of inflow of agricultural biogas digestate on bivalves' behavior. Environ Sci Pollut Res 28:67385–67393

Mitchell J (2020) PFAS contamination from US military facilities in mainland Japan and Okinawa. Asia-Pac J Jpn Focus 18(16):5457

Mizuno K, Terayama K, Hagino S, Tabeta S, Sakamoto S, Ogawa T, Sugimoto K, Fukami H (2020) An efficient coral survey method based on a large-scale 3-D structure model obtained by Speedy Sea scanner and U-net segmentation. Sci Rep 10:12416. https://doi.org/10.1038/s41598-020-69400-5

Morris RL, Heery EC, Loke LHL, Lau E, Strain EMA, Airoldi L, Alexanders KA, Bishop MJ, Coleman RA, Cordell JR, Dong YW, Firth LB, Hawkins SJ, Heath T, Kokora M, Lee SY, Miller JK, Perkol-Finkel S, Rella A, Steinberg PD, Takeuchi I, Thompson RC, Todd PA, Toft JD, Leung KMY (2019) Design options, implementation issues and evaluating success of ecologically-engineered shorelines. Oceanogr Mar Biol Ann Rev 57:169–228

Mowlem M, Beaton A, Pascal R, Schaap A, Loucaides S, Monk S, Morris A, Christopher L, Cardwell CL, Fowell SE, Matthew D, Patey MD, Lopez-Garcia P (2021) Industry partnership: lab on chip chemical sensor technology for ocean observing. Front Mar Sci 8:697611. https://doi.org/10.3389/fmars.2021.697611

Nagai K, Honjo T, Go J, Yamashita H, Oh SJ (2006) Detecting the shellfish killer *Heterocapsa circularisquama* (Dinophyceae) by measuring bivalve valve activity with a hall element sensor. Aquaculture 255:395–401

Nakaza E, Rahaman SMB, Kitamura Y, Pawlak G, Tsukayama S (2006) Cross-shore internal waves in Zanpa coastal region of Okinawa Island. J Waterway Port Coast Ocean Engineer 132(1): 36–46

Nishina A, Nakamura H, Park J-H, Hasegawa D, Tanaka Y, Seo S, Hibiya T (2016) Deep ventilation in the Okinawa trough induced by Kerama gap overflow. J Geophys Res Oceans 121:6092–6102

OECD (2019) Responding to rising seas: OECD country approaches to tackling coastal risks. OECD Publishing, Paris, 165 pp

Oh SJ, Moon S, Yoon YH, Kim DH, Kang IJ, Shimasaki Y (2021) Biomonitoring system to assist in early detection of oxygen–deficient sea water using shell valve movements of Pacific oyster (*Crassostrea gigas*). J Fac Agr Kyushu Univ 66:217–225

Okinawa Prefecture (2018) Current stats and problems of Okinawa tourism. Okinawa Prefecture, Naha, Okinawa, 8 pp. (In Japanese)

Okinawa Prefecture (2021) Status of Sewerage System Development in Okinawa Prefecture. https://www.pref.okinawa.jp/site/doboku/gesui/kokyo/seibi.html. Accessed 27 Apr 2022. (In Japanese)

Omori M (2019) Coral restoration research and technical developments: what we have learned so far. Mar Biol Res 15:377–409

Pang A-P, Luo Y, He C, Lu Z, Lu X (2020) A polyp-on-chip for coral long-term culture. Sci Rep 10:6964. https://doi.org/10.1038/s41598-020-63829-4

Quay AN, Fiske PS, Mauter MS (2022) Recommendations for advancing FAIR and open data standards in the water treatment community. ACS EST Eng 2:337–346

Rouf MA, Nakaza E (2014) Seasonal behavior of the internal tide and associated water temperature in Genka Bay, Okinawa, Japan. J Coast Res 30:102–110

Sakai K, Singh T, Iguchi A (2019) Bleaching and post-bleaching mortality of *Acropora* corals on a heat-susceptible reef in 2016. PeerJ 7: e8138. https://doi.org/10.7717/peerj.8138

Scholz S, Nichols JW, Escher BI, Ankley GT, Altenburger R, Blackwell B, Brack W, Burkhard L, Collette TW, Doering JA, Ekman D, Fay K, Fischer F, Hackermüller J, Hoffman JC, Lai C, Leuthold D, Martinovic-Weigelt D, Reemtsma T, Pollesch N, Schroeder A, Schüürmann G, von Bergen M (2022) The eco-exposome concept: supporting an integrated assessment of mixtures of environmental chemicals. Environ Toxicol Chem 41: 30–45

Schuijt LM, Peng FJ, van den Berg SJP, Dingemans MML, Van den Brink PJ (2021) (Eco) toxicological tests for assessing impacts of chemical stress to aquatic ecosystems: facts, challenges, and future. Sci Total Environ 795:148776

Shapiro O, Kramarsky-Winter E, Gavish A, Roman Stocker R, Vardi A (2016) A coral-on-a-chip microfluidic platform enabling live-imaging microscopy of reef-building corals. Nat Commun 7: 10860. https://doi.org/10.1038/ncomms10860

Shitaoka R, Ishibashi H, Takeuchi I (2021) Thermal tolerance of the hermatypic coral *Acropora tenuis* elucidated by RGB analysis and expression of heat shock proteins in coral and symbiotic dinoflagellates. Mar Pollut Bull 162:111812

Sutherland KP, Porter JW, Turner JW, Thomas BJ, Looney EE, Luna TP, Meyers MK, Futch JC, Lipp EK (2010) Human sewage identified as likely source of white pox disease of the threatened Caribbean elkhorn coral, *Acropora palmata*. Environ Microbiol 12: 1122–1131

Suzuki A, Iguchi A, Sakai K, Nojiri Y (2023) Succession of ocean acidification and its effects to reef-building corals in eastern Asia. In: Takeuchi I, Yamashiro H (eds) Coral reefs of eastern Asia under anthropogenic impacts. Coral reefs of the world, vol 17. Springer Nature Switzerland AG, Cham, pp 95–110

Takeda F, Mano H, Minamiyama M, Okamoto S (2018) Initial health and ecological risk assessment of chemicals in treated wastewater in Okinawa prefecture based on Japanese PRTR system. J Jpn Soc Civil Engineer Ser G Environ Res 74:25–34. (In Japanese with English abstract)

Takeuchi I (2023a) Introduction–The relevance of anthropogenic factors to coral reef conservation in the coastal areas of the East China Sea. In: Takeuchi I, Yamashiro H (eds) Coral reefs of eastern Asia under anthropogenic impacts. Coral reefs of the world, vol 17. Springer Nature Switzerland AG, Cham, pp 1–5

Takeuchi I (2023b) Anthropogenic stresses in coral reefs and adjacent ecosystem around the East China Sea. In: Takeuchi I, Yamashiro H (eds) Coral reefs of eastern Asia under anthropogenic impacts. Coral reefs of the world, vol 17. Springer Nature Switzerland AG, Cham, pp 111–135

Takeuchi I, Takayama K (2023) Development of a compact experimental system for ecotoxicological experiments on *Acropora* spp. In: Takeuchi I, Yamashiro H (eds) Coral reefs of eastern Asia under anthropogenic impacts. Coral reefs of the world, vol 17. Springer Nature Switzerland AG, Cham, pp 137–149

Takeuchi I, Yamashiro H (2017) Large *Porites* microatoll found by aerial survey at Sesoko Island, Okinawa, Japan. Coral Reefs 36:1317

Takeuchi I, Yamashiro H (2022) Bird-view of the reef flat off Sesoko Station on the east coast of Sesoko Island, Okinawa, Japan obtained by aerial survey. Galaxea J Coral Reef Stud 24:119–120

Tashiro Y, Takemura A, Fujii H, Takahira K, Nakanishi Y (2003) Livestock wastes as a source of estrogens and their effects on wildlife of Manko tidal flat, Okinawa. Mar Pollut Bull 77:143–147

The Ministry of the Environment, Japan (2017[2016]) The action plan to conserve coral reef ecosystem in Japan 2016–2020. To protect blessings from the coral sea. Nature Conservation Bureau, Ministry of the Environment, Japan, Chiyoda-ku, Tokyo, Japan. v + 48 pp.

The State of Queensland (2009) Reef Water Quality Protection Plan 2009 For the Great Barrier Reef World Heritage Area and adjacent catchments. The Reef Water Quality Protection Plan Secretariat, Brisbane, QLD, 30 pp

The State of Queensland and Commonwealth of Australia (2003) Reef Water Quality Protection Plan; for catchments adjacent to the Great Barrier Reef World Heritage Area, Queensland. October 2003. Great Barrier Reef Protection Plan Project, Department of Premier and Cabinet, Brisbane, QLD, 44 pp

Ventura D, Mancini G, Casoli E, Pace DS, Lasinio GJ, Belluscio A, Ardizzone G (2022) Seagrass restoration monitoring and shallow-water benthic habitat mapping through a photogrammetry-based protocol. J Environ Manag 304:114262. https://doi.org/10.1016/j.jenvman.2021.114262

Wang Y, Wang X, Li M, Dong J, Sun C, Chen G (2018) Removal of pharmaceutical and personal care products (PPCPs) from municipal waste water with integrated membrane systems, MBR-RO/ NF. Int J Environ Res Public Health 15:269. https://doi.org/10.3390/ijerph15020269

Wangpraseurt D, You S, Azam F, Jacucci G, Gaidarenko O, Hildebrand M, Kühl M, Smith AG, Davey MP, Smith A, Deheyn DD, Chen S, Vignolini S (2020) Bionic 3D printed corals. Nat Commun 11:1748. https://doi.org/10.1038/s41467-020-15486-4

Wear SL, Thurber RV (2015) Sewage pollution: mitigation is key for coral reef stewardship. Ann N Y Acad Sci 1355:15–30

Wei H, Pan D, Han H (2021) Electrochemical monitoring of marine nutrients: from principle to application. Trends Analyt Chem 138: 116242

West K, Van Woesik R (2001) Spatial and temporal variance of river discharge on Okinawa (Japan): inferring the temporal impact on adjacent coral reefs. Mar Pollut Bull 42:864–872

Wooldridge SA (2009) Water quality and coral bleaching thresholds: formalising the linkage for the inshore reefs of the great barrier reef, Australia. Mar Pollut Bull 58:745–751

WWF Japan (2010) Nansei Islands Biological Diversity Evaluation Project Report. WWF Japan, Tokyo, 215 pp
Ya M, Wu Y, Wu S, Li Y, Mu J, Fang C, Yan J, Zhao Y, Qian R, Lin X, Wang X (2019) Impacts of seasonal variation on organochlorine pesticides in the East China Sea and Northern South China Sea. Environ Sci Technol 53:13088–13097
Yamano H, Satake K, Inoue T, Kadoya T, Hayashi S, Kinjo K, Nakajima D, Oguna H, Ishiguro S, Okagawa A, Suga S, Horie T, Fukayama N, Hibiki A (2015) An integrated approach to tropical and subtropical Island conservation. J Ecol Environ 38:271–279
Yamashiro H, Wada N, Tang S-L (2023) Succession and spread of coral diseases and coral-killing sponges with special reference to microbes in Southeast Asia and adjacent waters. In: Takeuchi I, Yamashiro H (eds) Coral reefs of the world. Coral reefs of eastern Asia under anthropogenic impacts, vol 17. Springer Nature Switzerland AG, Cham, pp 71–82
Zayasu Y, Shinzato C (2016) Hope for coral reef rehabilitation: massive synchronous spawning by out planted corals in Okinawa Japan. Coral Reefs 35:1295
Zhang R, Du J, Dong X, Huang Y, Xie H, Chen J, Li X, Kadokami K (2021) Occurrence and ecological risks of 156 pharmaceuticals and 296 pesticides in seawater from mariculture areas of Northeast China. Sci Total Environ 792:148375